Books by Dr. Mousumi Boral

Pedagogy of Geography: Modern Theoretical Perspective
Pedagogy of Geography: Practicals and Practicum

Pedagogy of Geography

Practicals and Practicum

Mousumi Boral

M. Sc. (Geogr.), B.Ed. M.Ed.,
Ph. D. (Sc.), M. A. (Edu.)

SHIPRA

Rs. 995

ISBN: 978-93-91978-59-4

2024
First Published in India

Pedagogy of Geography: Practicals and Practicum

Published by:
SHIPRA PUBLICATIONS
LG 18-19, Pankaj Central Market
I.P. Ext., Patparganj, Delhi 110092, India
+91 11 47322068; 96500 28065, 9810522367
info@shiprapublication.com
www.shiprapublication.com

Foreword

In today's globalized world, geography is a vital subject that requires a comprehensive approach to teaching. The Pedagogy of Geography Teaching is an exceptional resource that covers both theoretical and practical aspects of the subject, offering teachers and learners a holistic view of geography. This book offers diverse teaching strategies that engage students and make learning geography more interactive and enjoyable, even for visually-impaired learners, making it an inclusive resource for all.

Additionally, Pedagogy of Geography Teaching equips students and teachers with the knowledge and skills necessary to understand and address contemporary issues effectively. With its emphasis on inclusivity and contemporary issues, this book is an essential resource for all geography educators. I strongly recommend it to anyone looking to improve their understanding and teaching of geography. The book's comprehensive coverage and focus on current trends and developments in the field make it an excellent choice for educators and learners alike.

Prof. Madhumita Bandyopadhyay
NIEPA
New Delhi

Preface

Throughout human history, most societies have sought to understand their place in the world, and the people and environments around them. Geography asks spatial questions—how and why things are distributed or arranged in particular ways on Earth's surface and how they develop and change over time. It looks at these different distributions and arrangements at many different scales. Geographers explore both the relationships between the physical properties of Earth's surface and the human societies spread across it and the varied cultural exchanges.

Whether geography is thought of as a discipline or as a basic feature of our world, developing an understanding of the subject is vehemently important. Thereby the approach to the study of geography has changed over several decades. Thinking geographically helps people to be aware and remain connected among and between places and important events.

This book makes an effort to broaden the teaching ambiance with relevant inclusions of the matter and also attempts to provide additional information to interested readers with comprehensive illustrations and justified analysis in simple and lucid language.

I would like to convey my heartfelt thanks to Priyanka Patra my geography student who has significantly helped me with her dedicated contributions. It is worth mentioning my younger son Shridipta who had been a constant companion in the making of this book. I owe him immensely for his relentless encouragement. Acknowledging the moral support and belief in my work, was invaluable inspiration received from my elder son Shrimanta.

Needless to say, I am grateful to all persons who have contributed their utmost to make it a success. It is hoped that the users will enjoy studying the book as much as the author has enjoyed writing it. Comments and suggestions for further improvement of the book are always welcome.

Kolkata
26 January, 2024

Mousumi Boral

Contents

PRACTICUM

PRACTICALS

1

Identification of Rocks

The scientific study of rocks is called *Petrology* which is an essential component of Geology. The subject matter consists of the origin, association, occurrences, mineral composition, chemical composition, texture, structure, and physical properties of rocks. *Petrography* deals with the descriptive part of the rocks and *Petrogeny* deals with the mode of formation of rocks.

Geologists study rocks because they contain clues about what the Earth was like in the past. A historical record of a planet can be discovered. Moreover, the past environment, pre-historic plant and animal species and climate can be studied. Moreover, topography can also be interpreted and the historical study of soil formation can be done.

Definition of a Rock

A rock is a naturally formed, non-living earth material consisting of minerals and other particulate matter. Primarily the three types of rocks are *Igneous, Sedimentary* and *Metamorphic*. Each type can be further classified into different types according to their mode of formation, mineral content, colour, structure, texture and fossil content.

Economic Significance of Rocks

1. It provides nutrients needed for plant growth.
2. Rocks are a storehouse of minerals.
3. They contain valuable deposits of coal, oil and natural gas
4. They provide building stones for the construction of roads, dams and other structures.
5. Rocks get weathered and ultimately lead to the formation of soil.

Definition of a Mineral

Whereas, minerals are defined as natural, inorganic solid substance which has a crystalline structure with a particular chemical composition. Whereas, the rock is composed of several minerals, and is generally classified based on the process of its formation.

Identification of Rocks

1. Colour of the rock is the effect of the composition
2. Texture is the result of crystals and grains present
3. Reaction to acids and alkalis

The other essentials needed for identifying rocks are hand-lens or magnifying glass, dilute hydrochloric acid, a nail, a knife and surely a streak plate. These enable one to view the presence of fossils, rock layers or foliations, glassy look, pores and interlocking crystals in the rocks.

Igneous Rocks

Igneous rocks have been formed by the solidification and cooling of magma. They may have crystals but have no stratification or fossils. Examples- Granite, Basalt and Dolerite.

Granite: Granites are coarse-grained, light-coloured, acidic and holocrystalline plutonic igneous rocks with a medium-specific gravity (2.63 to 2.75). They are hard and compact rocks composed essentially of quartz, feldspar and mica, usually with some ferromagnesian minerals. Feldspar is identified by its lath shape in which cleavage may be present, while quartz is identified by its vitreous lustre and mica by its flakes mafic. Both plagioclase and mafic minerals tend to form subhedral crystals while intergranular spaces are occupied by anhedral quartz. This typical interlocking of grains forms the usual granitic texture. Sometimes there is some micrographic intergrowth of quartz and feldspar. Some of these are porphyritic. Reading the larger-scaled structures, some have a bonded or layered structure. High-level granitic rocks may develop a drusy structure.

Basalt: Basalt makes up over 98 per cent of all volcanic rocks. It is a very fine-grained and dark-coloured mafic igneous rock with a typical volcanic texture, very high specific gravity and high hardness. It is very compact and is characterized by conchoidal fractures. Individual minerals cannot be identified by the naked eye. The characteristic minerals are calcic plagioclase, pyroxene and iron ore with or without olivine. They may be vesicular or the vesicles may be filled with secondary minerals forming amygdales and thereby giving rise to the amygdaloidal structure. The basaltic magma which erupted underwater forms pillow-like masses giving rise to basaltic pillow lava. The individual pillows have a glassy crust and a more or less crystalline interior which may show radiating contraction cracks.

Dolerite: Dolerites are the hypabyssal representatives of the gabbros and basalts, finer-grained than the former but coarser-grained than the latter. They occur mostly as dykes and sills and their texture is frequently ophitic or sub-ophitic. They are dark-coloured mafic igneous rocks with high specific gravity (2.64 to 3.12), high hardness and high compactness. Feldspar laths are randomly oriented forming criss-cross arrangement-the interspaces are filled up by ferromagnesian minerals in which traces of cleavage can be seen. As it is prone to weathering, local patches of reddish-brown tint can be often distinguished.

Sedimentary Rocks

Sedimentary rocks have been formed by the processes of deposition, cementation and solidification. They do not contain crystals and have stratifications and fossils. Examples-, Sandstone, Limestone and Conglomerate.

Sandstone: Sandstone is very fine to very coarse-grained, light-coloured, moderately compact and moderately hard sedimentary rocks with a rough feel and medium specific gravity. It is dominantly composed of detrital sand. Grains pack together against the grain to form a framework that provides that primary solid support in the deposit. The framework is cemented together in a matrix of ferruginous, siliceous, argillaceous and calcareous sediments. The texture is typically clastic-granular. Graded or cross-bedded structures may be present. Depending on grain provenance their main types are recognized as quartzose sandstone, arkose or feldspathic sandstone and lithic sandstone.

Limestone: Limestone is a rock whose primary constituents are composed dominantly of aragonite and calcite. They are very fine-grained, light-coloured with moderate specific gravity and very compact but of low hardness (easily scratched by a knife or even fingernail). The texture is typically non-clastic calcarenite, calcarenite and calcilutite. It strongly reacts with cold and dilute hydrochloric acid. A salient feature is that most of the limestones are particulate and bimodal-ie. they consist initially of relatively large carbonate particles mixed in various proportions with carbonate mud. Very often compositional layering characterized by different colour bands may be present.

Conglomerate: Conglomerates are clastic sedimentary rocks of rudaceous type. These are pebbly rocks of medium or fine size (over 2 mm in diameter). These are variegated in colour with moderate to high specific gravity and moderate compactness. Being mostly of fluvial, glacial and shoreline origin, these consist of typical mud-free gravels or their lithified equivalents. The texture is typically clastic as lithic clasts are the essential constituents. The pebbles, cobbles, gravels rock fragments, and granules are cemented together in a fine-grained matrix of ground mass of ferruginous (reddish brown), siliceous (light coloured), in a fine-grained matrix of ground argillaceous (muddy odour) and calcareous (reaction with dilute hydrochloric acid) sediments.

Metamorphic Rocks

Metamorphic rocks have been formed by the metamorphism processes wherein all igneous and sedimentary rocks undergo tremendous changes due to intense heat and pressure. They may be crystalline, foliated or non-foliated but possesses no fossils. Example- Marble, Quartzite and Schist.

Marble: Marbles are coarse-grained, light-coloured rocks with high compactness, low hardness and medium specific gravity. They are produced by the metamorphism of limestones and are, hence, dominantly composed of calcite. They strongly effervesce with dilute hydrochloric acid. Generally, they show an even-grained texture. Some show a crude foliation due to the segregation of minor silicate constituents or in some cases parallel alignment of lustrous flakes of graphite while some bear the imprint of stain in the form of lensoid outlines.

Quartzite: Quartzites are very hard granular rocks produced by the recrystallization of sandstone. They are light-coloured, fine to medium-grained rocks with high compactness, low to medium specific gravity, granoblastic texture and an overall vitreous lustre without any cleavage. They are non-foliated and composed largely of quartz.

Schist: Schists are most often associated with regional metamorphism in medium-grade metamorphic environments. They are medium-grade metamorphic, medium-grained, light-coloured rocks with low hardness, low compactness and medium specific gravity. They are dominantly composed of flaky minerals (mica) which are oriented in a preferred direction, imparting foliations in the rock. The grains of quartz and feldspar are interspaced with the foliation. The texture is typically schistose formed due to the parallel arrangement of flaky minerals. They break readily into parallel-sided slabs, due to the presence of phyllosilicates and the parallel alignment of amphibole prisms and fibres.

Rock Cycle of Transformation

The rock cycle is an idealized cycle of processes undergone by rocks in the earth's crust, involving igneous, sedimentary and metamorphic rocks interrelated with each other.

The Rock Cycle describes the transformations of rocks and subsequently the formation of new topographic structures on the dynamic Earth.

Identification of Rocks		
Igneous Rocks		
 Granite	 Basalt	 Dolerite
Sedimentary Rocks		
 Sandstone	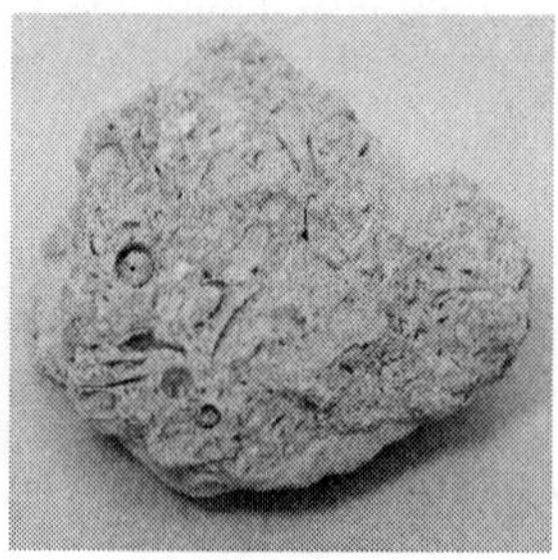 Limestone	 Conglomerate
Metamorphic Rocks		
 Marble	 Quartzite	 Schist

Identification of Minerals

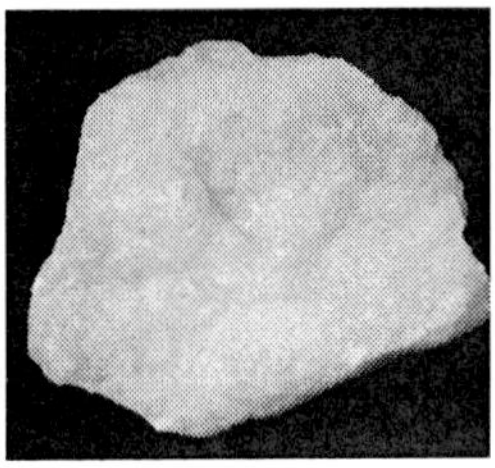

Quartz

Feldspar

Mica

Calcite

Pyroxene

Galena

Olivine

Hematite

Gypsum

2

Scale

The scale is the ratio between map distance and ground distance. Hence the map is simply the scale model of the real world. Map scales are represented in three basis and reversible forms—a statement, a numeric ratio and a graph and accordingly they are respectively called Statement scale, Ratio scale and Geographical scale. The scale is normally placed within a base, being associated with the north arrow positioned exactly middle, just above the scale bar. The location of the scale box depends on the page layout. The best cartographic presentation is decided by page borders, location, size, and shape of the outline map, available space, map index, key, or legends.

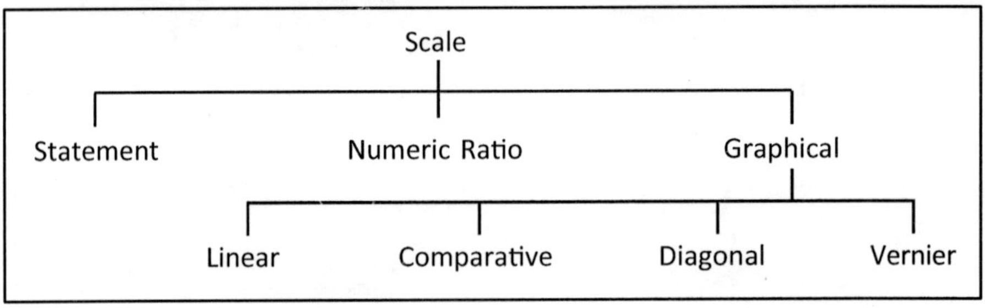

Statement Scale

A map scale may be in form of a simple statement in which the map distance is always expressed as a unit length. This is known as a statement scale and is written on a map as, 1 cm to 25 km or 1 inch to 50 miles. The value on the left-hand side of the statement indicates the map distance (cm/inch) and that on the right-hand side indicates the ground distance (km/mile).

Merits

(i) It is simplest to express.
(ii) It is not necessary to draw the scale like linear or graphical in the map.

Demerits

(i) A layman, ignorant of the various scale of measurements will not be able to read the map.
(ii) A reproduction of the original map by reduction or enlargement necessitates a determination of the scale all over again.

Numeric Ratio Scale

A map scale may also be expressed as a numeric ratio in which the numerator is the map distance and the denominator is the corresponding ground distance, both being expressed in the same unit of measurement. Therefore, it is a dimensionless fraction. It is also known as a ratio scale or representative fraction (R.F.), *e.g.* 1: 100000.

Merits

(i) The greatest advantage of a numeric ratio scale is that a map with a given R.F. can be used universally.

(ii) A statement scale or graphical scale in any system of measurement can easily be computed from R.F. and cartographically plotted as well.
Thus, 1: 100000 may be written as 1 cm to 100000 cm (CGS system)

$$\text{or, } 1 \text{ cm to} \frac{100000}{100000} \text{km}$$

$$\text{or, } 1 \text{ cm to } 1 \text{ km}$$

Demerits

(i) The scale will be changed by enlarging or reducing the map.

(ii) The previous knowledge about the unit of measurement is necessary to understand the representative fraction.

Graphical Scale

A map scale may also be cartographically represented by a linear graph or line. This is known as the graphical scale. This may be of four different forms—1. Linear Scale, 2. Comparative Scale, 3. Diagonal Scale, 4. Vernier Scale.

Linear Scale

This is the simplest form of graphical scale and is shown as a linear graph. The primary fundamental division (at least three) represents the ground distance in multiples of 1, 5 or 10. The primary division on the extreme left is subdivided into a suitable number of secondary divisions. The length of each secondary division depends on the fractional length up to which precision in measurement is desired.

Merits

(i) A linear scale helps to determine easily the idea of distance between two places on a map by visual observation or by a simple scale of measurement.

(ii) The ground distance can be easily measured from the map distance.

(iii) Enlargement or reduction does not change the measurement of the scale of the map as the scale also be reduced or enlarged consequently. This merit is not available in the case of the R.F. scale.

Demerits

(i) Drawing of a linear scale requires much time and may involve defects.

(ii) The respective unit of scale may not be acceptable all over.

(iii) Change in the unit of measurement of the scale may involve less of accuracy.

Problems

Example 1: Construct a linear scale to show 4 km in the primary division and 1 km in the secondary division and the given R.F. is 1: 200000.

Step 1: We know, 100000 cm = 1 km

$$\text{or , } 1 \text{ cm} = \frac{1}{10000} \text{km}$$

$$200000 \text{ cm} = \frac{1 \times 200000}{100000} \text{ km} = 2 \text{ km}$$

Step 2: The value of each primary division is 4 km.
∴ 2 km on the ground distance represented by 1 cm on the map distance.

1 km on the ground distance represented by $\frac{1}{2}$ cm on the map distance.

4 km on the ground distance represented by $\frac{1 \times 4}{2}$ cm on the map distance.

= 2 cm on the map distance.

Step 3: For the convenience of drawing the total length of scale will be—
2 cm on the map represented by 4 km on the ground.
2 × 7 cm on the map represented by 4 × 7 km on the ground.
14 cm on the map represented by 28 km on the ground.

Step 4: The value of each secondary division = 1 km.

$$\therefore \text{Total number of primary division} = \frac{\textit{Total ground length of scale}}{\textit{Ground length of primary division}}$$

$$= \frac{28\,km}{4} = 7 \text{ divisions.}$$

$$\therefore \text{Total number of secondary division} = \frac{\textit{Total ground length of each primary division}}{\textit{Ground length of each secondary division}}$$

$$= \frac{4\,km}{1} = 4 \text{ divisions.}$$

Methodology

- The map distance for each primary division is first evaluated from either statement or R.F.
- Select and fix the magnitude of ground distance for one primary division and secondary division.
- Based on the given problem or R.F., compute the map distance in cm or inch for the corresponding value of primary division.
- Draw a horizontal straight line.
- Divide the straight line with the help of a divider, diagonal scale, and perpendicular straight lines are drawn at each division with the help of set square.
- The first primary division from the left-side is drawn, then graphically sub-divided into secondary division with the desired precision.
- The drawing is then properly labelled and bordered.

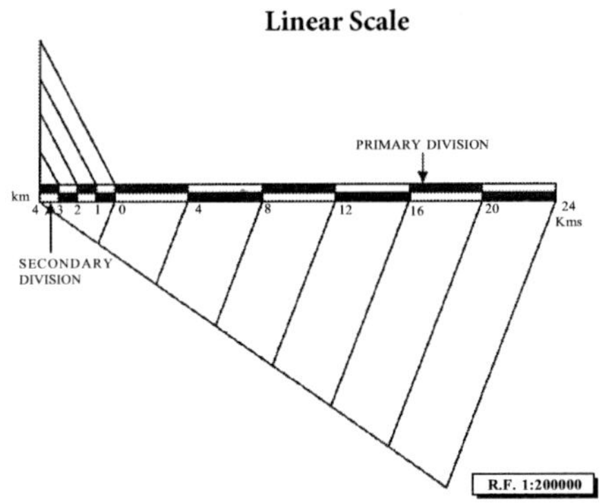

Figure 1: *A linear scale showing divisions in kilometres and metres*

A linear scale has been drawn on the basis of above calculation.

Example 2: Draw a linear scale to show 2 km in the primary division and 500 m in thesecondary division with R.F. = 1 : 98000.

Solution: R.F. = 1 : 98,000

$\therefore$ 1cm of scale represents 98,000 cm

$$\text{or, } \frac{98{,}000}{100{,}000} \text{ km } [\because 100{,}000 \text{ cm} = 1 \text{ km}]$$

$$\equiv 0.98 \text{ km}$$

or, 0.98 km = 1cm

$$\therefore 15 \text{ km} = \frac{15 \times 0.98}{100{,}000} = 14.7 \text{ km}$$

Since 14.7 km is not a round figure, take 15 km as a round figure and calculate the length of the line as:

14.7 km are represented by 15 cm

$$1 \text{ km is represented by } \frac{15}{14.7}$$

$$10 \text{ km is represented by } \frac{15 \times 10}{14.7} = 10.20 \text{ cm}$$

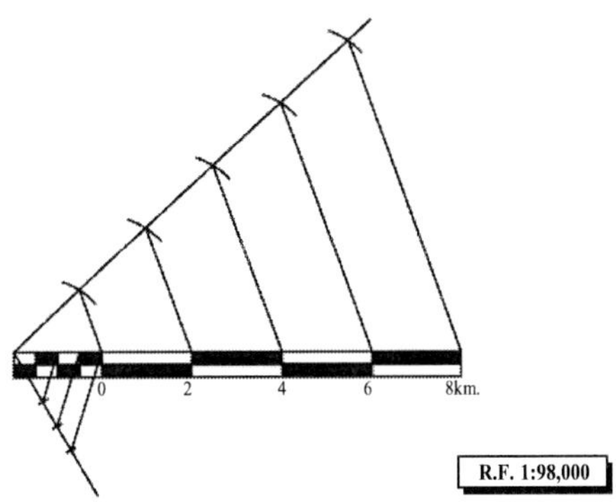

Figure 2: *A linear scale showing divisions in kilometres and metres*

Now a line of 10.20 cm is drawn which represents 10 km. It is divided into five primary divisions of 2 km each and then the first primary division is divided into four secondary divisions of 500 m each.

Example 3: Draw a linear scale to show 10 km in the primary division and 2 km in the secondary division when the given R.F. is 1 : 411,842.
Solution: R.F. = 1 : 411,842

$\therefore$ 1cm of scale represents 411,842 cm

$$\text{or, } \frac{411{,}842}{100{,}000} \text{ km } [\because 100{,}000 \text{ cm} = 1 \text{ km}]$$

$$\equiv 4.11 \text{ km}$$

$\because$ 4.11 km are represented by 1 cm

$\therefore$ 1 km is represented by $\frac{1}{4.11}$

$\therefore$ 50 km is represented by $\frac{1}{4.11} \times 50$ cm = 12.16 cm

Now a line of 12.16 cm is drawn which represents 50 km. It is divided into five primary divisions of 10 km each and then the first primary division is divided into five parts to show 2 km each. The given scale is drawn in the following section.

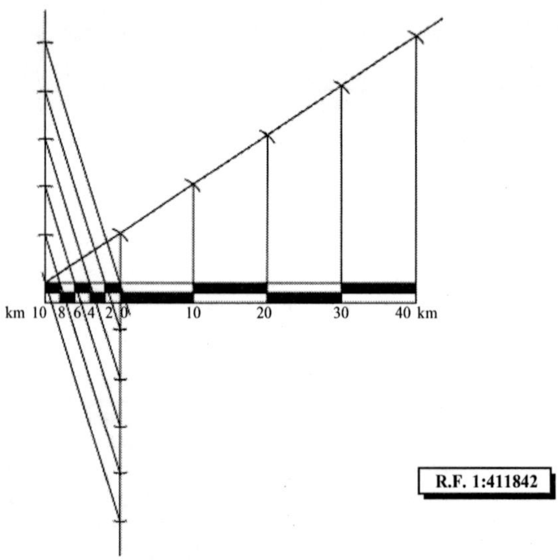

Figure 3: *A linear scale showing divisions in kilometres*

Example 4: Draw a linear scale to show 40 km and 10 km as primary and secondary divisions respectively when the R.F. is 1 : 2,000,000.
Solution: R.F. = 1 : 2,000,000

$\therefore$ 1cm of scale represents 2,000,000 cm

$$\text{or, } \frac{2,000,000}{100,000} \text{ km } [\because 100,000 \text{ cm} = 1 \text{ km}]$$

$$\equiv 20 \text{ km}$$

$\because$ 20 km is represented by 1 cm

$\therefore$ 1 km is represented by $\frac{1}{20}$ cm

$\therefore$160 km is represented by $\frac{1}{20} \times 160$ cm = 8 cm

Now a line of 8 cm is drawn which represents 160 km. It is divided into four primary divisions of 40 km each and then the first primary division is divided into four parts to show 10 km each as secondary division. The given scale is drawn in the following section.

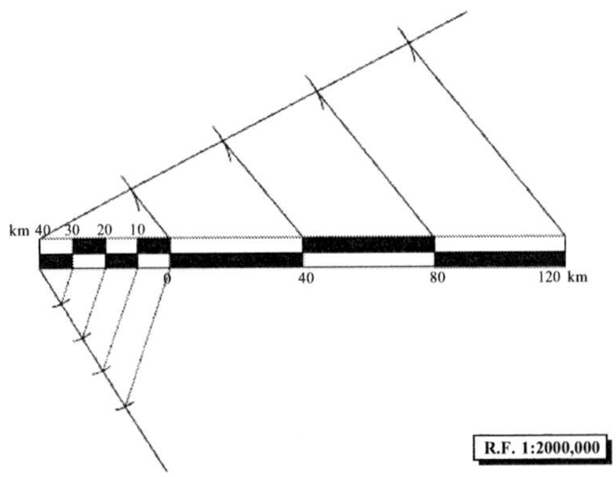

Figure 4: *A linear scale showing divisions in kilometres*

3

Temperature and Rainfall Graphs

A temperature and rainfall diagram is a composite diagram that represents the variation of temperature and rainfall of different months of a place in a year. Temperature can be diagrammatically represented with the help of a line graph (usually in red ink) in degrees Centigrade or Fahrenheit. The highest and lowest temperatures and the total rainfall in a year are depicted in the graph and thereby the climatic characteristics of the concerned place can be studied. The rainfall is usually graphically depicted as shaded columns (bar graphs usually in blue ink) for each month of a year like that of the temperature in cm/mm or in inches.

Meteorological stations, scattered all over the globe, are constantly gathering a wide range of data on temperature and rainfall for the construction of weather maps or temperature graphs. Climatic data mainly include the amount of rainfall collected through rain gauge and temperature which is measured with the help of various meteorologically designed thermometers.

Temperature is recorded daily which is laterally converted to the mean monthly temperature of a single month (according to the number of days of that month). The difference between the hottest and the coldest month is given as the annual range of temperature.

In the case of *rainfall*, daily records are added at the end of the month for the monthly rainfall amount. These monthly rainfalls are added at the end of the year (for 12 months) to find the annual rainfall and this annual amount can be divided by 12 months to obtain the mean annual.

Since the plotting of rainfall and temperature of a place for a year gives the learner, a fair idea about the weather of that particular place. This type of simple climate graph displaying yearly temperature and precipitation statistics for a particular location enables the learner to have an idea of the place. Most geographers worldwide have chosen the temperature-rainfall graph to identify the climatic region to which that particular place belongs and then can ensure regional planning.

Equatorial Climate

The equatorial region is hot and wet throughout the year. It is also known for its dense evergreen forest cover. It is situated between 5° to 10° North and South of the Equator in both hemispheres. More than 200 cm of annual rainfall and constant high temperature are the common features in this region.

Table 1: *Colombo (Sri Lanka)- 06°54'N, 79°25'E; Elevation -7 m*

Month	*January*	*February*	*March*	*April*	*May*	*June*	*July*	*August*	*September*	*October*	*November*	*December*	*Mean/ Annual*
Temp°C	26.1	26.4	27.2	27.8	28.0	27.2	27.2	27.2	27.2	26.7	26.1	25.8	26.9
Rainfall in mm	89	69	147	231	371	223	135	109	160	348	315	147	2344

Table 2: *Madang (New Guinea)- 5°14'S, 145°45'E; Elevation- 6 m*

Month	January	February	March	April	May	June	July	August	September	October	November	December	Mean/Annual
Temp°C	27.2	26.9	26.9	27.2	27.2	27.2	27.2	27.2	27.2	27.5	27.5	27.5	27.2
Rainfall in mm	307	302	378	429	384	274	193	122	135	254	338	368	3484

Highest temperature: 28 ° C
Lowest temperature: 25.8 ° C

Highest Rainfall: 429 mm
Lowest Rainfall: 69 mm

TEMPERATURE AND RAINFALL GRAPH

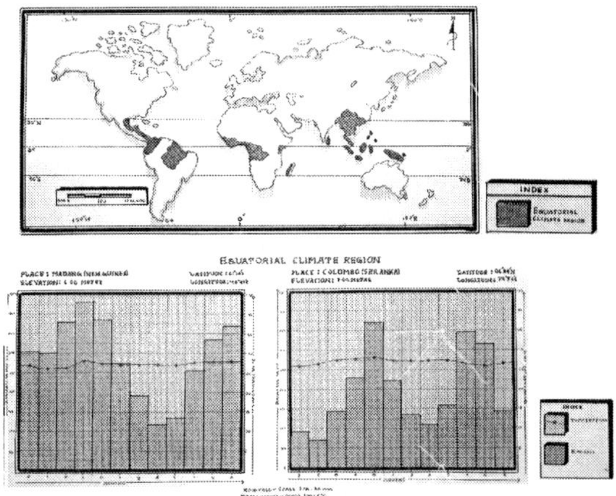

Interpretation

From the temperature values, it has been seen that there is a great uniformity of high temperatures throughout the year. There is hardly any variation. The annual range of temperature is not more than 30°C. In Colombo, the highest temperature is 28°C in the month of May. The lowest temperature found in the month of December is 25.8°C.

Precipitation is heavy and well distributed throughout the year. There are no months without rainfall. Instead, there are two periods of high rainfall (one is Apr/May and the other is Oct/Nov), which occur in the equatorial region. This double peak coincided almost with the equinoxes. The double peak is the characteristic of equatorial climates which are generally not found in other climatic regions. Hence it can be concluded that these areas are in the northern hemisphere of the Equatorial climate belt.

Tropical Monsoon Climate

Its latitudinal extension is 10° to 30° North and South of the Equator mostly near the Tropic of Cancer and the Tropic of Capricorn. Basically, extensive deciduous trees are found in the monsoon region and these trees shed their leaves during the dry season to conserve moisture.

Table 3: *Kolkata (India)- 22°34'N, 88°24'E; Elevation- 6 m*

Month	January	February	March	April	May	June	July	August	September	October	November	December	Mean/Annual
Temp°C	19.6	22.0	27.1	30.1	30.4	29.9	28.9	28.7	28.9	27.6	23.4	19.7	26.4
Rainfall in mm	12	28	34	50	134	290	331	334	253	127	27	4	1624

Table 4: *Papiti (Tahiti)- 17°32'S, 149°34'E; Elevation- 92 m*

Month	*January*	*February*	*March*	*April*	*May*	*June*	*July*	*August*	*September*	*October*	*November*	*December*	*Mean/ Annual*
Temp°C	29.6	26.9	26.9	26.9	25.8	25.3	25.3	25.0	25.3	25.8	26.4	26.7	26.3
Rainfall in mm	251	244	429	142	102	76	53	43	53	89	150	249	1881

Highest temperature: 30.4 ° C
Lowest temperature: 19.6 ° C

Highest Rainfall: 429 mm
Lowest Rainfall: 4 mm

TEMPERATURE AND RAINFALL GRAPH

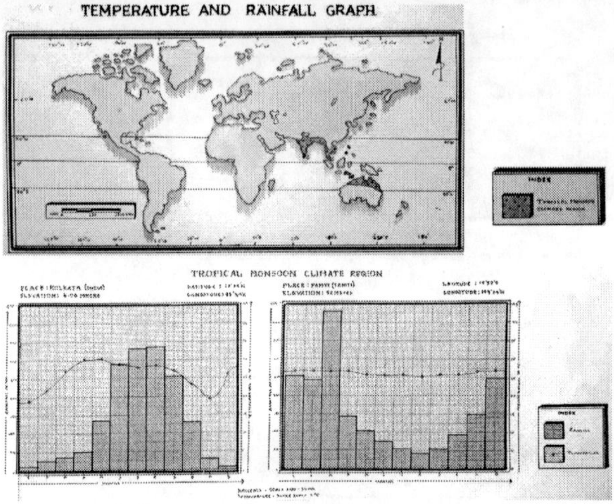

Interpretation

From the temperature values, it is observed that there is a great uniformity of temperature throughout the year in Papiti. But a small fluctuation in temperature has been observed in Kolkata in the year. The highest temperature has been observed in Kolkata during the month of May. It is because of the deficiency of rainfall that leads to the dry season which prevails during the months of April-May and December-January. Hence it is situated in the northern hemisphere in the ideal zone of Tropical Monsoon Climate.

The precipitation is varied all over the year in Kolkata. The highest amount of rainfall occurred in the month of July and August. But in Papiti, the rainfall is the same or more or less evenly distributed throughout the year with peaks in the December-January months. Situated at a 92-meter height and the highest rainfall recorded is 429 mm in March. But there is no month without rainfall.

From the above discussion, it can be interpreted that the uncertainty of rainfall and fluctuation of temperature have been observed throughout the year which are the main climatic characteristics of Tropical Monsoon Climate.

Hot Desert

In the tropics, these desertic regions are situated between 15° to 30° latitude in both hemispheres. Excess evaporation and less precipitation make these regions arid climatic zones. As a result, only the xerophytic vegetation thrives in these arid and semi-arid regions. These regions are mostly located in the western portions of the continents.

Table 5: *Aswan (United Arab Emirates, UAE)- 24°04'N, 32°53'E; Elevation- 116 m*

Month	*January*	*February*	*March*	*April*	*May*	*June*	*July*	*August*	*September*	*October*	*November*	*December*	*Mean/ Annual*
Temp°C	16.7	18.3	22.5	27.2	31.4	33.6	33.6	33.6	31.7	29.2	23.6	18.3	26.6
Rainfall in mm	2	3	3	3	3	3	0	0	0	2	2	2	23

Table 6: *Lima (Peru)- 12°05'S, 77°03'E; Elevation- 120 m*

Month	*January*	*February*	*March*	*April*	*May*	*June*	*July*	*August*	*September*	*October*	*November*	*December*	*Mean/ Annual*
Temp°C	23.3	23.9	23.6	24.7	19.4	17.2	16.7	16.1	16.9	18.1	19.4	21.1	20.03
Rainfall in mm	3	2	2	2	5	5	8	8	8	3	3	2	51

Highest temperature: 33.6 °C
Lowest temperature: 16.1 °C

Highest Rainfall: 8 mm
Lowest Rainfall: 0 mm

TEMPERATURE AND RAINFALL GRAPH

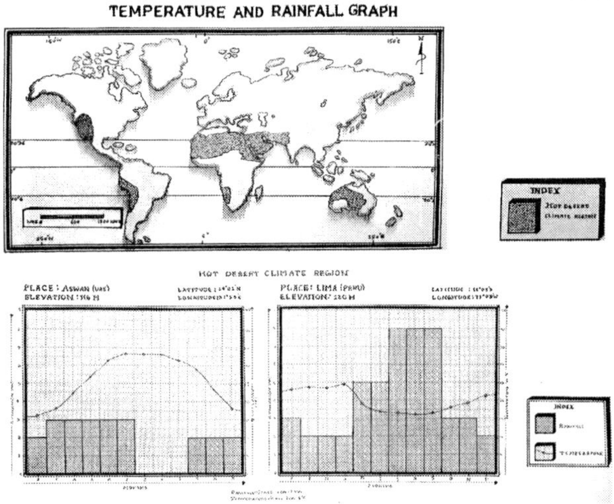

Interpretation

In Aswan, the climate is quite different from the climates of the equatorial or monsoonal regions. The highest temperature observed is 33.6°C and the lowest is 16.1°C. But the highest temperature was observed in the months of June, July, and August. It may be because of the negligible amount of rainfall. The amount of rainfall in these months is almost zero. In the other months, the amount of rainfall shows deficient trends. This is varied from 0 to 3 mm. Whereas Lima is situated in a hot climatic region so it receives rainfall with varying amounts from 2 mm to 8 mm in the year. The temperature remains almost consistent in all the months. But there is no month without rainfall.

From the above discussion, it can be concluded that the high temperature and negligible/low rainfall are the striking characteristics of the hot desert climatic regions.

Mediterranean Climate

It is situated between 30° to 45° N and S in the temperate climatic zone. This region bears a stark specialty than other regions with a distinct climatic and vegetation type. It mainly encompasses areas around the Mediterranean Sea but in the southern hemisphere, there are few regions that resemble such climatic types.

Table 7: *Marsais, (France)- 43°10'N, 05°23'E; Elevation- 79·6 m*

Month	*January*	*February*	*March*	*April*	*May*	*June*	*July*	*August*	*September*	*October*	*November*	*December*	*Mean/ Annual*
Temp°C	7.5	6.9	8.1	10.0	13.1	16.7	20.0	22.2	21.9	19.2	14.7	10.6	14.2
Rainfall in mm	48	38	46	51	48	25	15	23	66	94	79	56	589

Table 8: *Santiago (Chile)- 33°27'S, 70°42'W; Elevation- 522 m*

Month	*January*	*February*	*March*	*April*	*May*	*June*	*July*	*August*	*September*	*October*	*November*	*December*	*Mean/ Annual*
Temp°C	20.5	20.0	18.0	15.3	11.4	8.6	8.9	10.3	12.2	14.7	17.2	23.3	15.0
Rainfall in mm	3	3	5	13	63	84	76	56	30	15	8	5	361

Highest temperature: 23.3 ° C
Lowest temperature: 6.9 ° C

Highest Rainfall: 84 mm
Lowest Rainfall: 3 mm

TEMPERATURE AND RAINFALL GRAPH

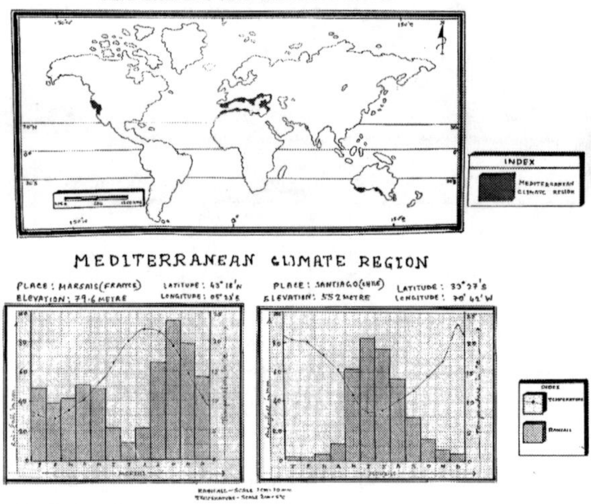

Interpretation

From the temperature values of Marsais (France), it is observed that the temperatures are highest in the months of August- September & lowest in January–February, and December, whereas, in Santiago (Chile), in the southern hemisphere lowest recorded temperature is in the June–July.

Whereas, the rainfall recordings are highest in the October–November, and December months and the lowest in July–August in Marsais in the Northern hemisphere. In the Southern Hemisphere, the opposite situation occurs, the highest rainfall is in June–July, and August months like in the Tropical Climatic region. But the stirring difference is that the temperature is lowest in these months. Moreover, rainfall is lowest in the month of January–February, and December.

So, from the above discussion, it can be said that the striking characteristics of winter rainfall primarily indicate the Mediterranean climatic type.

4

Drawing of Isolines

Isotherms, Isobars, Isohyets and Contours

Isolines

Isolines are lines of equal values [Iso = equal; lines = lines = isolines]. They mark the stage or state of uniformity. These lines of great importance are drawn or constructed to show the uniformity in the distribution of temperature (air), pressure (air), rainfall, relief (features) and other geographical aspects. Major isolines which are generally used are (i) isotherms, (ii) isobars, (iii) isohyets and (iv) contours.

Method of Drawing Isolines

Isolines are drawn by calculations based on specific points of interaction. The specific points are available from data collected from weather stations or survey offices. Weather offices supply data on air temperature, air pressure and rainfall of different stations or places. Specific spot heights or heights of the land are collected by means of land surveying or from survey offices. These data are plotted on a piece of paper or on maps. These maps or plotted sheets of paper are given to draw isolines.

The main problem in drawing isolines is finding out the location of the position of the said lines to their actual site. It can be achieved by following the *method of interpolation.*

Finding out the Position of Isoline

Interpolation is a method by which the position of a point is determined from the interaction of two other points located nearby.

Suppose, a point is to be determined between points A and B. The value of A is 80 and the value of B is 120. Now finding out the value of C to be located in between the two points. By the simple mathematical method, it can easily be determined (80 + 120) ÷ 2 = 100. Therefore, the value of C is 100. In the same manner, an iso-point can be easily determined.

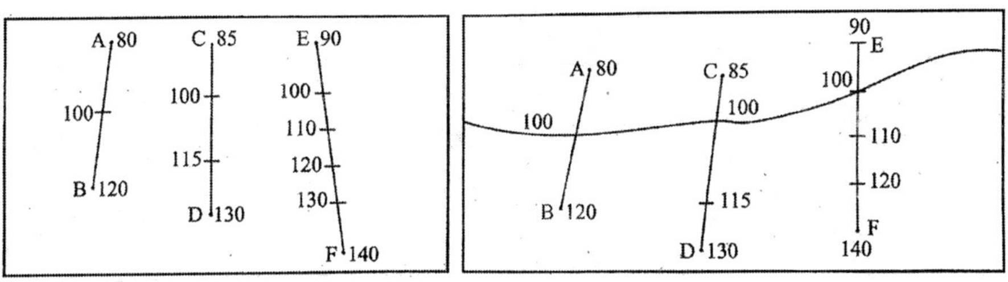

Figure 1: *Locating the exact iso-point (100) between two points of different values*

A graphical method of finding out iso-points and drawing out of isoline is shown in Figure 1.

Problem: To draw the given isolines on the practice paper with 50, 100 and 150 in Figure 2.

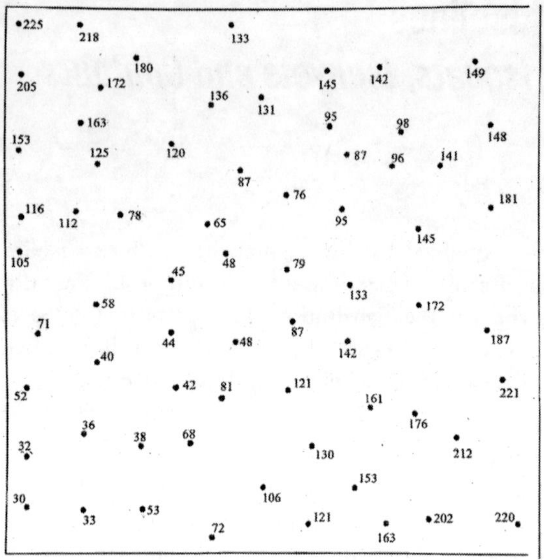

Figure 2: *Values to Draw Isolines (for practice)*

Isotherms

It is known that the distribution of temperature is not equal all over the earth's surface. The places which record equal temperatures in a particular season are joined by an imaginary line on the map which is known as the Isotherm. Isotherm means (Iso = equal, therm = temperature) equal temperature.

Definition

Isotherms are imaginary lines drawn on a given map joining places of the same average temperature for a specified period, supposing them to be at sea level (reduced to sea level).

Method of Drawing Isotherms

The following steps are:

(i) Conversion of temperature to the sea level location: All places on the earth's surface are not situated at the same sea level. Some places are adjacent to the coastal part, lying almost at sea level, while others are located at higher altitudes on mountains, hills and, plateaus. For their altitudes, differences in temperature occur. If isotherms are drawn according to the temperature of those places at different altitudes, a number of complications may arise for a simple interpretation.

To avoid such complications and eliminate the effects of altitudes, temperatures are reduced to sea-level temperature i.e., as if all the places are located at the same sea level. Thus, isotherms are drawn with the temperatures which are made reduced to sea level. At every 100 metres rise (ascent), the temperature reduces at the rate of 0.6°C (1°F at every 300 feet height).

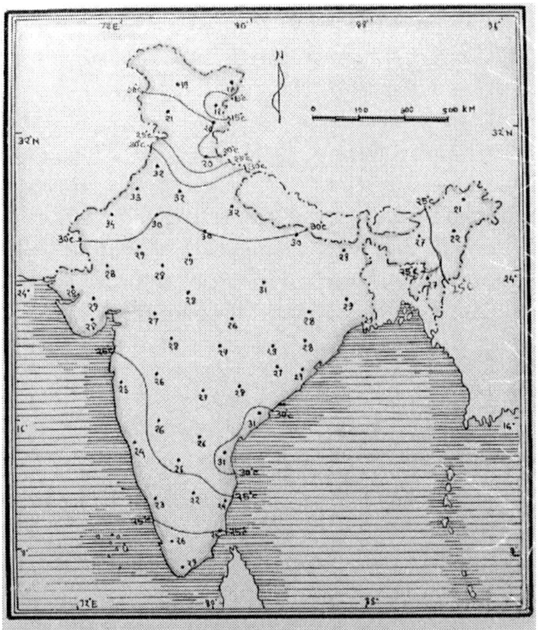

Isotherms drawn for the summer season are 15°C, 20°C, 25°C & 30°C

Map No 1: *Isotherms showing Summer Temperatures over India*

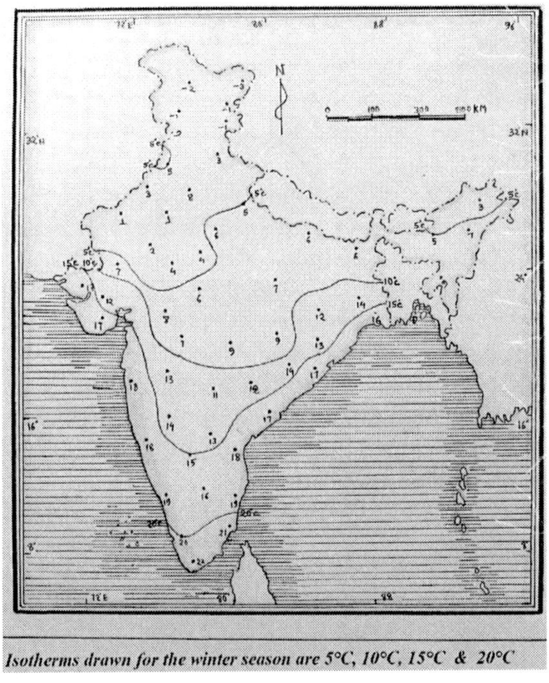

Isotherms drawn for the winter season are 5°C, 10°C, 15°C & 20°C

Map No 2: *Isotherms showing Winter Temperatures over India*

Accordingly, a reduction in temperature is to be made. If a place is at an altitude of 600 metres and its actual temperature is 20°C, its reduced temperature at sea level would be 20°C + 6°C = 26°C (because temperature decreases at the rate of 0.6°C for every 100 metres ascent).

(ii) Plotting the temperature values or data on a map: Thus, having prepared the temperature data reduced at sea level, students should proceed for plotting them on a map according to their placing of respective stations.

(iii) Drawing of isothermal lines: It is a simple act of joining the points of stations having equal temperatures shown on the map. In order to do this, the students need to follow the same method of drawing contour lines.

Precautions in Drawing Isotherm lines

The students need to:

1. Examine the location and distribution of the points, marking the values of temperature of the different places given on a map.
2. Join the places with a line having the same temperature (according to the method following for drawing contours).
3. Observe at first and then draw carefully so that a line does not pass through or cross any other line, having a different temperature.
4. Remember that an isotherm line never crosses another isotherm line.

Problem: Map No. 1 shows the summer temperature in India. Four isotherms are to be drawn at 5°C intervals like 15°C, 20°C, 25°C and 30°C.

Interpretation: These are quantitative areal maps where quantities are indicated by lines of equal value known by a multiplicity of such terms as isopleth (*iso* meaning equal and, *pleth* meaning lines). Wright (1944) proposed that isograms be used for all lines of quantity with two subdivisions- the isometric lines (*metron* meaning measurement) that represents a constant value or intensity pertaining to certain areas through which it passes.

Isotherms are lines which join places of equal temperature conditions. From the value of temperature of Map nos. 1 and 2 it becomes very clear for, the identification of summer maps or winter maps. The temperatures during summer are very high in contrast to the winter temperatures which are low. The most heated regions during the summer are north-west part of India in the states of Punjab, Rajasthan, and Delhi and more or less, all over central India, the temperature conditions remain quite high. By closely observing the isotherms drawn in the map of India showing winter conditions, it becomes clear that the lowest temperatures are recorded in the northern part of India in the states like Jammu and Kashmir, Delhi, Punjab, Rajasthan and, Haryana. The southern part of India remains comparatively warm because of the presence of the Bay of Bengal, Indian Ocean and the Arabian Sea, and also its southerly position in terms of latitudes.

Isobars

It is known that the distribution of pressure is not equal all over the earth's surface. The places on the map which record equal pressure in a particular season are joined by imaginary lines known as the Isobars. The instrument used to measure atmospheric pressure is called the barometer. Isobars mean (Iso-equal, bar-barometric pressure). These lines are on a weather map of constant barometric pressure drawn on a given reference surface. The isobaric pattern on a constant-height surface is extremely helpful in weather forecasting as temperature and wind are in close association with pressure conditions. Regions of low pressure at sea level tend to be areas of bad weather, especially in winter.

Definition

Isobars are imaginary lines drawn on a given weather map joining places of equal atmospheric pressure for a specified period, taking them to be at reduced sea level. A map with the isobar marked 1004 represents an area of high pressure, while the isobar marked 976 represents an area of low pressure.

Features of Isobars

1. The isobars are drawn in a similar manner to the isolines.
2. The numbers measure the atmospheric pressure in millibars.
3. Usually, isobars are drawn at intervals of two or four millibars (one thousandth of a bar).
4. The closer the isobars are together the pressure gradient is steeper and it is windier in the region.
5. If the lowest number is in the middle circle, this is a low pressure or depression which may render wind and rain.
6. If the highest number is in the middle circle, it is a high-pressure or anticyclone which results in dry and sunny weather conditions.

Problem: Map No. 3 shows the overall pressure conditions in India. Three isobars are to be drawn like 996 mb, 1000 mb, and 1004 mb.

Interpretation: The map depicts general atmospheric pressure conditions over Indian territory. The general trend of the isobars shows low-pressure conditions towards the interior locations and relatively high-pressure areas adjacent to the coast except in the

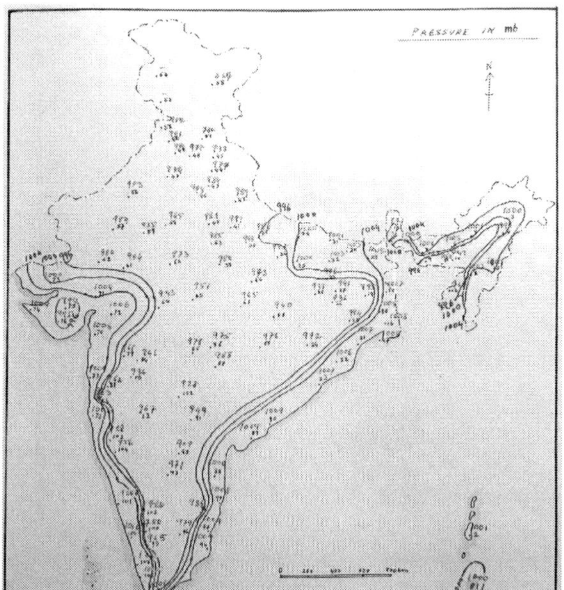

Map No 3: *Isobar Map of India*

north-eastern region where the wedge of high pressure is towards north-east. Moreover, the islands also exhibit an increasing trend of pressure. It exhibits the period which is at end of summer when low-pressure conditions exist on the interior of the landmass and high over the adjacent outer regions.

Isohyets

The isohyets maps are the quantitative maps which are indicated by the lines of equal value. The isohyets maps represent the constant value or intensity pertaining to certain area through which it passes. These isohyets are imaginary lines which join the places of equal rainfall area.

From the values of rainfall, it becomes very easier to the identification of distribution or trend of rainfall during summer or winter seasons. In these tropical regions, the rainfall is very high during the monsoon season and less during the winter season. The major rainfall occurs from mid-June to mid-September in West Bengal, Uttar Pradesh, Odisha, Punjab, Meghalaya, Maharashtra etc. Tamil Nadu region experiences rainfall twice a year i.e., summer monsoon (South-west monsoonal winds) and winter (north-east monsoonal winds) seasons. The amount of rainfall decreases from east to north-west ward. There are few rain shadow regions situated in India like one situated in south-eastern coastal India and another in the north-east region.

Problem: Map No. 4 shows the overall rainfall trend in India. Four isohyets are to be drawn like 60 cm,120 cm,180 cm, and 240 cm.

Interpretation: The map depicts the rainfall trend over the Indian subcontinent. The isohyets exhibit a decreasing trend of rainfall as one moves from the south, south-eastern regions towards the western and north-western regions. Moreover, the amount

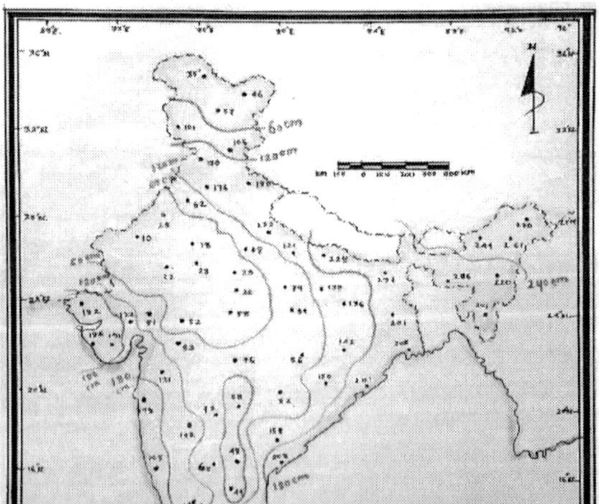

Isohyets drawn for the rainy season are 60cm, 120cm, 180cm & 240cm

Map No 4: *Isohyets showing Rainfall over India*

of rainfall is also low towards northern India as precipitation may be in the form of snow. Comparatively, high rainfall is recorded in north-eastern India. The isohyet map shows the temperature, pressure and wind conditions over the area.

Contours

Land areas on the earth's surface rise at different levels and heights above sea level. Hence, they have different altitudes. By their different elevations, they form different types of landforms, such as mountains, hills, valleys, plateaus, and plains. Land slope that rises from the sea level is known as altitude. In fact, altitude is the vertical distance above mean sea level, measured in feet or metres.

Definition

A contour may be defined as an imaginary line showing the same height or altitude on a map. Thus, the meandering lines which show the altitude of land on a map are called contours. On a map, land-heights are usually joined to draw a series of contour lines. A map showing the height of land by contours is called a *contour map*. Contour lines are lines showing the same height on the map. These are not straight lines, but are zigzag, winding or serpentine in nature. Values of their altitudes are marked within the line at definite intervals.

During surveying, spot heights are noted in the survey book, marking points with their respective heights (values). With the help of these heights, contours are drawn on a map according to certain intervals, such as 10, 20, 25, 50, 100, 200, 250 (metres or feet) and so on, as required or surveyed.

Method of Drawing Contours

1. Careful observation, examination and concentration of the mind in drawing contour lines at the definite interval (as directed) inf the location with spot heights are essential.
2. Drawing from the lower contour lines of the lowest spot heights towards the higher contours of the highest spot heights is required to avoid complications.
3. It should be noted that a contour never ends in the middle of a map. It starts from one end and then terminates at last on the other end or starts from one point and then goes on around the different directions and meets again, at last, forming a circular or elongated shape.
4. A contour never cuts away or crosses another contour. Sometimes, it may touch/merge just another contour to show the steepest gradient as in the case of waterfalls or high/steep cliffs/mountains.
5. After drawing the contours, it is advisable to write their respective values in numerical form. Values should be given in writing either within the line or at the ends. In the case, of circular or semicircular contours, values should be given in writing within the lines.

The most careful issue in drawing a contour line is determining the location of points (in between spot heights) which are required but not given as spot heights or there is no marking of them. But in drawing, this line is needed to find out their position in between two spot heights. The points are to be found out from the positions of other spot heights. Thus, it is a matter of finding out their positions (Refer Figure 1)

According to the given map, a judicious choice of scale is to be made. To avoid miscalculations and misconceptions arbitrary markings need to be avoided. All the points are to be found out by proper calculation according to the given scale through the following method of interpolation of contour. In fact, interpolation is a method of finding out a position in between two poles *i.e.*, two points (spot heights) as in the case of contour drawing in which spot heights are given as points showing heights or altitudes. In determining or finding out a position of a point, according to the scale between two spot heights, the calculation should be made accurately and contours are then to be drawn through these points with the help of other spot heights.

By measuring the linear distance between the two points of given values, the spot value of the contour directed can be identified by calculations, placed and then drawn between them.

In the same way, a contour of 600 metres is to be drawn. In this case, at first, an examination of the places through which this line may enter is needed. The 600m contour line of this area (Figure 3) may proceed through a suitable point between 670 m and 550 m spot heights. The necessity is there to find out the 600 m. By minutely calculating the actual linear distance and distributing the difference (670-550) m on the line the exact point of drawing the contour of 600 m can be plotted.

Example 1: Examine carefully how are contour lines of 100, 200, 300 and 400 metres drawn on the given diagram in Figure 3 and observe their positions.

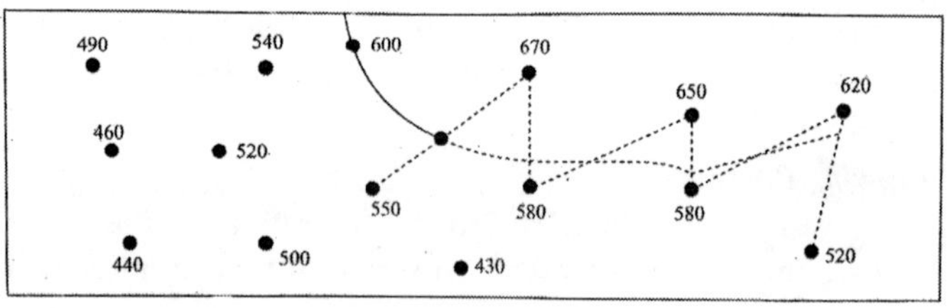

Figure 3.

Problem: To draw contours on the given sheet of paper with 100, 200, 300 and 400 metres in the given Figure 4.

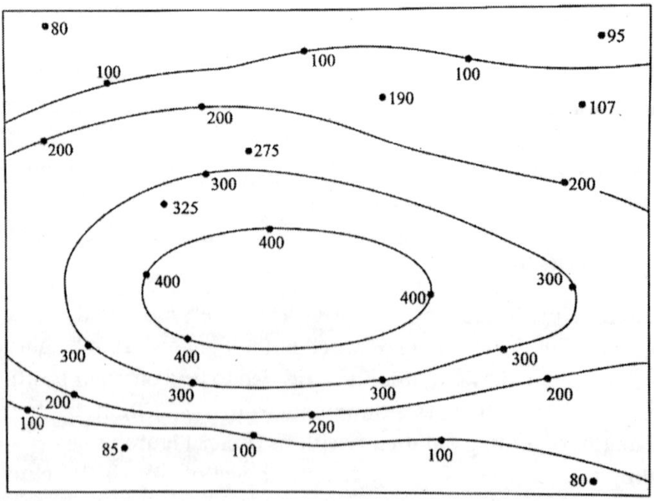

Figure 4.

5

Meteorological Instruments

To observe and record weather conditions and various elements of weather, different types of meteorological instruments are needed. The following are the weather instruments—

1. Six's Minimum and Maximum Thermometer
2. Hygrometer and
3. Fortin's Barometer.

Six's Minimum and Maximum Thermometer

- *Air Temperature Recording*: The warmth or coldness of the air is called air temperature. The sun is the source of all heat and energy in the atmosphere as well as in the earth. The atmosphere *i.e.* the air is not directly heated by the sun. At first, the surface of the earth is heated up by the sun. Then the heat of the earth's surface is radiated to the layers of the atmosphere by the processes of conduction, convection and radiation.
- *Measurement of Temperature*: The temperature or the degree of the heat of air may be high (*i.e.* hot) or it can be very low (*i.e.* cold). It is measured with the help of an instrument known as Thermometer.

A Six's Minimum and Maximum Thermometer is employed to know the maximum and minimum temperatures of the day. A self-recording instrument known as a Thermograph is used for measuring temperature.

There are two common scales used for graduating a thermometer: [i] Centigrade Scale, [ii] Fahrenheit Scale.

The temperature of melting-point of ice is recorded as ± 0° in the Centigrade Scale and 32° in the Fahrenheit Scale. The temperature, at which water boils, is taken as 100° in the Centigrade Scale and 212° in the Fahrenheit Scale. It is, therefore, there are 100 divisions between the melting point of ice and the boiling point of water in the Centigrade Scale. Whereas in the Fahrenheit Scale there are 180 divisions between these two points.

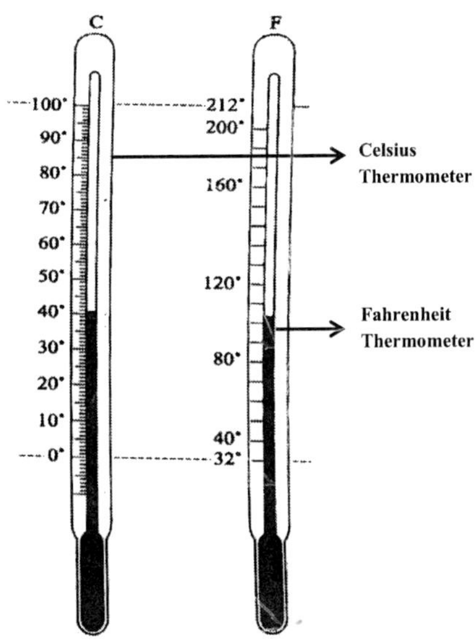

Figure 1: *Celsius and Fahrenheit Thermometer*

1. *Conversion Method from °F to °C Formula:*

   $\because$ Temperature in °C = Temperature in (°F-32) $\times \frac{5}{9}$

   $\therefore$ If the temperature of a place is 90°F, the temperature in °C

   = Temp. (90-32) $\times \frac{5}{9}$ = 32.2°C

2. *Conversion Method from °C to °F Formula:*

   $\because$ Temperature in °F = Temperature in (°C $\times \frac{9}{5}$) + 32

   $\therefore$ If the temperature of a place is 17°C, the temperature in °F

   = Temp. (17 $\times \frac{9}{5}$) + 32

   = 30.6 + 32 = 62.6°C

It is a thermometer designed to record both the maximum and minimum temperature of a day. The indexes can be made up and down with help of a small horse-shoe magnet to set them on the mercury surface.

At first, the thermometer has to be set before use. It can be done by moving each index up or down with the help of a magnet until they reach the mercury surface just in contact.

The reading at the lower ends of the indexes I_1 and I_2 register the minimum and maximum temperature of the day respectively. Then indexes are set again on the mercury surface by the magnet for reading of the next day.

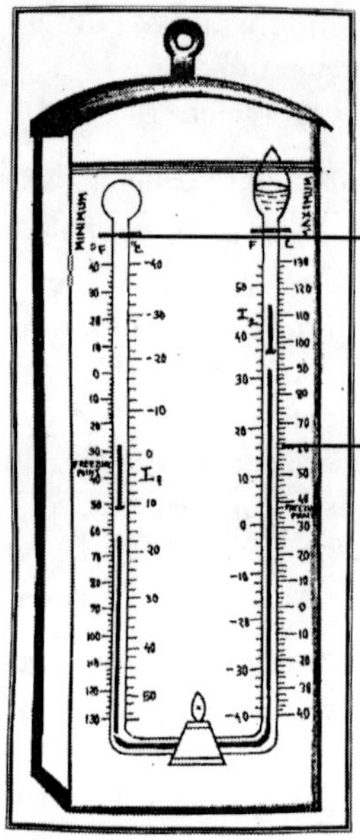

Figure 2: *Six's Minimum and Maximum Thermometer*

- *Temperature Recording Method*

The following table shows how to record temperatures (°C):

Table 1: *Diurnal Temperature Recording*

Date: 06.11.2022 and 07.11.2022 Place: Gandhinagar, Gujarat Time: 8.30 a.m.

Date	*Maximum Temperature*	*Minimum Temperature*	*Difference Temperature*	*Average Temperature*
05.11.2022	36°C	17°C	19°C	26.5°C
06.11.2022	36°C	18°C	18°C	27°C

Record taken by: India Meteorological Department, Meteorological Centre, Ahmedabad 382475

- *Terms of Temperature Data Calculation:*

Temperature varies from day to day and from time to time. The highest temperature of the day is called daily maximum temperature. The air gets maximum temperature in the afternoon at about 2 p.m. It is for the afternoon is hotter than noon-time. The lowest temperature of the day is called the daily minimum temperature. The minimum temperature is found at about 4 a.m. in the early morning, when the earth becomes cool due to rapid radiation.

1. *Mean daily temperature*: The average of maximum and minimum temperature recorded during a day is known as mean daily temperature. To get the mean daily temperature, the total of maximum temperature and minimum temperature is to be divided by 2.
   Example: On 5th November 2022 the maximum temperature of Gandhinagar was 36°C and minimum temperature was 17°C. Therefore the daily mean temperature on 5th November 2022 was:
   $= \frac{\text{Maximum} + \text{Minimum}}{2} = \frac{36+17}{2} = 26.5°\text{C}$
2. *Diurnal or daily range of temperature*: The difference between maximum and minimum temperature recorded in a particular day at a place is called Diurnal or Daily range of temperature.
   Example: If the maximum temperature of a place is 36°C and minimum temperature was 17°C, the diurnal or daily range of temperature is: (36° - 17°)C = 19°C
3. *Mean monthly temperature:* The average temperature of a month is known as mean monthly temperature. To get it, the total of daily temperature for all the days of a month is divided by the number of days of that month.
4. *Mean annual temperature*: The average of the monthly mean temperature of a year is known as the mean annual temperature. To get it, the total of mean monthly temperature of all the months of a year is divided by the number of months *i.e.* 12 months.
5. *Annual range of temperature*: The difference between the mean temperature of the hottest month and the mean temperature of the coldest month is called the mean annual range of temperature.
   Example*:* If the mean temperature of the hottest month is 35°C and the mean temperature of the coldest month is 17°C, the annual range of temperature is: (35°C - 17°C) = 18°C.

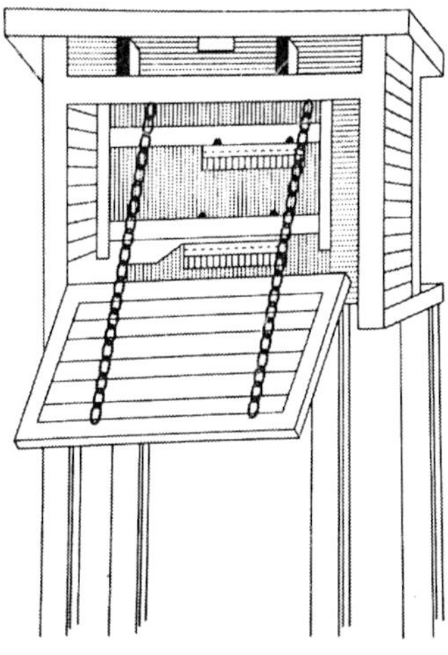

Figure 3: *Stevenson Screen*

Stevenson Screen

To obtain correct reading, the weather stations use the Stevenson Screen. It is manufactured in such a manner that the shade temperature of the air can be measured.

The Stevenson Screen consists of a wooden box, painted white and kept to stand at a height of about a metre placed away from building and trees. The four sides of the box are louvered to allow the free entry of the air. The roof has an intervening air space to prevent direct sunshine and radiant heat. One side is hinged so as to serve as a door. The instruments are kept inside the box supported on a frame. The instruments which are kept in the box are: 1. The Maximum and Minimum Thermometer, 2. Ordinary Temperature Thermometer and 3. Dry and Wet Bulb Thermometer (Hygrometer) etc.

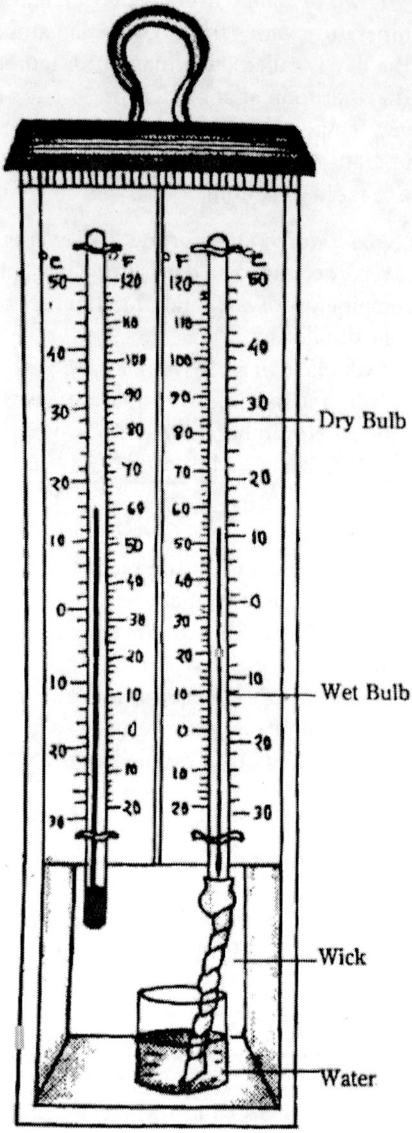

Figure 4: *Dry and Wet Bulb Thermometer*

Hygrometer or Dry and Wet Bulb Thermometer

To measure the humidity of air, an instrument called Hygrometer is employed. It consists of two thermometers: a dry bulb thermometer and a wet bulb thermometer. The wet bulb thermometer is covered with a piece of muslin or cotton cloth. It is kept constantly moist by a wick dipping into a small vessel containing water.

The dry bulb thermometer records the temperature of the atmosphere. The wet bulb thermometer measures a lower temperature due to cooling effect caused by the evaporating of water from the vessel. The rate of evaporation depends upon the relative humidity of the surrounding air and for this, fluctuation in temperature occurs in the wet bulb reading.

If the air remains saturated, evaporation would be little and the thermometer will record almost the same reading. When the relative humidity will be low, more evaporation will occur and the difference in temperature between the two thermometers will be high. (i) a small difference in temperature indicates high relative humidity; (ii) a large difference denotes low relative humidity and (iii) same reading in the two thermometers gives the indication of saturation. Actual amount of relative humidity is calculated by the difference of temperature from Standardise Table.

A table is given here to show how the amount of Relative Humidity varies at different temperatures:

Table 2: *Relative Humidity Table*

Dry Bulb Reading	*Depression of Wet Bulb (°C)*																		
(°C)	*0.5*	*1.0*	*1.5*	*2.0*	*2.5*	*3.0*	*3.5*	*4.0*	*4.5*	*5.0*	*5.5*	*6.0*	*6.5*	*7.0*	*8.0*	*9.0*	*10.0*	*11.0*	*12.0*
-5	85	71	58	44	30	17	4												
4	87	73	61	48	35	53	12												
3	88	75	64	52	40	29	18	7											
2	89	77	67	56	45	34	24	14	3										
1	90	79	69	59	49	39	30	20	10	1									
0	90	81	71	61	52	44	34	25	16	7									
+1	90	81	73	64	55	47	38	29	20	13	4								
2	91	82	73	64	57	49	41	33	24	17	9	1							
3	91	83	74	65	57	49	43	35	28	29	14	7							
4	92	83	75	67	59	51	43	35	32	25	18	11	4						
5	92	84	76	68	61	53	46	38	31	24	21	15	8	2					
6	92	85	77	70	62	55	48	41	34	27	20	14	12	6					
7	93	85	78	71	64	57	50	44	37	30	24	17	11	5					
8	93	86	79	72	65	59	52	46	39	33	27	21	15	9					
9	93	86	80	73	67	60	54	48	42	36	30	24	18	12	12	3			
10	93	87	81	74	68	62	55	50	44	38	33	27	21	16	15	6			
11	94	87	81	75	69	63	68	52	46	41	45	35	30	24	19				
12	94	88	82	76	70	65	59	54	48	43	37	32	27	22	21	12	4		
13	94	88	83	77	71	66	60	55	50	45	40	35	30	25					
14	94	89	83	78	72	67	62	57	52	47	42	37	32	27					
15	94	89	84	78	73	68	63	58	53	48	42	39	34	30	28	20	13	4	
16	95	89	84	79	74	69	64	59	55	50	43	41	37	32					
17	95	90	83	80	75	70	65	61	56	52	47	43	39	34					
18	95	90	85	80	76	71	66	62	57	53	49	45	40	36	35	27	20	13	6
19	95	90	86	89	76	72	67	63	59	54	50	46	42	38					
20	95	91	86	81	77	73	68	64	60	56	52	48	44	40	38	30	24		
21	95	91	86	82	78	73	69	65	61	57	53	49	45	42	39	32	26	19	13
22	95	81	87	82	78	74	70	66	62	58	54	50	47	43					
23	96	91	87	83	79	75	71	67	63	59	55	52	48	45					
24	96	91	87	83	79	75	71	68	64	60	57	53	49	46	43	37	31	26	21
25	96	92	88	84	80	76	72	68	65	61	58	54	51	47	44	38	33		
26	96	92	88	84	80	76	73	69	66	62	59	55	52	49					
27	96	92	88	84	81	77	73	70	66	63	59	56	53	50	47	41	36	31	26
28	96	92	88	84	81	77	73	71	63	65	61	58	55	53					
29	96	93	90	86	82	78	74	72	68	66	63	60	57	54					
30	96	93	90	86	82	79	76	73	70	67	64	61	58	55	50	44	39	35	30
33	96	93	90	86	83	80	77	74	71	68	66	63	60	57	52	47	42	37	33
35	96	93	90	87	83	80	77	74	71	70	66	64	60	59	54	50	44		
36	97	93	90	87	84	81	78	75	72	70		64		59	54	50	45	41	36
39	97	94	91	88	95	82	79	76	74	71		66		61	56	52	47	43	39
40		94		88						72		66		61	56	52	48		

- *Determination of Relative Humidity:*

In order to determine the Relative Humidity of a place, we have to record the temperature readings of the dry bulb and wet bulb. Then it is needed to calculate the difference of these two readings. Now from the table the relative humidity of a place can be determined.

Example 1

Temperature of dry bulb reading is 11°C
Temperature of wet bulb reading is 10.5°C
∴ Difference of the two readings = 0.5°C (Depression of the wet bulb)
∴ From the table it is found that the relative humidity of the place is 94% (Humidity is high)

Example 2

∵ Temperature of dry bulb reading is 11°C, and wet bulb reading is 4°C
∴ The difference of the reading is 7°C
∴ It is found from the table that the relative humidity of the place is 19% (Very dry).

Table 3: *Table for Determination of Relative Humidity*

Date: ________________ Place: ________________ Time: ________________

No. of Reading (N)	*Temperature of Dry Bulb Reading (°C) (a)*	*Temperature of Wet Bulb Reading (°C) (b)*	*Difference (a—b) is the Depression of the Wet Bulb*	*Mean Difference*	*Relative Humidity (RH) in % to be Determined on the Basis of Given Table*	*Remarks if Any*
1						
2						
3						

Records taken by:

Fortin's Barometer

A barometer means an instrument for measuring atmospheric pressure. Fortin's barometer is an improved barometer. It is light and portable. It consists of a vertical glass tube filled with mercury. It is closed in the upper end. The open end of this tube is inverted and immersed into a cistern of mercury. This cistern has a flexible bottom which can be adjusted by a screw. With the help of the screw, the zero level is adjusted. Above this level, the column of mercury is measured which indicates the air pressure. An ivory index is fixed at the zero level.

The glass-tube is fixed within a brass-tube on which a graduated scale is inscribed. The mercury level in the tube can be seen through a slit. There is a sliding Vernier (scale) on it for ensuring accurate reading. Vernier can easily be adjusted by screw. There is an attached thermometer on the barometer. It helps to make a temperature correction.

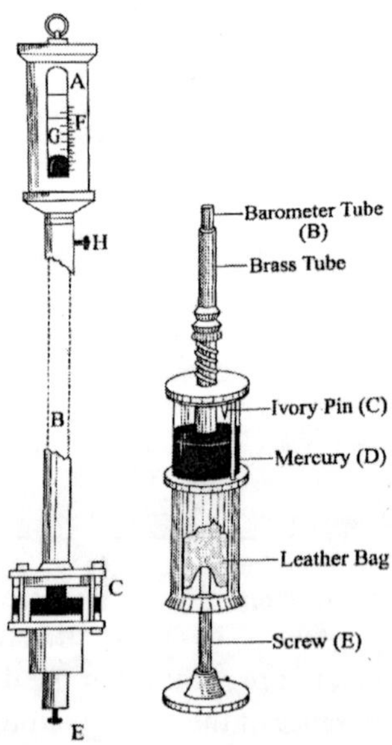

Figure 5: *Fortin's Barometer*

- *Reading of Barometer*

In order to have a reading of Barometer, it is needed to see whether the ivory pointer C just touches the mercury surface in the reservoir D or not. If it does not, then adjustment is needed. Then there is gap between the tip of the pointer and its image formed by the shining mercury surface. If, on the other hand, the tip dips into the mercury, a dimple will be formed there. Hence adjusting the mercury surface by the screw E so that the ivory pointer C just touches it as needed. This happens when the tip of the pointer just coincides with its image in the mercury. It is known as zero-adjustment, because by this adjustment the surface of mercury in the reservoir comes in level with zero-mark of the scale F.

Then, by turning the screw H, raise the Vernier G to a certain height above the level of mercury in the tube and then slowly lower it.

Now with the eye at the level of the convex surface of the mercury meniscus and slowly sliding down the Vernier till its lower edge is tangential to the convex mercury meniscus is needed. To do this adjustment correctly, a milk-white plate is attached on the wooden board behind the Vernier. As soon as the Vernier is adjusted correctly, the milk-white plate (being obstructed completely by the Vernier) will disappear from there. It is known as Vernier adjustment.

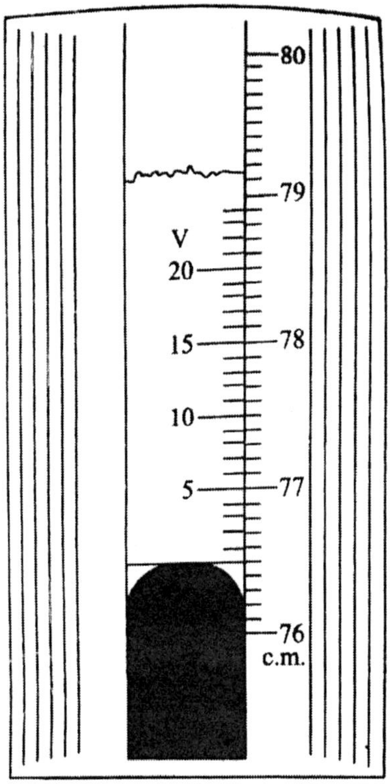

Figure 6: *A Portion of the Barometer showing the Vernier Scale and Reading*

Then reading can be made. Read the Main scale and the Vernier scale avoiding parallax. Repeat the observations at least thrice; now the mean height of mercury column is found. The temperature from the thermometer attached with the barometer can also be noted.

- *Determination of Vernier Constant and Reading*

It is necessary to determine the Vernier Constant (VC); because it is different instrument. Suppose, if there are 10 small divisions in main scale, and if the 10 small divisions of Vernier scale are equal to 9 small divisions of main scale, then Vernier Constant would be:

$$1 \text{ divisions of Vernier Scale} = \frac{9}{10} \text{ divisions of main scale}$$

$$\therefore \text{VC} = 1 - \frac{9}{10} \times 0.1 \text{ (small division of main scale)}$$

$$= \frac{10-9}{10} \times 0.1$$

$$= \frac{1}{10} \times 0.1$$

$$= \frac{1}{100} = 0.01\text{cm}$$

After determining the VC, proceeding to have the reading of main scale is needed. Suppose, the reading of the main scale is 75.6cm; but the height of the mercury is somewhat higher, then need to determine it from Vernier scale. Here in this case, the Vernier scale touches the main scale at 13. Hence the Vernier reading would be 13×0.01 = 0.13. It should now be added to the main scale reading. Therefore, the atmospheric pressure of this is 75.6 + 0.13 = 75.73cm or 757.3mm

To make accurate reading, taking reading thrice is needed and the average is considered. It is shown in the following table:

Table 4: *Calculation of Atmospheric Pressure*

Calculation Table To Measure The Atmospheric Pressure

Date:_______________ Place:_______________ Time:_______________

No. of Reading	*Main Scale Reading (a)*	*Vernier Constant*	*Vernier Reading*	*Vernier Reading × V.C (b)*	*Total Reading (a + b)*	*Mean Height (cm)*	*Mean Temperature*
1	75.6		13	0.01×13 = 0.13	75.6+0.13=75.73		
2	75.6	0.01	14	0.01×14 = 0.14	75.6+0.14=75.74	75.73	29°C
3	75.6		13	0.01×13 = 0.13	75.6+0.13=75.73		

Records taken by:

∴ At 29°C temperature, atmospheric pressure is:

$$\frac{75.73 + 75.74 + 75.73}{3}$$

$$= \frac{227.20}{3} = 75.73 \text{ cm or } 757.3 \text{ mm}$$

- *Method of Transferring mm (millimetre) to mb (miillibar)*

Given, 1000 dynes = 1 millibar

And 1000 millibars = 1 bar

1 bar = 29.53″ or 76 cm or 760 mm

∵ 760 mm = 1 bar

$$\therefore 1 \text{ mm} = \frac{1}{760} \text{ bar}$$

$$\therefore 757.3 \text{ mm} = \frac{1 \times 757.3}{760} = \frac{757.3}{760}$$

$$= 0.996 \text{ bar} = 0.996 \times 1000 = 996 \text{ mb}$$

This should be noted that the atmospheric pressure has an inverse relation with atmospheric temperature. The higher the temperature, the lower is the pressure. The general rule is that a rising thermometer shows a falling barometer, whereas a falling thermometer shows a rising barometer.

The atmospheric pressure is shown in the map by means of millibar (mb). It is shown in the weather map as isobars. Pressure gradients determine the direction and velocity of wind.

Rain Gauge

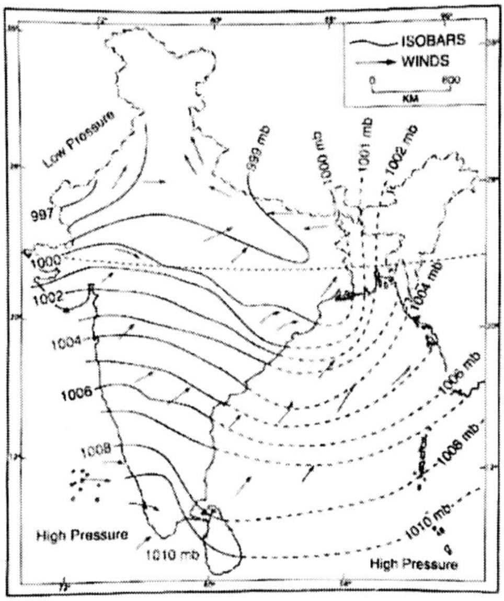

Figure 7: *Millibars are shown in the Weather Map of India*

The rainfall of a place is measured by an instrument called rain- gauge. It consists of 2 parts, a collecting can and a measuring glass. The collecting can A is placed in an upright cylinder opened at the top containing a funnel BCD which drains into the can below. Any rain which falls over the mouth of cylinder runs down in the can is collected. At fixed intervals, the water is emptied out and the amount is measured in the graduated glass E. Though the rainfall is measured in inches, India has adopted a metric unit and the unit of measurement is mm. When we say, the mean annual rainfall of a place is 50 mm, we mean that if all the rain that falls upon that particular place during a normal year collected without the least amount being lost by evaporation or flowing away. The amount of water would cover the area to the depth of 50 mm. the line joining places having same mean monthly or yearly rainfall is called isohyets.

Wind Vane

The instrument widely used for measuring wind direction is wind vane or weather cock. As wind direction is always blocked by trees and tall buildings wind-vanes or cocks need to be erected in an exposed position to get a five direction. It is made up of 2 parts – A

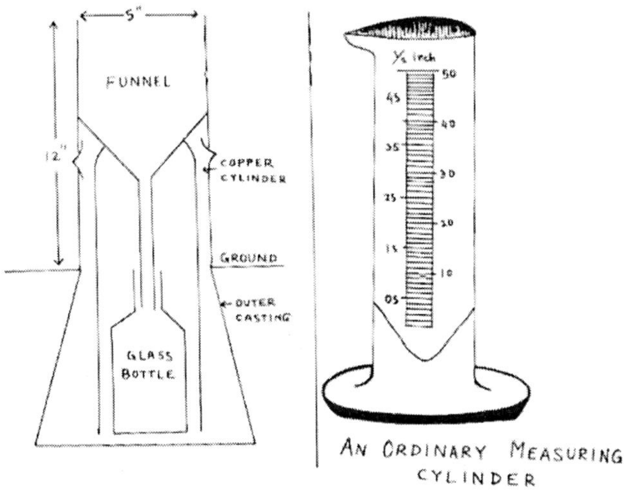

Figure 8: *Rain Gauge*

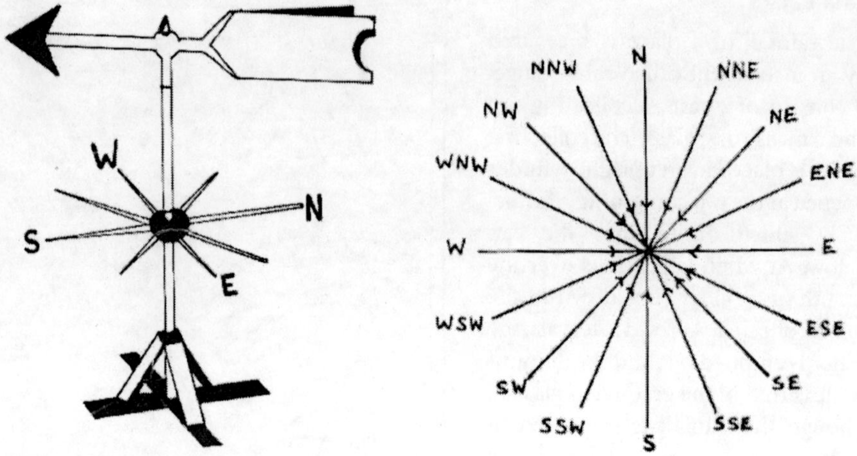

Figure 9: *Wind Vane and Wind Directions*

Table 5: *Table Showing Various Directions of Wind and Related Codes*

Code No.	*Description*
00	Calm
02	North–North East
05	North East
07	East–North East
09	East
11	East–South East
14	South East
16	South–South East
18	South
20	South–South West
23	South West
25	West–South West
27	West
29	West–North West
32	North West
34	North–North West
36	North
39	Variable

and B. A is one part and i.e., the arrow or vane on the top which is free to move with the prevailing wind. The other part (B) with the 4 compass points is stationary and shows in which direction is wind moving. Winds are always named from the direction they blow. A south-west wind is one that blows from the south-west.

6

Cartograms

Cartograms are the diagrammatic representation of numerical information and statistical data. The word 'carto' means diagram and 'gram' means making or composing; thus, it means 'diagram making'. According to the characteristics of the data it needs to be suitably represented by an appropriate graphical method and thereby meet the purpose of research and study. The authenticity of the data will lead to the accuracy of the graphical representation and interpretation.

Simple Bar Diagram

Simple Bar or Column diagrams consist of a series of columns or bars proportional in length in accordance to the represented quantities. If the width of the bars remains the same, they are called a one-dimensional diagram. When the data with only one component is available and each bar represents a single value, simple bar diagrams are drawn. A bar graph can help to identify the trend of data better than in a tabulated form.

Importance

- Comparison: Simple Bar Graph is quite an effective tool used for the purpose of convenient and quicker comparison among observations based on a statistical parameter.
- Convenience: It is easily understandable as information can be retrieved by observing the chart at a glance.
- Help in Decision Making: The bar graph can help to make a quick summary of a large set of data in a visual form, which may help to take any crucial decision.
- The trend of Data: A bar graph can help to identify the trend of data better than in a tabulated form.

Table 1: *The Calculation for Representing the Total Population of Different Districts of West Bengal, 2011*

Sl No.	*Districts of West Bengal*	*Total Population (m = in million)*	*Population Reduced to Scale (1 cm ≡ 3 million)*
1.	Darjeeling	1846823 = 1.84 m	0.61
2.	Jalpaiguri	3872846 = 3.87 m	1.29
3.	Cooch Behar	2819086 = 2.82 m	0.94
4.	Uttar Dinajpur	3007134 = 3.01 m	1.00
5.	Dakshin Dinajpur	1676276 = 1.68 m	0.56
6.	Malda	3988845 = 3.99 m	1.33
7.	Murshidabad	7103807 = 7.10 m	2.37
8.	Birbhum	3502404 = 3.50 m	1.17
9.	Burdwan	7717563 = 7.71 m	2.57
10.	Nadia	5167600 = 5.17 m	1.72
11.	North 24 Parganas	10009781 = 10.01 m	3.34
12.	Hooghly	5519145 = 5.52 m	1.84

Sl No.	Districts of West Bengal	Total Population (m = in million)	Population Reduced to Scale (1 cm ≡ 3 million)
13.	Bankura	3596674 = 3.60 m	1.20
14.	Purulia	2930115 = 2.93 m	0.98
15.	Howrah	4850029 = 4.85 m	1.62
16.	Kolkata	4496694 = 4.50 m	1.50
17.	South 24 Parganas	8161961 = 8.16 m	2.72
18.	Paschim Medinipur	5913457 = 5.91 m	1.97
19.	Purba Medinipur	5095875 = 5.09 m	1.70

Source: State Statistical Handbook, 2014, Bureau of Applied Economics and Statistics, Government of West Bengal

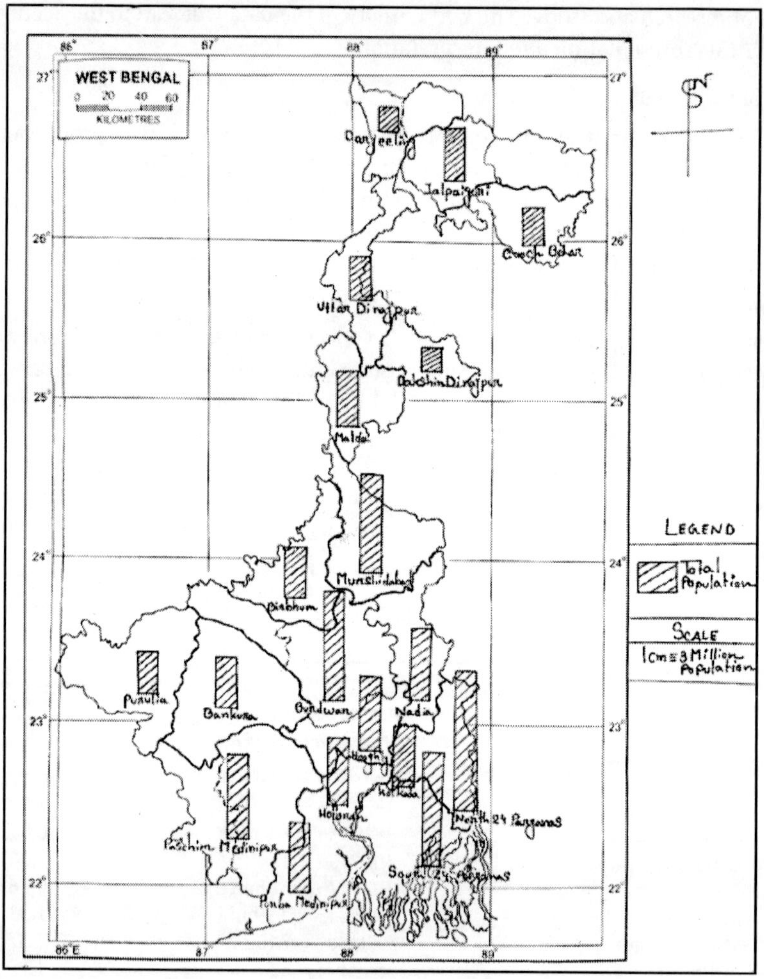

Figure 1: *Simple Bar Diagram showing the Total Population of Different Districts of West Bengal (2011)*

Source: Developed by the author.

Interpretation

According to the given data on the population of West Bengal (2011), it is seen that the simple bar graphs represent the distribution of the district-wise population of West Bengal.

From the map, it is seen that the highest populated district is North 24 Parganas (10.01 million). Bankura, Howrah, Hooghly, Purulia, Kolkata, Nadia, and Murshidabad consist of a medium range of population (4 to 6 million population). The least populated districts are Darjeeling, Cooch Behar, Malda, Jalpaiguri, Birbhum, etc. Dakshin Dinajpur has the lowest share of the population (1.68 million) among all the districts of West Bengal.

In short, it can be said that the total population of the northern districts is much lower than the southern districts of West Bengal. The northern districts consist of rocky and hilly terrain, infertile soil, and forested zones and this has resulted in lesser-developed infrastructure and other civic facilities in comparison to other regions. Thereby, these districts represent a low population. Whereas the southern districts possess fertile low-lying agricultural tracts, varied industrial plants, favourable transport and communication network, well-developed infrastructure, etc. This has resulted in the development of the tertiary sector of the economy and the generation of employment which has ultimately accentuated the growth of the population in these regions.

Multiple Bar Diagram

In a *multiple bar diagram or multiple bar graph*, two or more sets of interrelated data are represented using two sets of horizontal bars. Multiple bar diagram facilitates comparing or exhibiting the relationship between two or more two sets of data. The multiple bar graphs are drawn similar to simple bar graphs but in this case, for two bars two different shades, colors, or dots are used to represent two distinct sets of data.

The basic principle of drawing a multiple bar diagram is that the length of the bars is proportional to the value of the representable item. Depending on the nature of the data, multiple bar diagrams are drawn when different bars of proportionate lengths are drawn side by side to represent the constituents. Multiple bar graphs are also one-dimensional diagrams.

Importance

- Comparison: It can be applied to contrast two or more two sets of data.
- Establishing Relationship: Through the multiple bar diagram relationship between two or more phenomena inherited can be established.
- Information Retrieve: A single diagram provides information about two or more variables just side by side.
- Simplicity: It looks visually appealing and is very simple to understand.

Table 2: *The Calculation for Representing Male-Female Population of Different Districts of West Bengal (2011)*

		Total Population		*Population Reduced to scale (1 cm = 20,00000)*	
Sl No.	*Districts of West Bengal*	*Male*	*Female*	*Male*	*Female*
1.	Darjeeling	937259	909564	0.47	0.45
2.	Jalpaiguri	1983064	1889782	0.99	0.94
3.	Cooch Behar	1451542	1367544	0.73	0.68
4.	Uttar Dinajpur	1551066	1456068	0.78	0.73

Sl No.	Districts of West Bengal	Total Population		Population Reduced to scale (1 cm = 20,00000)	
		Male	Female	Male	Female
5.	Dakshin Dinajpur	857199	819077	0.43	0.41
6.	Malda	2051541	1937304	1.03	0.97
7.	Murshidabad	3627564	3476243	1.81	1.74
8.	Birbhum	1790920	1711484	0.90	0.86
9.	Burdwan	3966889	3750674	1.98	1.88
10.	Nadia	2653768	2513832	1.33	1.26
11.	North 24 Parganas	5119389	4890392	2.56	2.45
12.	Hooghly	2814653	2704492	1.41	1.35
13.	Bankura	1838095	1758579	0.92	0.88
14.	Purulia	1496996	1433119	0.75	0.72
15.	Howrah	2500819	2349210	1.25	1.17
16.	Kolkata	2356766	2139928	1.18	1.07
17.	South 24 Parganas	4173778	3988183	2.09	1.99
18.	Paschim Medinipur	3007885	2905572	1.50	1.45
19.	Purba Medinipur	2629834	2466041	1.31	1.23

Source: State Statistical Handbook, 2014, Bureau of Applied Economics and Statistics, Government of West Bengal

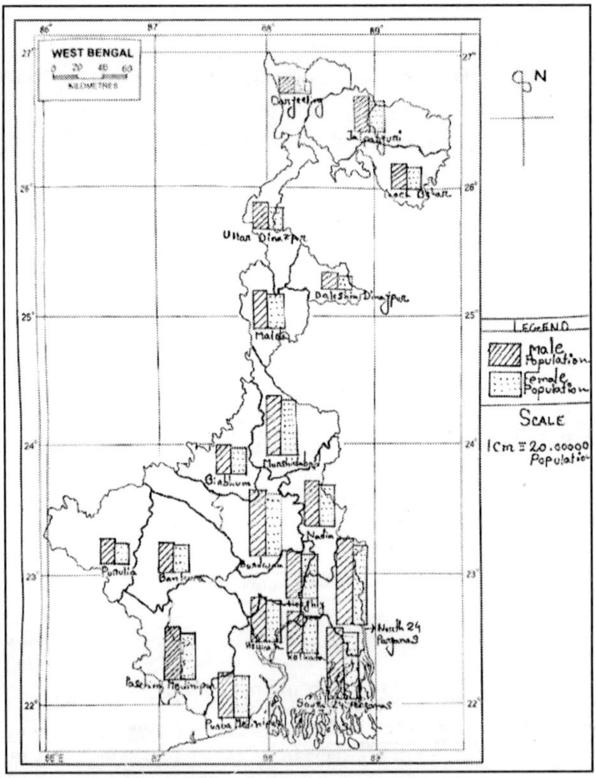

Figure 2: *Multiple Bar Diagram showing the Male-Female Population of Different Districts of West Bengal (2011)*

Source: Developed by Author

Interpretation

The multiple bar diagrams have been drawn with the district-wise male and female populations of West Bengal. It can be observed that the trend of the male population is higher in all of the districts. The male-female difference is more in the districts of Burdwan, Howrah, Kolkata, Bankura, North 24 Parganas, etc. The male population had moved to these urban sites in search of jobs which has led to the stark male-female difference as their families remain in the rural native land. Comparatively, the difference is low in Hooghly, Paschim Medinipur, Birbhum, and Murshidabad where the availability of agricultural land, agro-based industries, and other small scale-industries generate employment for the population. Again, the mountainous, and Uttar and Dakshin Dinajpur districts exhibit a parity between male and female populations.

Compound Bar Diagram

Compound Bar Diagrams are drawn when the data comprises more than one component within a total and each bar is proportionately divided to show the constituents as well as the total. Each category is distinguished by different shades. A suitable scale for all the bars should be selected for the drawing and appropriate presenting the components.

Importance

- Division in totality: This bar diagram is used to represent data in which the total magnitude is divided into different components.
- Retrieve information: Compound Bar Diagram helps to retrieve information not only one component rather its sub-parts at a glance.
- Visual impression: It is very easy to understand for any layperson as the number of elements is proportionate to the height of the bar and within each bar, the sub-parts are shown in different shades/colours or hachures.
- Easy comparison: A comparison of elements existing in different areas can be done using this diagram.
- Summarise large data set: It helps to summarise huge data sets with multiple sub-parts and reaching to any conclusion.

Table 3: *The Calculation for Representing Rural-Urban Population of Different Districts of West Bengal (2011)*

		Total Population			*Population Reduced to Scale (1cm = 3 Million)*		
Sl No.	*Districts of West Bengal*	*Rural*	*Urban*	*Total*	*Rural*	*Urban*	*Total*
1	Darjeeling	1118860	727963	1846823	0.37	0.24	0.62
2	Jalpaiguri	2812495	1060351	3872846	0.94	0.35	1.29
3	Cooch Behar	2529652	289434	2819086	0.84	0.10	0.94
4	Uttar Dinajpur	2644906	36228	2681134	0.88	0.01	0.89
5	Dakshin Dinajpur	1439981	236295	1676276	0.48	0.08	0.56
6	Malda	3447185	541660	3988845	1.15	0.18	1.33
7	Murshidabad	5703115	1400692	7103807	1.90	0.47	2.37
8	Birbhum	3052956	449448	3502404	1.02	0.15	1.17
9	Burdwan	4639264	3078299	7717563	1.55	1.03	2.57
10	Nadia	3728727	1438873	5167600	1.24	0.48	1.72
11	North 24 Parganas	4277619	5732162	10009781	1.43	1.91	3.34
12	Hooghly	3390646	2128499	5519145	1.13	0.71	1.84
13	Bankura	3296901	299773	3596674	1.10	0.10	1.20

Sl No.	Districts of West Bengal	Total Population			Population Reduced to Scale (1cm = 3 Million)		
		Rural	Urban	Total	Rural	Urban	Total
14	Purulia	2556801	373314	2930115	0.85	0.12	0.98
15	Howrah	1775885	3074144	4850029	0.59	1.02	1.62
16	Kolkata	0	4496694	4496694	0.00	1.50	1.50
17	South 24 Parganas	6074188	2087773	8161961	2.02	0.70	2.72
18	Paschim Medinipur	5190771	722686	5913457	1.73	0.24	1.97
19	Purba Medinipur	4503161	592714	5095875	1.50	0.20	1.70

Source: State Statistical Handbook, 2014, Bureau of Applied Economics and Statistics, Government of West Bengal

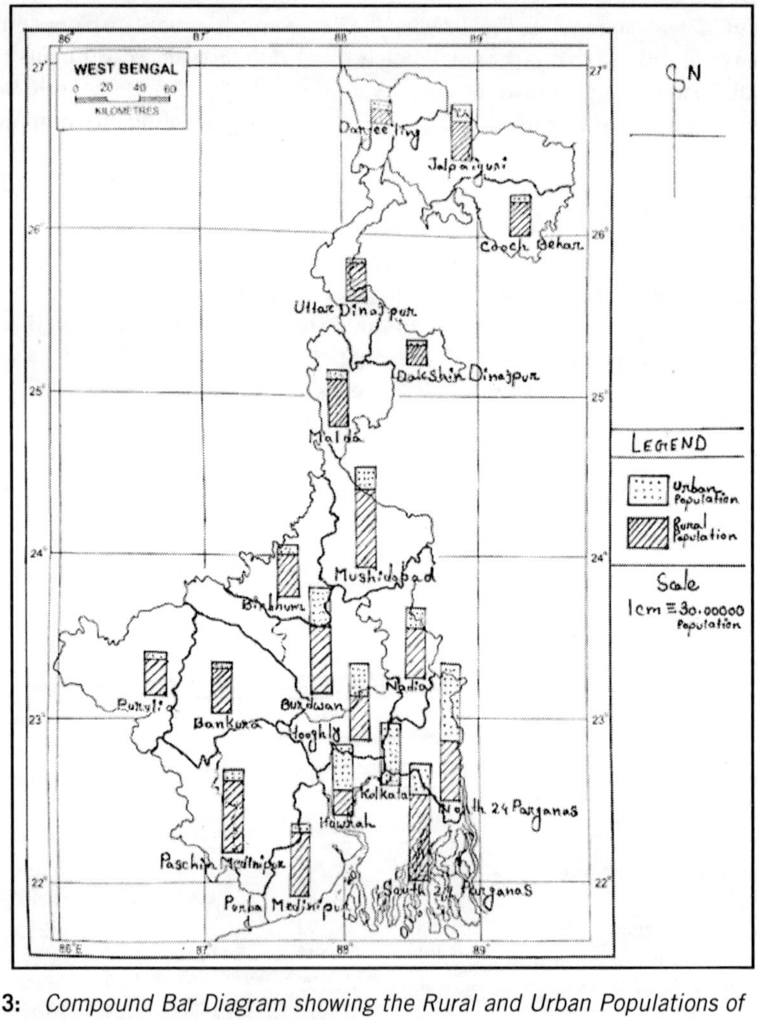

Figure 3: *Compound Bar Diagram showing the Rural and Urban Populations of Different Districts of West Bengal (2011)*

Source: Developed by Author

Interpretation

The compound bar graphs have been drawn with the district-wise rural and urban populations of West Bengal. The trend of the rural population is high in most of the districts of West Bengal.

Kolkata records no rural population. Infact this megacity is the seventh-most populous city in India(2011). Other districts like North 24 Parganas, Howrah exhibit high urban population in comparison to the rural total. The southern districts possess a greater share of the urban population and bear the facilities of diversified occupational structure, infrastructure, transport, communication and other services.

In the northern districts, especially Darjeeling, Uttar Dinajpur, Dakshin Dinajpur, etc. rural population is much higher than the urban population. Most of the people are engaged in agricultural activities, possess lesser diversity in occupational opportunities and the civic facilities are in the stage of developing.

Pie Diagram

Pie diagrams are represented by proportional circles which are further divided into sectors to present associated items into these sectoral parts. To calculate the radii of the proportional circles the area of the total individual value is taken to represent the circle i.e. πr^2. Thus the radius can be obtained by dividing the total value by π and then finding the square root. Sometimes to simplify the calculation only the square roots are calculated since π is common. The radii are then reduced to scale and drawn for suitable comparison.

To divide each circle into its sectors, the total value is taken to represent 360° and the various sectors are calculated in their angular degrees to the total. To give a visual effect the sectors are then distinctly coloured or shaded.

Importance

- Easy to understand: It is a simple picturesque representation that can be easily understood.
- Comparison: Through this diagram not only proportionality of any element can be represented but also its inherited sub-division or sector wise analysis can be derived.
- Decision Making: A pie diagram helps to take any critical decision as a long set of data with diverse sub-divisions/sectors can be represented in visual form.

Interpretation

Pie graphs have been drawn to show the type of total workers engaged in economic activities of six selected districts of West Bengal (2011). In terms of total workers, South 24 Parganas has the highest number of total workers and Purulia has the lowest. In this data series, the workers are subdivided into 3 broad divisions, those are: i) Agricultural workers, ii) Non-agricultural workers, iii) Handloom or handicraft or footloose industry workers. After representing the pie diagram in respective districts, it is seen that the share of agricultural workers is much lower than non-agricultural-sector workers in all the districts of West Bengal. It can be observed that the share of the workers at the non-agricultural sector is strikingly highest in all the six districts as urbanisation is gradually taking place. The agricultural-sector shows the least share in comparison to the Handloom/ handicraft or footloose industry with the exception of Jalpaiguri, Paschim Medinipur, and Purulia. These three districts possess agricultural land which results in considerable workers engaged in this sector. As Jalpaiguri is in the hilly terrain, the handloom/ handicraft/ Footloose industry has a very low or meagre percentage of workers. Whereas, with the changing land-use pattern, all the districts

Table 4: *The Calculation for representing the Type of Total Workers engaged in Economic Activities of Six selected Districts of West Bengal (2011)*

		Type of Workers						*Area in Degree*		
Sl No.	*Districts*	*Agricultural Sector*	*Non-Agricultural Sector*	*Handloom/ Handicraft/ Foot Loose Industry*	*Total Workers*	*Pie Radii*	*Radius in Scale (1 cm = 400 units)*	*Agricultural Sector*	*Non-Agricultural Sector*	*Handloom/ Handicraft/ Foot Loose Industry*
1	Jalpaiguri	46249	265717	4415	316381	317.34	0.79	52°37′	302°22′	5°01′
2	Nadia	39426	300877	51698	392001	353.24	0.88	36°13′	276°19′	48°28′
3	Howrah	15577	309397	52854	377828	346.79	0.87	14°50′	294°48′	50°22′
4	South 24 Parganas	29395	468686	41487	539568	414.43	1.03	19°37′	312°42′	27°41′
5	Paschim Medinipore	76212	261754	14713	352679	335.05	0.84	77°48′	267°11′	15°01′
6	Purulia	21239	108835	7135	137209	208.99	0.52	55°44′	285°33′	18°43′

Source: Economic Census of West Bengal, Government of India, 2012-13.

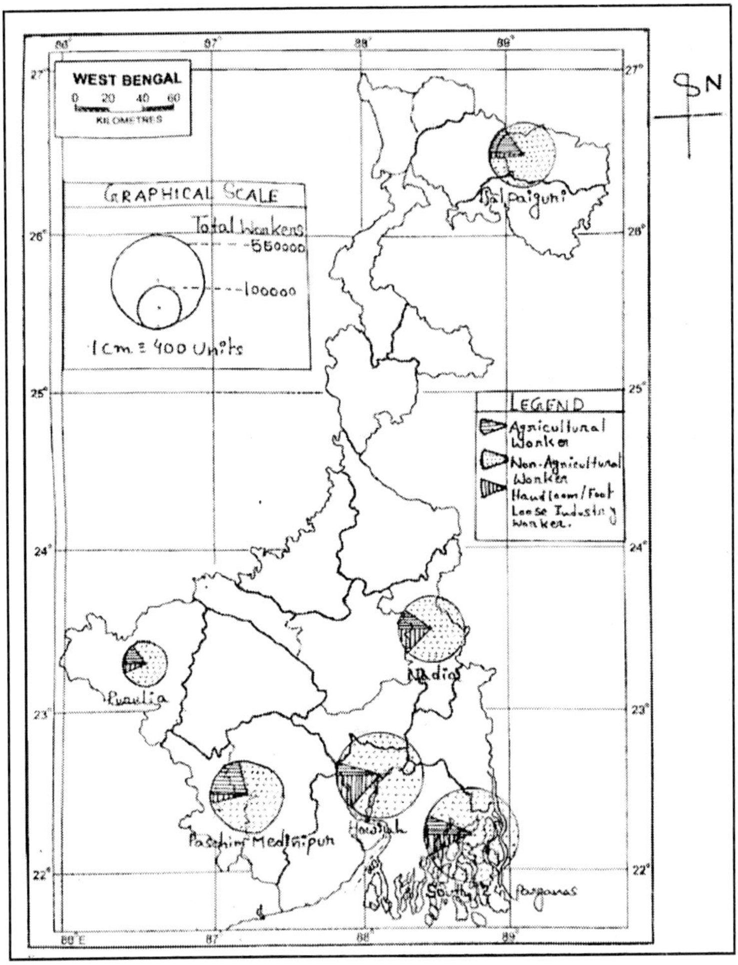

Figure 4: *Pie graphs showing the Type of Total Workers engaged in Economic Activities of Six selected Districts of West Bengal (2011)*

Source: Developed by author

draw huge workers at non-agricultural sector which encompasses trade and commerce, business, transport and communication, manufacturing, construction, and other services. In the future, the scenario of the workers engaged in the Handloom/ handicraft or footloose industry might change with governmental intervention.

Choropleth

Choropleth is a statistical thematic map that gives easy visualization of the range of intensity of the variable or element. The word "Choropleth" has been derived from two Greek words – "Choros" meaning 'area/ region' and "Plethos" meaning 'multitude'. Thus, the literal meaning of choropleth is "quantity in the area". Technically, choropleth is a quantitative areal map that shows the spatial distribution of intensity or density of any element with the help of a system of graded shading or colour drawn within the boundaries of the administrative units.

According to Monk House and Wilkinson (1826), "if the isopleth is the chief tool to a climatologist, the choropleth may be said to be the chief tool for human geographers." This method is primarily used to show the density of commodities per unit area, the average measure of percentages, etc., for example, the density of population, male-female ratio (sex ratio), percentage of agricultural lands under different crops to total land, or percentage of irrigated (types used) land to cultivated land or other physical and cultural geographical elements.

Importance

- Spatial distribution of elements: Through choropleth mapping, the spatial distribution of any element can be represented effectively. The intensity of shading is directly proportional to the density of the element.
- Easy to understand: These maps are developed using shading variation proportional to the distribution of any element so that it is easily understandable and one can retrieve information from the map.
- Comparative distribution: This is a simple method that gives comparative information about the distribution of an element over two or more places at a glance.
- To represent an abrupt change in element: Shading method is especially suited to those elements in which there is an abrupt change. For example – the density of the population.
- Resolve misconception: The misconception that commonly arises in the drawing of an isopleth map is the grouping of a unit under one average value, which can be resolved through the choropleth technique as the density zones are based on administrative units, reveals the average distribution of the element of the particular administrative region.

Table 5: *The Calculation of Choropleth for representing the Density of Population in Different Districts of West Bengal, 2011*

Sl. No.	*Name of Districts*	*Area in sq. km.*	*Total Population 2011*	*Density/sq.km. (Calculated)*	*Density/sq.km. (Actual)*
1.	Darjeeling	3149	1846823	586.479	586
2.	Jalpaiguri	6227	3872846	621.944	622
3.	Cooch Behar	3387	2819086	832.325	832
4.	Uttar Dinajpur	3140	3007134	957.686	958
5.	Dakshin Dinajpur	2219	1676276	755.420	755
6.	Malda	3733	3988845	1068.536	1069
7.	Murshidabad	5324	7103807	1334.299	1334
8.	Birbhum	4545	3502404	770.606	771
9.	Burdwan	7024	7717563	1098.742	1099
10.	Nadia	3927	5167600	1315.915	1316
11.	North 24 Parganas	4094	10009781	2444.988	2445
12.	Hooghly	3149	5519145	1752.666	1753
13.	Bankura	6882	3596674	522.620	523
14.	Purulia	6259	2930115	468.144	468
15.	Howrah	1467	4850029	3306.087	3306
16.	Kolkata	185	4496694	24306.454	24306
17.	South 24 Parganas	9960	8161961	819.474	819
18.	Paschim Medinipur	9368	5913457	631.240	631
19.	Purba Medinipur	4713	5095875	1081.238	1081

Source: State Statistical Handbook, 2014, Bureau of Applied Economics and Statistics, Government of West Bengal

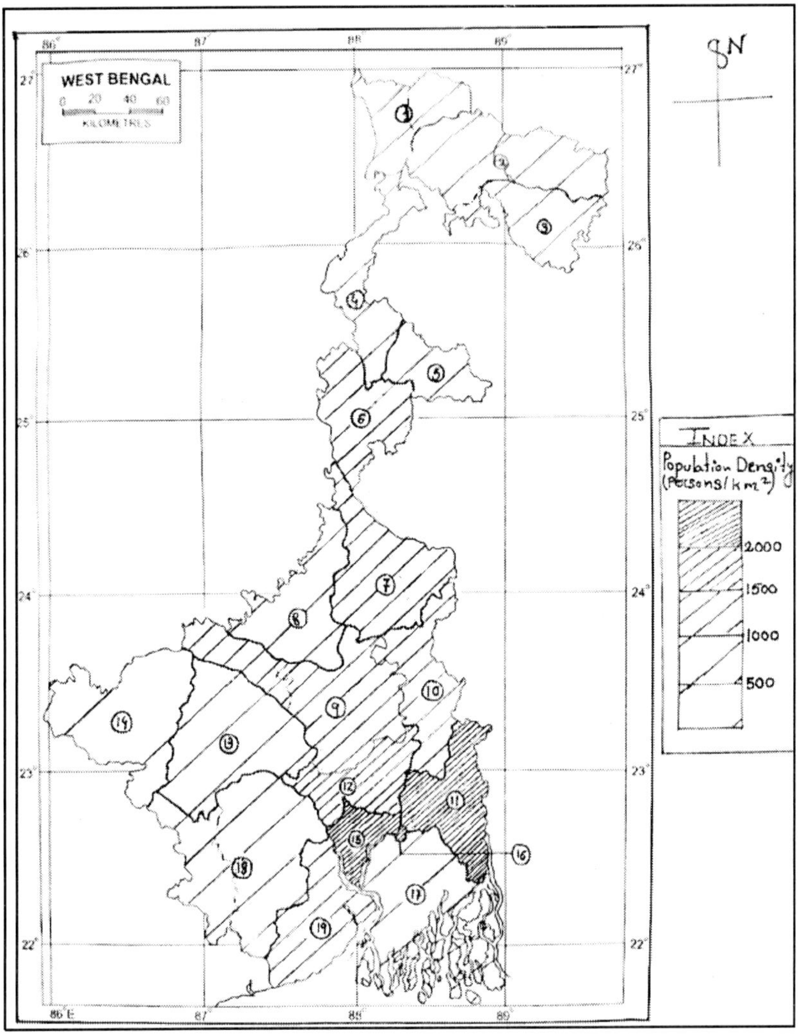

Figure 5: *Choropleth Map showing Population Density of Different Districts of West Bengal*

Source: Developed by the author

Calculation for the Division of Classes/Class Intervals with Population
Highest density = 24306.45
Lowest density = 468.14
Class Range (Highest-Lowest) = (24306.45-468.14) = 23838.31
Number of classes = 5
Class Interval will be = 23838.31 / 5 = 4767.662

For a more or less, equal representation of the districts in all the Class Intervals, it is judicious to take Class Intervals at 500 instead of 5000.

∴ Class Intervals taken are: i) <500, ii) 500-1000, iii) 1000-1500, iv) 1500-2000, v) >2000

Table 6: *Name of Districts Based on Serial No. Placed on Map*

Darjeeling (1)	Jalpaiguri (2)	Cooch Behar (3)	Uttar Dinajpur (4)	Dakshin Dinajpur (5)	Malda (6)	Murshidabad (7)
Birbhum (8)	Burdwan (9)	Nadia (10)	North 24 Parganas (11)	Hooghly (12)	Bankura (13)	Purulia (14)
Howrah (15)	Kolkata (16)	South 24 Parganas (17)	Paschim Medinipur (18)	Purba Medinipur (19)		

Table 7: *Name of Districts under Different Population Density Classes*

Class Division (Based on Population Density)	*Density Zones*	*Name of Districts*
>2000	High	Kolkata, Howrah, North 24 Parganas
1500-2000	Medium	Hooghly
1000-1500		Malda, Murshidabad, Burdwan, Nadia, Purba Medinipur
500-1000	Low	Darjeeling, Jalpaiguri, Cooch Behar, North Dinajpur, South Dinajpur, Birbhum, Bankura, Paschim Medinipur, South 24 Parganas
<500		Purulia

Source: Developed by Author Based on Population Density Map of West Bengal, 2011

Interpretation

Population density is a measurement of population per unit of land area. The density of the population varies based on the geographical location, the scope of occupation, civic services for better living, amenities of education, available infrastructure, urbanization trend and stability of the political regime, etc.

The 5 distinct zones under the Population Density mapping of West Bengal can be broadly classified under 3 distinct density zones, namely Low (Below 1000), Medium (Between 1000 to 2000), and High (Above 2000) Density Zone.

High population density is noticed in 3 districts of West Bengal, namely Kolkata, Howrah, and North 24 Parganas. Among them the highest and distinct density has been recorded in Kolkata, more than 24000 persons per sq. km. Kolkata, the capital of West Bengal by the side of River Ganga is one of the populous metropolitan areas possessing all the urban civic facilities to its maximum. This is the main hub with business, education, administration, and medical facilities. Howrah and North 24 Parganas are two adjacent districts of Kolkata which are also very developed and hold a high concentration of population.

Relatively medium density of population has been observed in Hooghly, Malda, Nadia, Purba Medinipur, Burdwan and Murshidabad. With the availability of fertile agricultural land and agro and ancillary-based industries, a comparatively appreciable amount of the population is found in these regions. The primary sector provides huge employment and to some extent the secondary and the developing tertiary sectors.

The districts like Purulia, Bankura, Darjeeling, North Dinajpur, South Dinajpur have low population primarily due to the harsh physical location, climate, and infertile soil. South 24 Parganas, Paschim Medinipur, Jalpaiguri, Cooch Behar, and Birbhum are developing amidst unfavourable geographical conditions. The basic transport and communication system, civic amenities, and infrastructural outlay are yet to be developed to their maximum which would then surely draw a greater population.

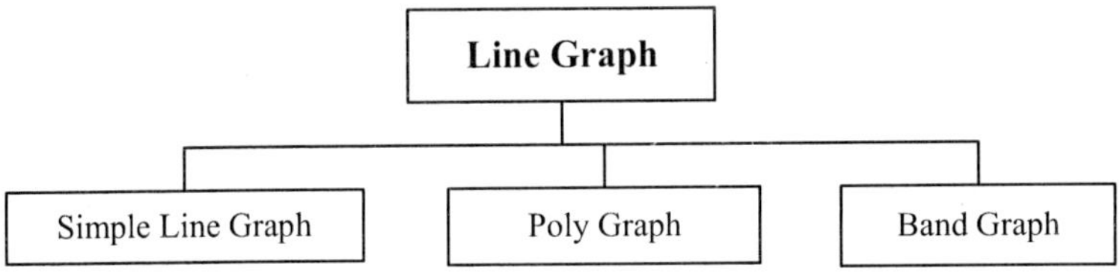

Figure 6: *Different Types of Line Graph*

With the help of choropleth mapping in different districts of West Bengal, the difference in regional developmental conditions truly reflects on the density distribution.

Line Graph

A *line graph* may be defined as a graph represented by a line or a series of lines. It precisely represents some statistical data and information. There are different types of line graphs. These may be grouped into the following three types:

Simple Line Graph

In a simple line graph, a single graphical line is drawn according to the value put forwarded by statistical data.

Methods of Drawing Line Graph

1. Drawing a line on the bottom of a graph (at abscissa) and dividing it into equal parts as many as one requires to show the divisions of days/months/years/or other units.
2. On the two definite ends of the line (abscissa), erecting two vertical lines (on the ordinates) as shown in the figure and dividine them to show the scale (of temperature, rainfall, amount of production, population growth, or other items given on the statistical table).
3. Now plotting or marking the determining coordinates along X-axis and Y-axis on the graph by points.
4. Then joining the points by a straight line (curve line) to show the line graph (curve).

Table 8: *The Calculation of Simple Line Graph for representing the Urban Population of India (1901- 2011)*

Year	*Urban Population in Million*	*Population Reduced to Scale 1 cm ≡ 5 Millions*
1901	2.5	0.5
1911	2.6	0.52
1921	2.8	0.56
1931	3.3	0.66
1941	4.4	0.88
1951	6.2	1.24
1961	7.9	1.58
1971	10.9	2.18
1981	15.9	3.18
1991	21.4	4.28
2001	28.5	5.7
2011	37.7	7.54

Source: Census of India, 2011

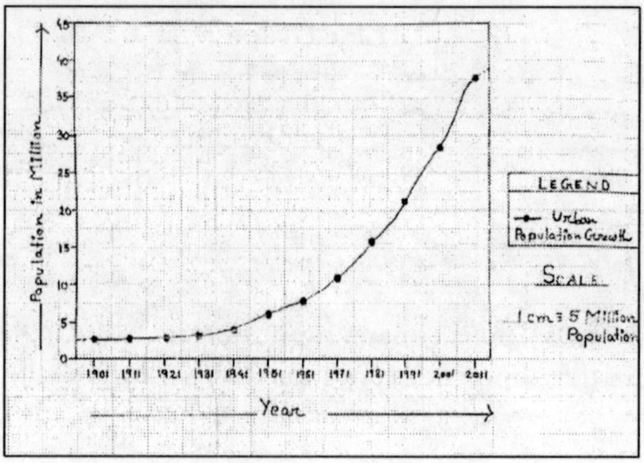

Figure 7: *Simple Line Graph showing Urban Population Growth of India (1901-2011)*
Source: Developed by Author

Interpretation: The simple line graph represents the statistical data in a single graphical line. After drawing the line graph showing the growth trend of urban population of India from 1901 to 2011 it is studied that in the pre-independence period it was very negligible, but in post-independence especially, after 1991 the growth of the urban population escalated tremendously. Due to the availability of urban civic facilities for education, services, infrastructure, administration, business and commerce, and medicine, the urban population has increased in India. The standard of living has changed, job opportunities extended, and a good communication network system has developed. With the growth of numerous urban centers, several challenges have also cropped up. With globalization, Indian urbanization would change immensely in the twenty-first century.

Polygraph

In a *polygraph*, there are a series of line graphs. They are known as multiple-line graphs. In this, more than one series of data can be represented in a single graph. Through this graph statistical data can not only be represented simultaneously but also, they can be compared.

Methods of Drawing Polygraph

1. The method of drawing a polygraph is the same as drawing of the line graph.
2. In the poly graph two or more than two variables are presented through more than one line graph.

Interpretation: The polygraph is such a graph that can place more than one series of data within a single graph. In the present study, the polygraph is showing the rural and urban population growth in India from 1901 to 2011 with two sets of line graphs separately. After plotting two data sets it is studied that the growth rate of the rural population is much higher than the urban population. The rural population is rising positively but its trend is negative (in 1991-2001 growth of the rural population is 11.3 million whereas in 2001-2011 growth of the rural population is 9.1 million). In contrast, the urban population is rising positively and its trend is also positive (in 1991-2011 growth of the urban population is 7.1 million whereas in 2001-2011 growth of the urban population increased to 9.2 million). In India, the trend of rural population growth exhibits a diminishing trend in comparison to the urban population.

Table 9: *The Calculation of Polygraph for representing the Rural-Urban Population of India (1901- 2011)*

Year	Population in Million		Population Reduced to Scale (1 cm ≡ 10 Million)	
	Rural	Urban	Rural	Urban
1901	21.3	2.5	2.13	0.25
1911	21.6	2.6	2.16	0.26
1921	22.3	2.8	2.23	0.28
1931	24.6	3.3	2.46	0.33
1941	27.5	4.4	2.75	0.44
1951	29.9	6.2	2.99	0.62
1961	36	7.9	3.6	0.79
1971	43.9	10.9	4.39	1.09
1981	52.4	15.9	5.24	1.59
1991	62.9	21.4	6.29	2.14
2001	74.2	28.5	7.42	2.85
2011	83.3	37.7	8.33	3.77

Source: Census of India, 2011

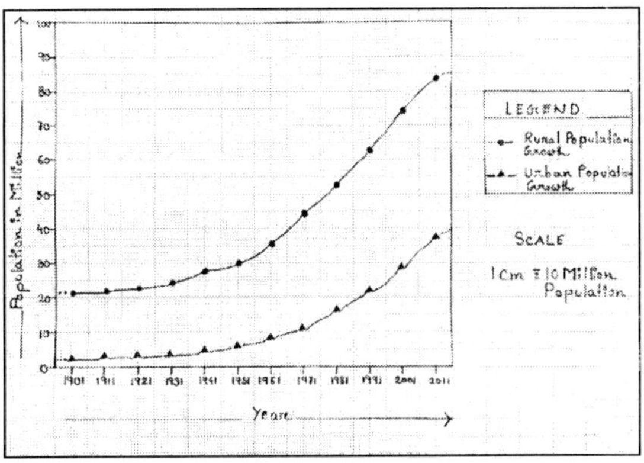

Figure 8: *Polygraph showing the Rural-Urban Population Growth in India (1901-2011)*

Source: Developed by Author

Band Graph

Band graph is also known as a compound line graph. In this graph, in between a series of line graphs, different shades are made between two-line graphs so that separations are made clearly. These may show the range and high-low value series of the variables studied.

Methods of Drawing Band Graph

1. The same method of drawing line graphs and polygraph are followed.
2. Then marking different shades in between the regions of line graphs so that differences can be easily made and layers can distinctively be shown.

Interpretation: A band graph is similar to a polygraph but the only difference is that in between line graphs separate shades are given to denote the area of separation. Herein, it is tried to represent the growth of the population of India from 1901 to 2011 in three distinct aspects i.e. total population growth, rural population growth, and urban

Table 10: *The Calculation of Band Graph for representing the Rural-Urban Population of India (1901-2011)*

	Population in Million			*Population Reduced to Scale (1 cm ≡ 10 Million)*		
Year	*Urban*	*Rural*	*Total*	*Urban*	*Rural*	*Total*
1901	2.5	21.3	23.8	0.25	2.13	2.38
1911	2.6	21.6	24.2	0.26	2.16	2.42
1921	2.8	22.3	25.1	0.28	2.23	2.51
1931	3.3	24.6	27.9	0.33	2.46	2.79
1941	4.4	27.5	31.9	0.44	2.75	3.19
1951	6.2	29.9	36.1	0.62	2.99	3.61
1961	7.9	36	43.9	0.79	3.6	4.39
1971	10.9	43.9	54.8	1.09	4.39	5.48
1981	15.9	52.4	68.3	1.59	5.24	6.83
1991	21.4	62.9	84.3	2.14	6.29	8.43
2001	28.5	74.2	102.7	2.85	7.42	10.27
2011	37.7	83.3	121	3.77	8.33	12.1

Source: Census of India, 2011

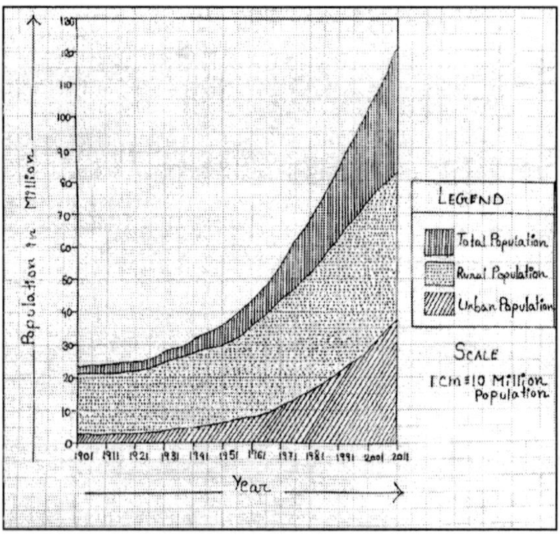

Figure 9: *Band Graph showing the Trend of Population Growth in India (1901-2011)*

Source: Developed by Author

population growth. After plotting all the sets of data, and giving shades it is seen that all three graphs exhibit a rising trend in India. Rural-urban migration has occurred for employment and a better standard of living opportunities. The presence of transport, communication, infrastructure, and services is leading to urban sprawl. With time these develop into cities and townships and attract a greater population and these regions tender all-inclusive facilities for the community.

Flow Maps/Charts

The Flow chart is a combination of graph and map. It is drawn to show the flow of commodities or people between the places of origin and destination. It is also called a Dynamic Map. Passenger and vehicle movement is represented through flow charts. In flow diagrams/charts the frequency of traffic flow between two areas is shown by the breadth of the line joining the two points (areas) on the map. The thickness of the line, the more will be the amount of traffic/passenger flow implying a developing region and *vice versa*. Generally, metropolia, industrial regions, busy cities, and centres of business and commerce show flow diagrams with thick lines. Sometimes flow diagrams are shown by parallel lines. In this case, the frequency of traffic flow is shown by the number of lines. The more the amount of flow of passengers or more will be the number of parallel lines and *vice versa*.

Example: Four stations A, B, C, and D are shown the flow of vehicles are shown between these stations. The daily number of vehicles is as follows:

Road	*Length in km.*	*Length in cm. 1 cm = 8 km*	*Direction*	*Number of*	*Vehicles in Scale 1 cm = 50 vehicles*
A-B	65	8.13	East	126	2.4
B-C	65	8.13	South-East	36	0.72
B-D	77	9.63	North-East	48	0.96

Interpretation

The flow diagram shows the maximum number of vehicles from A toward B which is 126. But the minimum number of vehicles moving from B to C is 36. The number of vehicle and passenger movement directly represent the level of development or the availability of facilities. Several planning agencies (government and NGOs) prepare flow charts to show the problems and prospects of that region. Consequently, developmental measures can be taken up and the regional disparities can be studied.

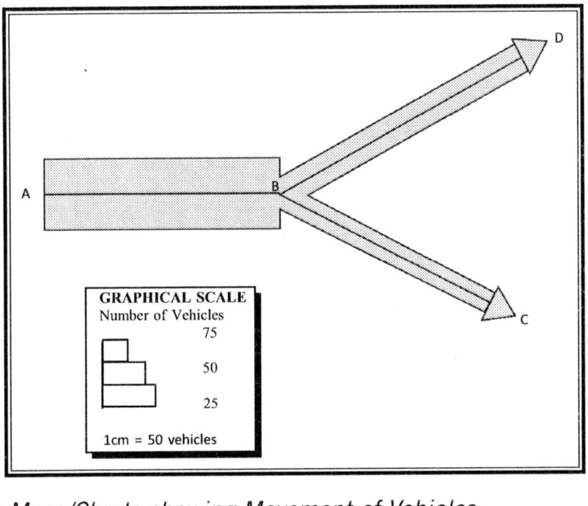

Figure 10: *Flow Maps/Charts showing Movement of Vehicles*

Star Diagram

A star diagram is used for organizing the characteristics of a single topic. A central space is used for displaying the topic, with each 'point' of the star listing some fact, attribute, or trait about the topic.

A Star diagram is also known as a clock diagram or vector diagram or direction diagram. Wind Rose is an example of a Star diagram. The wind rose shows the direction and intensity of wind for any particular region. From a point, straight lines are drawn in various directions. According to a given number of days, corresponding to each direction expressed in proper scale. Then these points are joined. A number of calm days, if any, are written in a circle at the centre.

Example: Below is the given temperature and rainfall of the months in a year to draw a Star diagram.

Month	*Jan.*	*Feb.*	*Mar.*	*April*	*May*	*June*	*July*	*Aug.*	*Sept.*	*Oct.*	*Nov.*	*Dec.*
Temperature in 0°C	28.9	27.8	25	20	15.6	12.2	11.7	14.4	18.3	22.8	26.1	27.8
Rainfall in cms.	4.2	3.3	2.8	1.0	1.5	1.2	0.8	0.8	0.8	1.7	3.0	3.8

Interpretation

On the basis of the above data, a Star diagram is drawn. From the diagram, it has been observed that the temperature is high during the month of January, February, March, November, and December, and the rainfall has also occurred about 4.2 cm in the month of January. But the rainfall and temperature are less in the month of June. So, it can be said that the given data is recorded from a region situated in the southern hemisphere as in this hemisphere the temperature is high during the month of December.

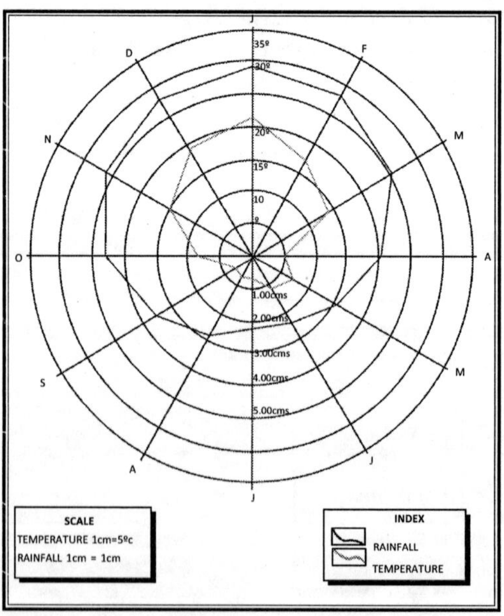

Figure 11: *Star Diagram showing Temperature and Rainfall*

7

Chorochromatic Map

The term "Chorochromatic" can be applied to a special category of the thematic map, where the quantity and values of the statistical data are not considered. This category of map is often called a "Colour patch map"(Monkhouse and Wilkinson, 1971), as the term colour is used in its widest interpretation of tints or shades. The purpose of this chorochromatic map is to identify and differentiate the different regions of a map based on colours. It can be commonly used in geography studies and also for many other studies related to land- use. Within the boundaries of the respective administrative unit, spatial patterns of the ground surface are outlined and then displayed with colour symbols such as tints or colours.

Importance

- *Land-use pattern-related information:* The chorochromatic maps are also known as land-use maps as they represent the quantity of land utilization and distribution of different natures of land-use and other types of commodities. These maps can be used for regional planning,
- *Use of conventional colours, shades, and signs*: Functional chorochromatic maps or morphology of urban centres are prepared through the selection of some conventional colours or signs. These may be yellow for cultivation lands, green for forest, brown for fallow, red for settlements, metalled roads, un-metalled roads, cart tracts, railways including both meter gauge and broad gauge, tanks, and hillocks, which can be shown on the maps, through their respective conventional signs. The areal distributions of different elements on these maps are clearly shown by using different conventional colouring, shading, and tinting.
- *Easy to understand:* As conventional signs, colours, and shades are used, maps look attractive and informative and can be easily comprehendible by any map reader.
- *Easy to depict the data:* The most decided advantage of drawing these maps is that cumbersome calculations for the gathered data are unnecessary. These maps are qualitative and non-quantitative in nature.
- *Minute Variation:* Chorochromatic maps can be used to show minute differences in the regional distribution of several elements in any region by use of tints and symbols which can be used for regional planners.

Problem: Draw a Chorochromatic Map of Land-Use of the Shanrpur Mouza; P.S. – Ghatshila from the given data.

Table 1: *Data Table for Drawing of Chorochromatic Map*

1. H	20. P	39. P	58. Pond (W)	77. P	96. P	115. P	134. H	153. V	172. H	191. V
2. H	21. H	40. P	59. V	78. P	97. P	116. P	135. H	154. H	173. P	192. V
3. V	22. V	41. P	60. V	79. H	98. P	117. P	136. H	155. H	174. P	193. V
4. H	23. V	42. P	61. RL	80. V	99. P	118. P	137. H	156. H	175. V	194. V
5. R	24. P	43. P	62. H	81. R	100. P	119. P	138. H	157. H	176. P	195. V
6. V	25. P	44. P	63. H	82. W	101. P	120. H	139. H	158. V	177. P	196. P
7. H	26. P	45. P	64. H	83. V	102. P	121. H	140. H	159. H	178. H	197. P
8. R	27. P	46. H	6.5. H	84. H	103. R	122. H	141. V	160. H	179. P	198. P
9. H	28. P	47. PP	66. H	85. H	104. R	123. H	142. V	161. V	180. H	199. V
10. H	29. P	48. H	67. R	86. H	105. V	124. P	143.V	162.V	181. P	200. V
11. H	30. P	49. H	68. P	87. R	106. V	125. P	144. P	163. V	182. P	201. V
12. RL	31. P	50. H	69. V	88. V	107. P	126. P	145. V	164. V	183. P	202. V
13. P	32. P	51. P	70. H	89. H	108. P	127. P	146. H	165. V	184. P	203. V
14. P	33. P	52. P	71. G	90. V	109. P	128. P	147. H	166. Nullar (W)	185. P	204. V
15. P	34. P	53. H	72. P	91. H	110. P	129. P	148. H	167. P	186. P	205. V
16. V	35. P	54. V	73. V	92. V	111. H	130. V	149. H	168. V	187. P	206. V
17. P	36. P	55. H	74. V	93. H	112. R	131. V	150. H	169. V	188. P	207. V
18. P	31. P	56. G	77. P	94. H	113. H	132. R	151. V	170. P	189. W	208. V
19. P	38. P	57. G	76. P	95. H	114. H	133. V	152. V	171. V	190. Mill (H)	
1. H	20. P	39. P	58. Pond (W)	77. P	96. P	115. P	134. H	153. V	172. H	191. V
2. H	21. H	40. P	59. V	78. P	97. P	116. P	135. H	154. H	173. P	192. V
3. V	22. V	41. P	60. V	79. H	98. P	117. P	136. H	155. H	174. P	193. V
4. H	23. V	42. P	61. RL	80. V	99. P	118. P	137. H	156. H	175. V	194. V
5. R	24. P	43. P	62. H	81. R	100. P	119. P	138. H	157. H	176. P	195. V
6. V	25. P	44. P	63. H	82. W	101. P	120. H	139. H	158. V	177. P	196. P
7. H	26. P	45. P	64. H	83. V	102. P	121. H	140. H	159. H	178. H	197. P
8. R	27. P	46. H	6.5. H	84. H	103. R	122. H	141. V	160. H	179. P	198. P
9. H	28. P	47. PP	66. H	85. H	104. R	123. H	142. V	161. V	180. H	199. V
10. H	29. P	48. H	67. R	86. H	105. V	124. P	143.V	162.V	181. P	200. V
11. H	30. P	49. H	68. P	87. R	106. V	125. P	144. P	163. V	182. P	201. V
12. RL	31. P	50. H	69. V	88. V	107. P	126. P	145. V	164. V	183. P	202. V
13. P	32. P	51. P	70. H	89. H	108. P	127. P	146. H	165. V	184. P	203. V
14. P	33. P	52. P	71. G	90. V	109. P	128. P	147. H	166. Nullar (W)	185. P	204. V
15. P	34. P	53. H	72. P	91. H	110. P	129. P	148. H	167. P	186. P	205. V
16. V	35. P	54. V	73. V	92. V	111.H	130. V	149. H	168. V	187. P	206. V
17. P	36. P	55. H	74. V	93. H	112. R	131. V	150. H	169. V	188. P	207. V
18. P	31. P	56. G	77. P	94. H	113. H	132. R	151. V	170. P	189. W	208. V
19. P	38. P	57. G	76. P	95. H	114. H	133. V	152. V	171. V	190. Mill (H)	

KEYS : H - House, V- Vacant, P – Paddy, R – Road, RL – Railway, PP – Petrol Pump, G – Garden, W – Water body

Chorochromatic Map
Illustrating Land-Use Pattern of Shanrpura Mouza P.S. Ghatshila

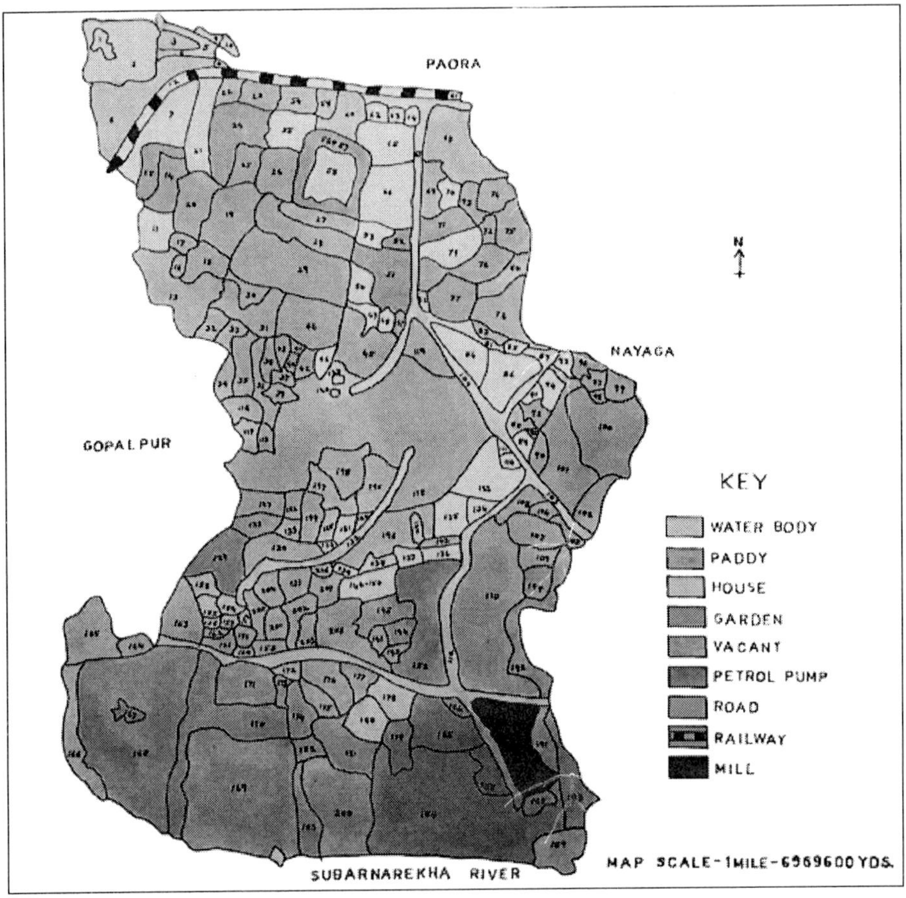

8

Weather Map

Introduction

This weather map of Sunday 16th May 1954 was observed at 17:30 hrs. I.S.T. (1200 hrs. G.M.T.). It is seen that summer prevails over the Indian sub-continent.

Temperature

With the increase in the heat of the land, the cold weather season changes to warmer and warmer weather. The high pressure of the north-western sub-continent which is the characteristic of the cold weather season gradually, changes to the low pressure of the hot weather. In summer the average temperature is about 32° to 33° C. From the isotherms drawn it can be stated that the temperature is about (32° + 8°) = 40 °C. In the southern states of India, low temperature prevails due to the maritime influence.

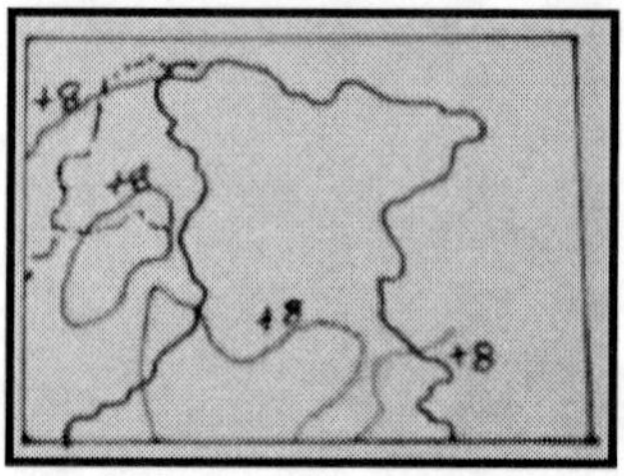

30° N — 35° N
70° E — 80° E

A rough hand sketch

Pressure

Due to the rising temperature in the interior landmass, the pressure is low in the 994 mb, 996mb, and 998 mb isobars are predominant. Comparatively, in the south, 1006 mb, and 1004 mb isobars are seen due to the lesser high temperature prevailing there.

Another important trend of the isobars is that it runs parallel to the border of southern India. There is a low-pressure trough in central India – due to the formation of a local thunderstorm. Moreover, low pressure exists in the north-western frontier provinces due to the western disturbances. These cellular-like pressure zones in a region are termed as bar high when high pressure exists and bar low when low pressure exists.

Another important feature is that there is a steep pressure gradient in the south-western coastal region because isobars are narrowly spaced. Moreover, a steep pressure gradient exists over the land. Whereas, isobars are widely spaced on the sea which

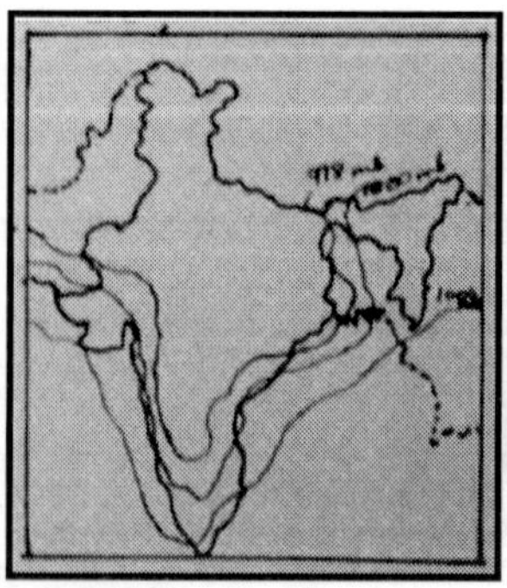

37° 6′ N — 8° 4′ N
68° 7′ E — 97° 25′ E

A rough hand sketch

results in a gentle gradient. But in the Sind, Punjab, and Baluchistan portions the isobars are relatively closer and therefore the pressure is quite moderate.

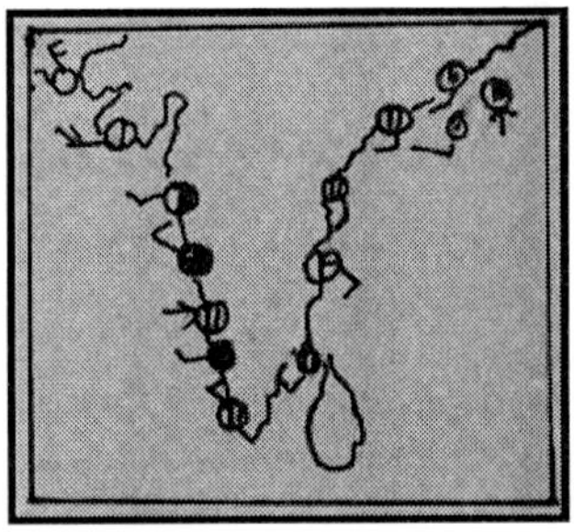

5° N — 22° N
67° E — 90° E

A rough hand sketch

Wind

As high pressure exists over the sea and low pressure on the land, wind direction is obviously directed from the sea towards the land. In some places, due to local disturbances and intense heating, the wind-direction changes. Velocity primarily depends on the gradient of pressure. Since the gradient of pressure was gentle the wind velocity almost throughout the middle sub-continent was also less. Calm to light conditions prevails over the west and north-west India, Uttar Pradesh, Bihar, Bengal, and Assam. But, on the western coast, due to the steep gradient, there is a higher velocity than 50 knots.

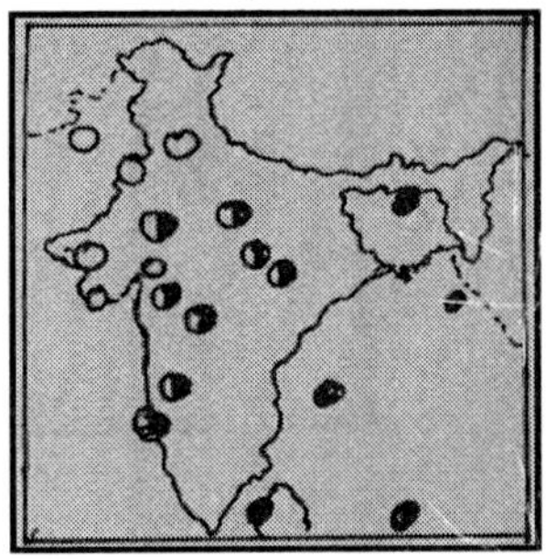

37° 6' N — 8° 4' N
68° 7' E — 97° 25' E

Cloud Cover

In northern India, the sky was almost clear of clouds except locally, at a few places, where the cloud varied from $^{1}/_{8}$ to ½. One notable exception was eastern India where at Dibrugarh and Tezpur the sky was overcast. But in the south, mainly in the western coastal plain, the sky was more than ½ clouded.

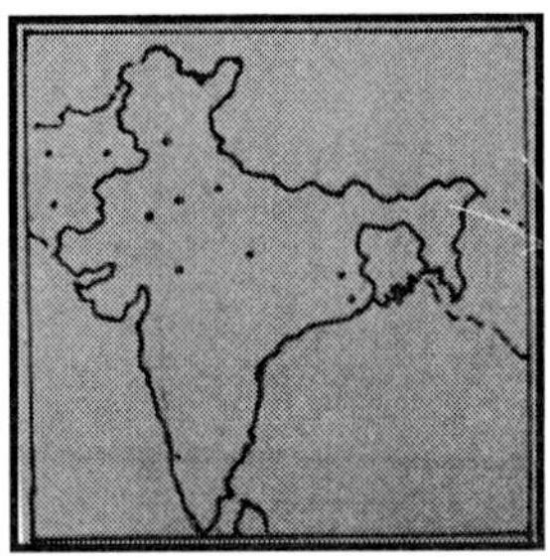

37° 6' N — 8° 4' N
68° 7' E — 97° 25' E

High or medium clouds prevailed over Assam, and in Madhya Pradesh. The Peninsular had chiefly high or medium clouds except in the western coastal plain.

There is no fog or mist which is recorded. But thunderstorms are reported in the area near Ceylon.

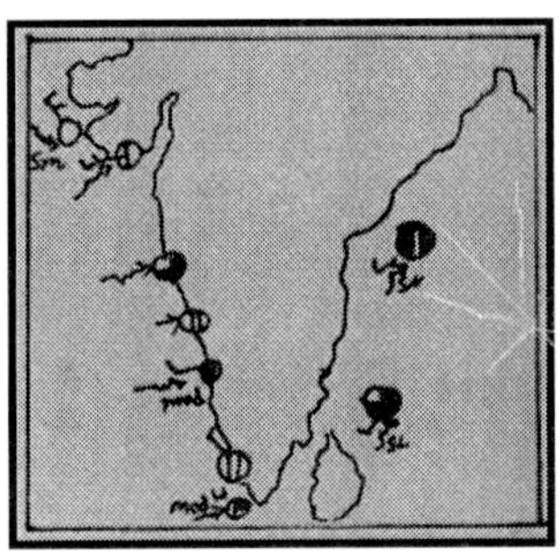

5° N — 22° N
67° E — 90° E

A rough hand sketch

Precipitation

Due to the intense heating and high temperature, there is no possibility of

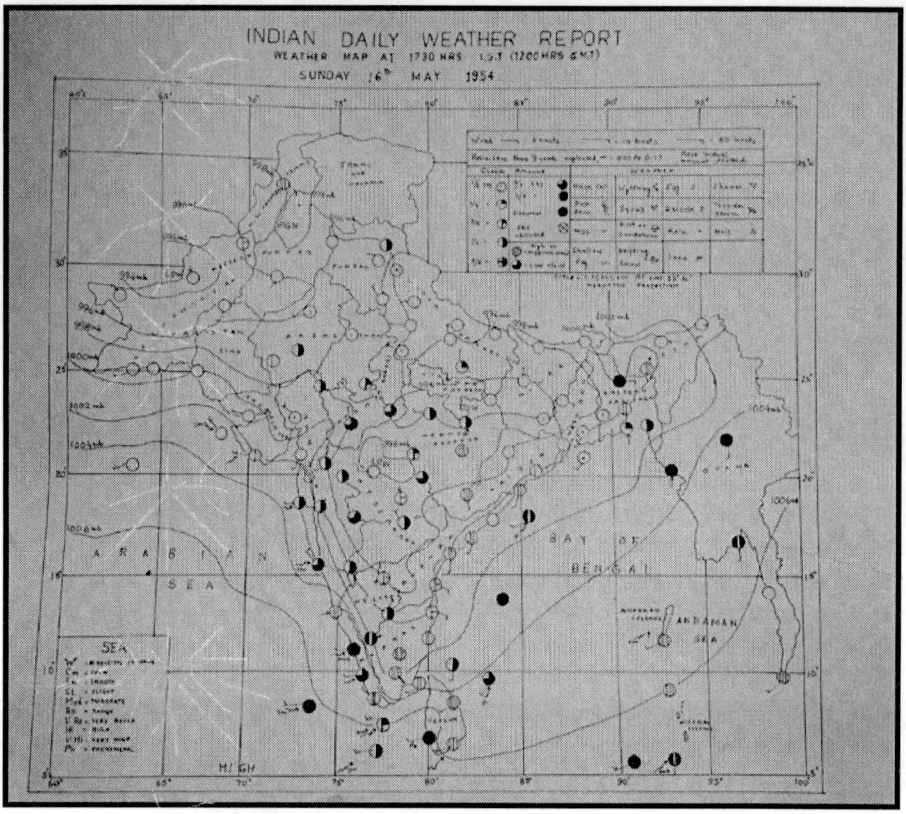

rainfall. Only local thunderstorms in the form of 'Kalbaishakhi' bring rainfall to give relief from the hot dry unbeatable summer.

Sea Conditions

The sea is mostly calm with moderate waves prevailing over it. The waves are directed from the south-south-west to the north-east. As the waves are moderate, the sea is naturally calm and not turbulent.

9

Map Projections

Cylindrical Equal-Area Projection

Introduction: The cylindrical equal-area projection has been derived by projecting the surface of the globe with parallel rays on a cylinder touching it all the equator. The circles of latitudes and longitudes both are projected as straight lives intersecting, one another at right angles. The area between two parallels is made equal to the corresponding surface on the sphere at the cost of great distortion in shape towards higher latitudes. This is why it is an equal-area projection.

Graphical Construction: A circle from the centre O with the radius of the reduced sphere is drawn. OE is produced to Q making EQ equal to the true length of the equator. The tangent TM at E is drawn which should be parallel to PP', from the middle of EQ, draw the central meridian NS right-angled to the EQ, Oa', Ob', Oc', and Od' are drawn in accordance to the given interval. From a', b', c' lines are drawn parallel to the line EQ. All the meridians should be parallel to the tangent, equal in length, and according to the interval given. In this way, the TGLM required graticule is obtained (Refer Figure 1).

Problem: Prepare a graticule for Africa on a scale of 1:50,000,000 at an interval of 10 with latitudinal extent – 40° N to 40° S and longitudinal extent – 20° W to 60° E.

Properties: The scale along the Equator and other parallels become increasingly exaggerated and along meridians, it is progressively diminished. This leads to pronounced deformation of shape in high latitudes. This projection is suitable for Equatorial regions and for low latitudes, where both shape and area are more or less correctly represented.

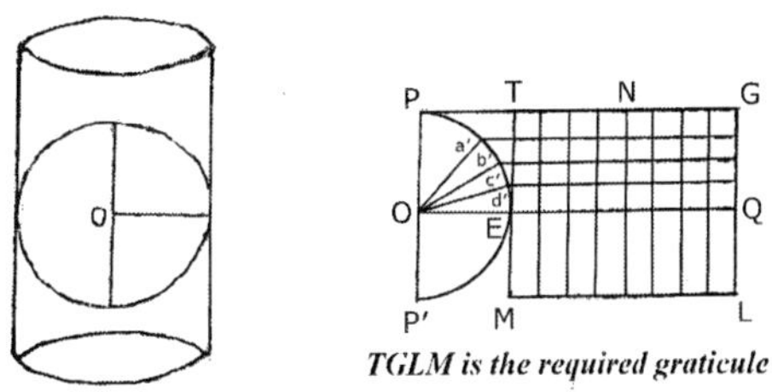

TGLM is the required graticule

Figure 1.

Simple Conical Projection with One Standard Parallel

Introduction: In simple conical projection with one standard parallel the cone is supposed to touch the sphere along the Central Meridian which is truly represented by the graticules. Other parallels are drawn at the true distances from it but the scale along these points is joined with the common centre to show the meridians.

Graphical Construction: The reduced radius is calculated and a circle is drawn. 10 and 50 angles are marked off with the help of a protractor.

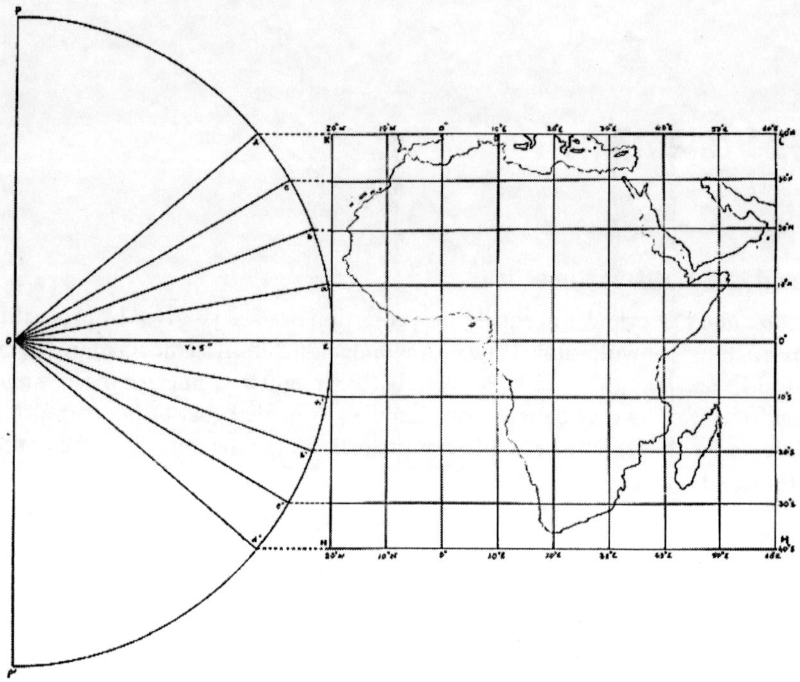

Figure 2: *Cylindrical Equal–Area Projection*

A tangent AV to the circle is drawn at A so VA is the projected radius of the 50 N latitude. ∠rOQ = 10° is the given interval between the latitudes. Therefore, Qr is the true distance between the two parallels at 10° intervals. Now, with centre 0 and radius Qr a semi-circle is drawn which bisects OA at b. From b, a line parallel to OQ is drawn, and the line VA meeting OV at point a is also drawn. Thus, ab is the longitudinal distance between two meridians at an interval of 10° along the standard parallel. Then VO in the centre of the paper is drawn with centre V and radius VA an arc of 50° N is drawn. From the standard parallel, points for the parallels are marked along the central meridian. Similarly, along the arc of 50° N, the distance 'ab' is marked for the spacing of meridians is done. Straight lines from V passing through these points are drawn and thus, the complete graticule is obtained (Refer Figure 3).

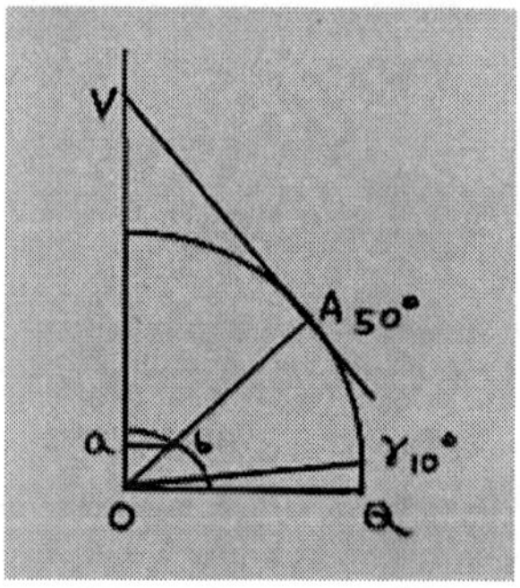

Figure 3.

Problem: Prepare a graticule for Europe on a scale of 1:70,000,000 at an interval of 10° with latitudinal extent – 10° N to 90° N and longitudinal extent – 170° W to 50° W.

Properties: The parallels are concentric circles and the meridians are straight lines and have the correct scale along them. The projection may be completed for one

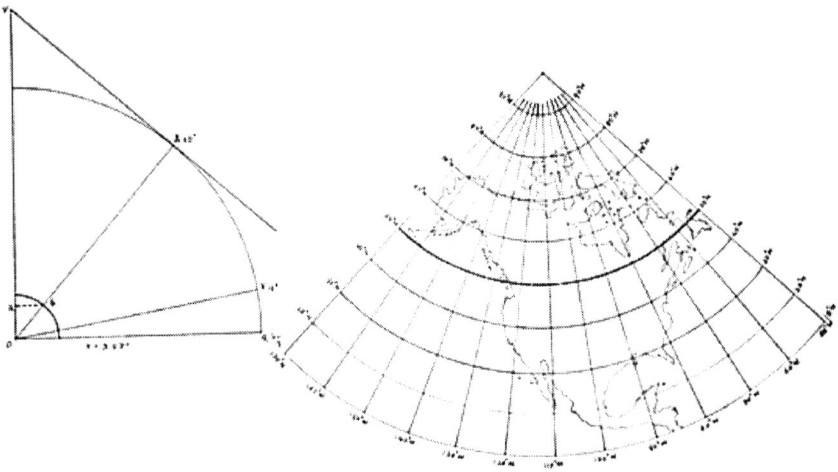

Figure 4: *Simple Conical Projection with One Standard Parallel*

complete hemisphere but suitable for temperate mid-latitude regions, which can be accurately represented.

Polar Zenithal Equal Area Projection

Introduction: This projection was designed by J.H. Lambert in 1772. Like other zenithal projections in this system also the meridians are straight lines, drawn at true angular distances radiating from the pole and parallels are concentric circles. The circles of latitudes become closer away from the pole. Their spacing is so adjusted as to make it an equal area projection. This is commonly used for polar areas.

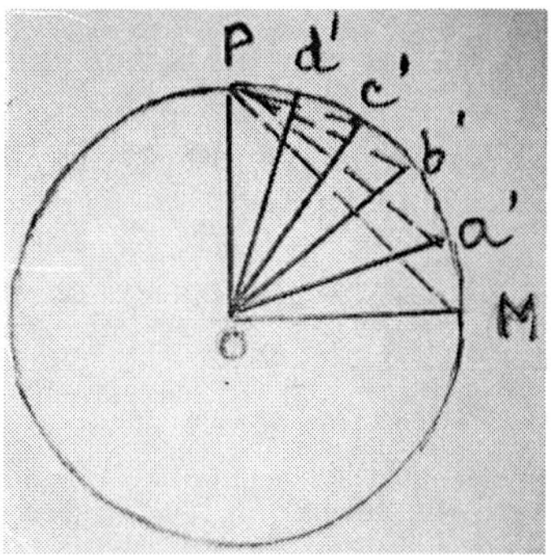

Figure 5.

Graphical Construction: With centre O and with the measurement of the reduced sphere, a circle is drawn to represent the reduced sphere. Through O, a vertical line to meet the circumference at P in the north is drawn to represent the polar point of the reduced sphere. At 15° interval, the angles are marked off from the Equator to the Pole. P is joined to the point of intersection of the arms of the angles of the circumference (Refer Figure 5).

With the centre O and radii each equal to these radiating lines circles to represent the different latitude from 0° to 75° N is drawn. Through 0°, a vertical line / Central Meridian is drawn and on either side of it points at 15 intervals are marked. From the centre to these points, radiating lines are drawn to represent the longitudes.

Problem: Prepare gratitude for Arctic Region on a scale of 1:125,000,000 at an interval of 15° with latitudinal extent - 0° to 90° N and longitudinal extent of 180° E to 180° W.

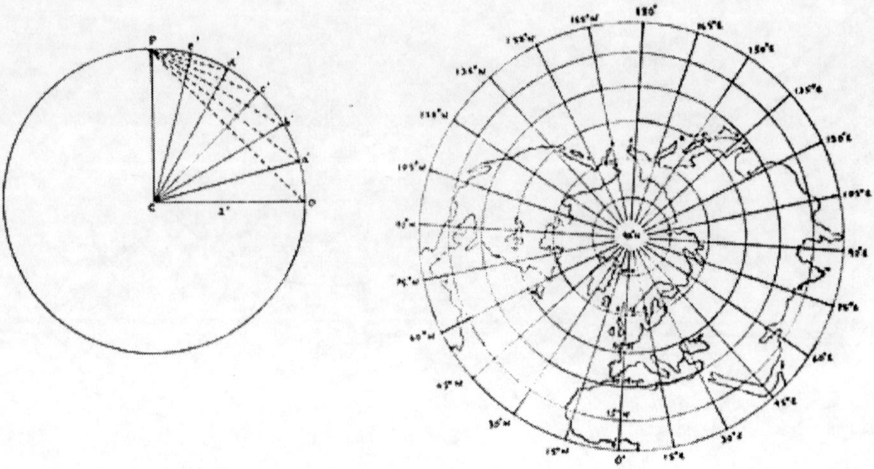

Figure 6: *Polar Zenithal Equal–Area Projection*

*Properties***:** This projection is suitable for polar regions, but distances are observed at the edge of the map. The meridians show the correct bearing or azimuths of all the points from the centre of the map. As the area is maintained, there are distortions in the shape of the continents. But in this projection, the world can be represented in hemispheres only, with the pole as the centre.

10

Topographical Maps

Topographical maps, also known as general-purpose maps, are drawn at relatively large scales. These maps show important natural or physical and cultural or non-physical features such as relief, vegetation, water bodies, cultivated land, settlements, and transportation networks, etc. Other regional maps are drawn by National Atlas and Thematic Mapping Organisation (NATMO).

The Survey of India (SOI) prepares the topographical maps of India for the entire country. These topographical maps are drawn in the form of a series of maps at different scales. Hence, in the given series, all maps employ the same reference point, scale, projection, conventional signs, symbols, and colours.

The topographical maps in India are prepared in two series, i.e.

1. India and Adjacent Countries Series and
2. The International Map Series of the World.

1. *India and Adjacent Countries Series:* Topographical maps under India and Adjacent Countries Series were prepared by the Survey of India till the coming into existence of adjoining countries was abandoned and the Survey of India confined itself to preparing and publishing the topographical maps for India as per the specifications laid down for the International Map Series of the World. However, the Survey of India for the topographical maps under the new series retained the numbering system and the layout plan of abandoned India and Adjacent Countries Series.

   The topographical maps of India are prepared on 1:10,00,000, 1:250,000, 1:1,25,000, 1:50,000, and 1:25,000 scales providing a latitudinal and longitudinal coverage of 4°×4°, 1°×1°, 30′×30′, 15′×15′, 5′×7′30″ respectively. The numbering system of each one of these topographical maps is shown in Figure 1.
2. *The International Map Series of the World:* Topographical maps under the International Map Series of the World are designed to produce standardized maps for the entire World on a scale of 1:10, 00,000, and 1:250,000.

Reading of Topographical Maps

The study of topographical maps is simple. It requires the learners to get acquainted with the legend, conventional signs, and colours on the sheets. The conventional signs and symbols depicted on the topographical sheets are shown in Figure 5. The whole of India and adjacent countries bounded by Latitude 0° to 40°N and Longitude 60°E to 100°E were subdivided into uniform sections of 4°×4° dimensions from 1 to 137, implying that the fundamental basis of these topographical maps are the well-defined latitudes and longitudes of the Earth.

These 4°×4° topographical sheets are drawn on Lamberts Conical Orthomorphic Projection to the scale of 1:10, 00,000 (or 1inch to 16 miles) and are also known as 'Million Sheets' or 4′ sheets or 1M Maps. The present political or international boundary organization, there are 137 1M sheets covering all the major parts of India.

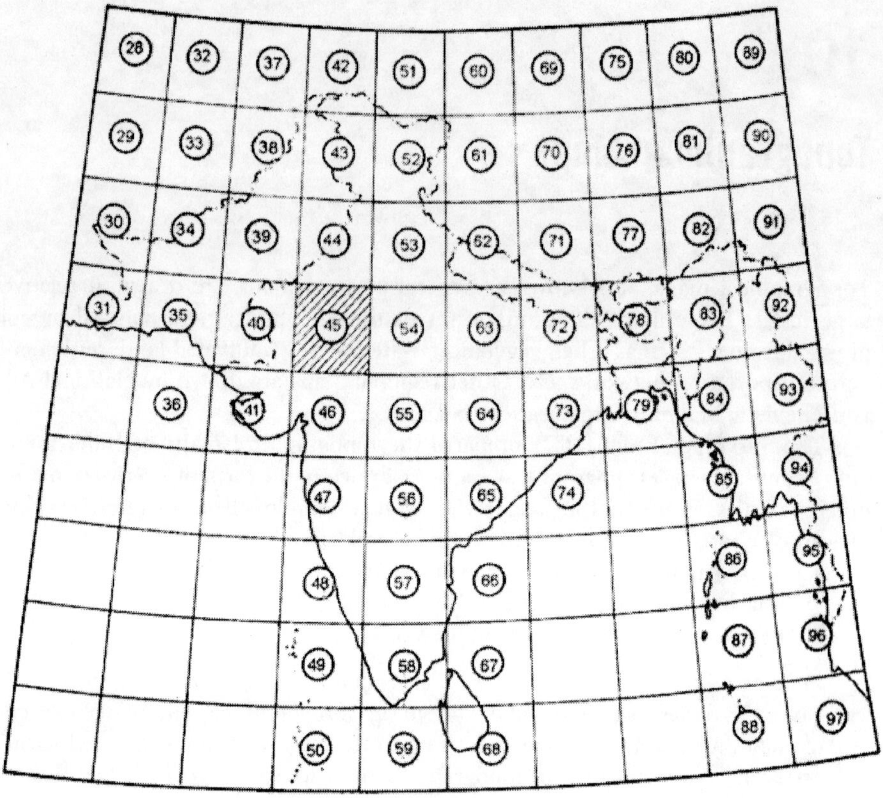

Figure 1: *The Map Series Covering the Indian Subcontinent*

Each of these 1M sheets, Toposheet No. 55 is further divided into 16 uniform divisions of 1°×1° dimensions and are designed by 16 letter symbols of English alphabets from A to P as depicted in Figure 2a. These 1°×1° sheets are drawn to the scale of 1:250,000 (or 1 inch to 4 miles) and are referred to as 'quarter-inch sheets' or 'degree sheets'. (Refer to Table No 1)

Topographical Map 55D/7

Table 1: *The Different Sheets of Indian Topographical Maps (According to Scale)*

Sheets	*Scale (Metric System)*	*Scale (FPS System)*	*Dimensions*	*Contour Intervals*	*Examples*
Million Sheet or 1M Map	$1:1,000\times10^3$	1 inch to 16 miles	4°×4°	—	55
Quarter-inch or Degree Sheet	$1:250\times10^3$	1 inch to 4 miles	1°×1°	100m/250ft/500ft	55D
Half degree or Half inch Sheet	$1:100\times10^3$	1 inch to 2 miles	30′×30′	50m/100ft/250ft	55D/NW
Quadrant Sheet or Inch Sheet	$1:50\times10^3$	1 inch to 1 miles	15′×15′	20m/50ft/100ft	55D/7
Special	$1:25\times10^3$	—	5′×7′30″	10m	55D/7/5

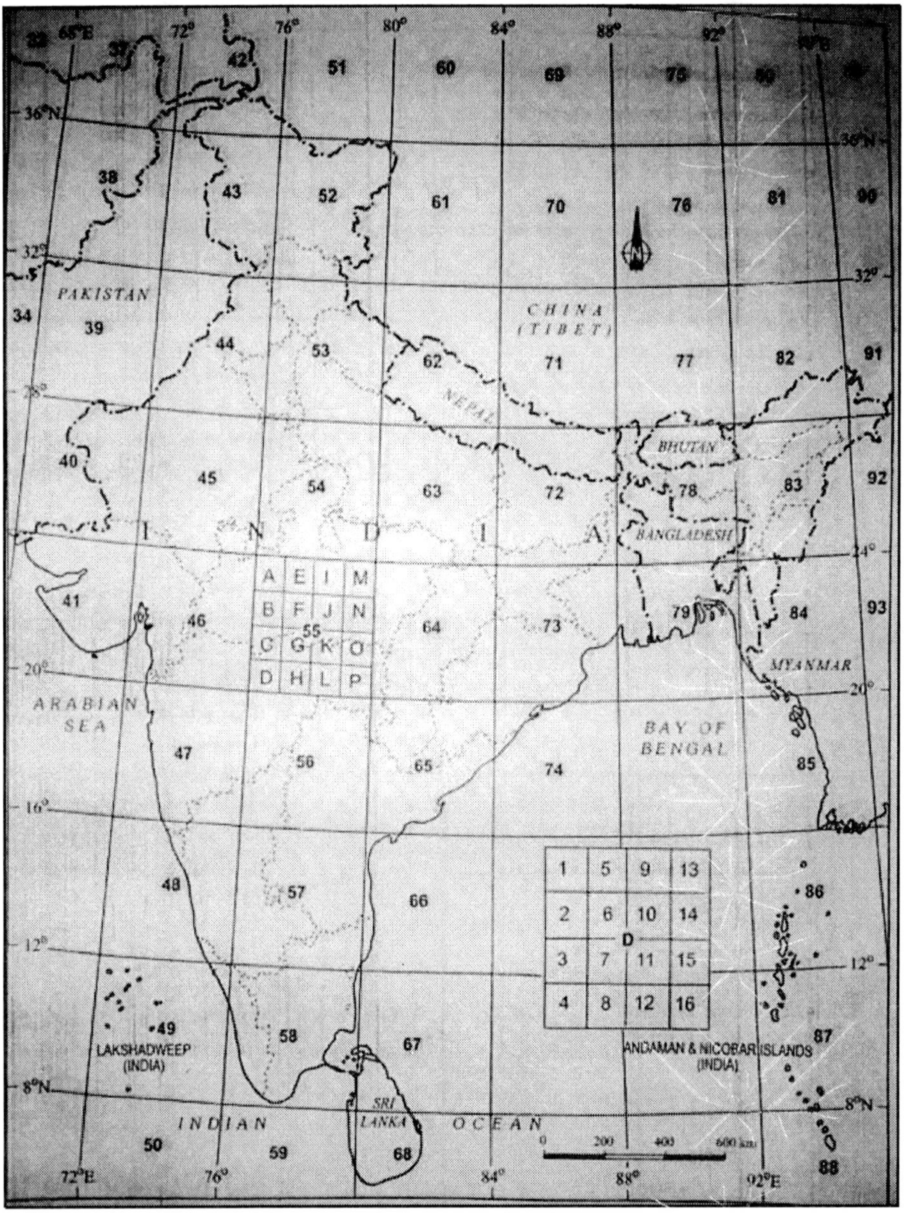

Figure 2a: *Reference of Topographical Sheets Published by Survey of India (SOI)*

Further, each of these degree sheets, for example, D is subdivided into 16 parts of 15′×15′ dimensions and is numbered from 1 to 16. These squares are drawn to a scale of 1inch to 1mile, commonly marked as 1:50,000, but being 1:63,360 to be exact in the ratio form. Each of the squares is identified by an index number; for example, the one shown hatched in Figure 2b, has the index number 55D/7.

Each of the 1:50,000 sheets is further subdivided into four 1:25,000 scale sheets of $7^1/_2{}' \times 7^1/_2{}'$ dimensions and the four squares are labeled NW, NE, SW, SE as shown

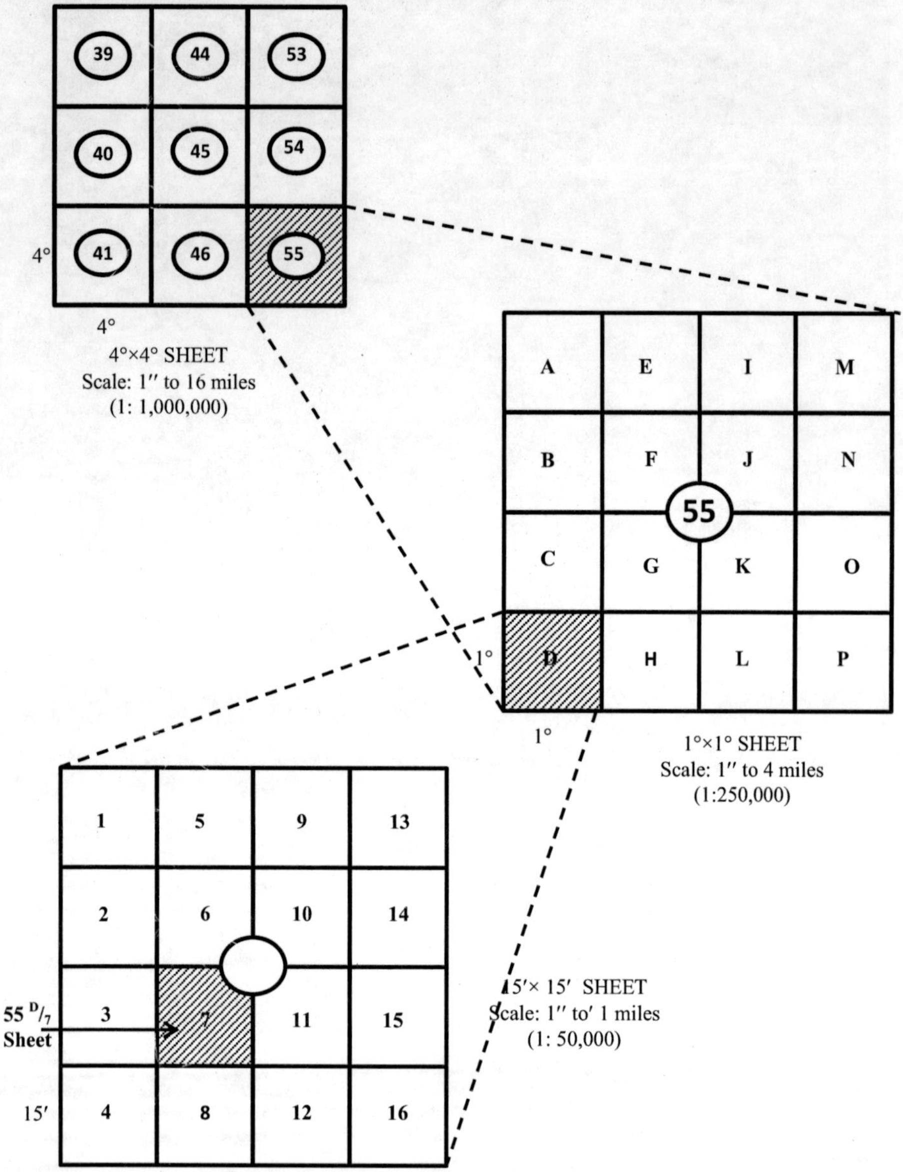

Figure 2b: *The Different Subdivision of the Square 45 with the Classification of the Scale*

in Figure 3a, and identified appropriately. Alternatively, the sheet is divided into 6 divisions (5′ latitude × $7^1/_2$′ longitude) and numbered 1 to 6 as depicted in Figure 3b.

Grid Reference

The topographical maps bear the national grid of squares drawn to the scale of 2cm = 1 km, with each square having a side measuring 1 km. These are the sheets most commonly used for various purposes and are of vital practical importance.

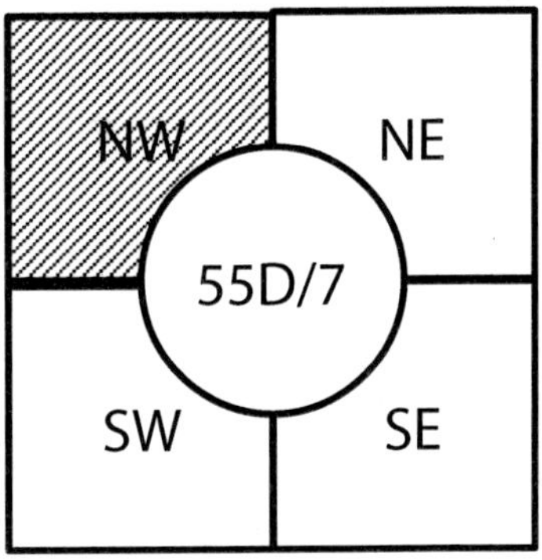

Figure 3a: *1:25,000 Topographical Maps (7½' latitude×7½' longitude)*

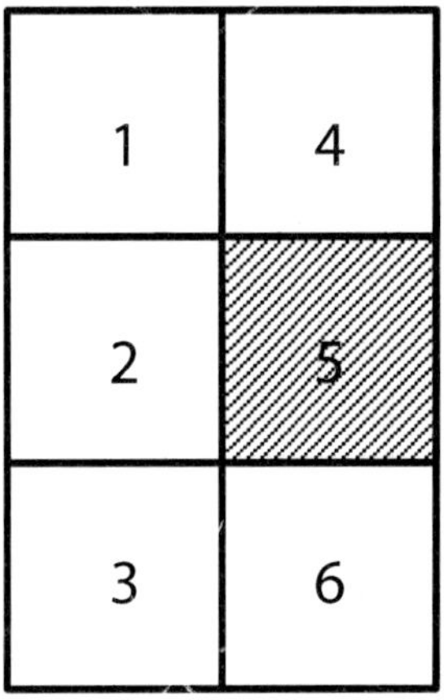

Figure 3b: *Old-layout Topographical Maps (5' latitude×7½' longitude)*

Figure 3a and 3b: *The Detailed Division of a Square*

With reference to any of these sheets, the vertical lines are called Eastings since they are numbered from West to East. Similarly, the horizontal lines are called Northings because they are numbered from South to North (Figure 4.)

This network of horizontal and vertical lines, or the Eastings and Northings, is called the Grid. It is also identified as the Grid Reference.

Principles of Map Interpretation

The topographical map gives a representation of the area on paper, using conventional signs and symbols, in order to describe the area of land use represented on the map. The objective of map interpretation is to bring out the relationship between various physical and cultural features.

Physical Features: The contour lines provide the basis for the study of the relief features of an area. The location of the highest point and lowest point and their height depict the general slope of the area which can also be inferred from the direction of major rivers in the area. Contrasts in slope may be examined with reference to the spacing of contour lines and physical features as valley or lowland, upland, ridge, hill, cliffs, rock outcrops, sand hill, ravines, etc. may be inferred.

Drainage Features: The main river and the directions in which they flow, whether the rivers are perennial or not, and the presence of waterfalls, rapids, etc., may also be noted. The presence of dams, canals, and distributaries may indicate that the rivers are used for irrigation.

The presence of non-perennial streams shows that the region probably receives low rainfall. The location of tanks, wells, springs, marshes, and swamps, and other drainage features bring out the relation between the relief and drainage features.

Land Use: The type of natural vegetation such as forest, grass, or scrub gives an indication of the climatic conditions of the area. The relation between physical features and land use may also be examined. As, the area probably gets low rainfall the distribution of wastelands, and sparse vegetation with palmyra and cactus is widespread. A large area

Topographical Map

State: West Bengal — Toposheet No.:73 M/10

Districts: Barddhaman and Birbhum

EASTINGS

NORTHINGS

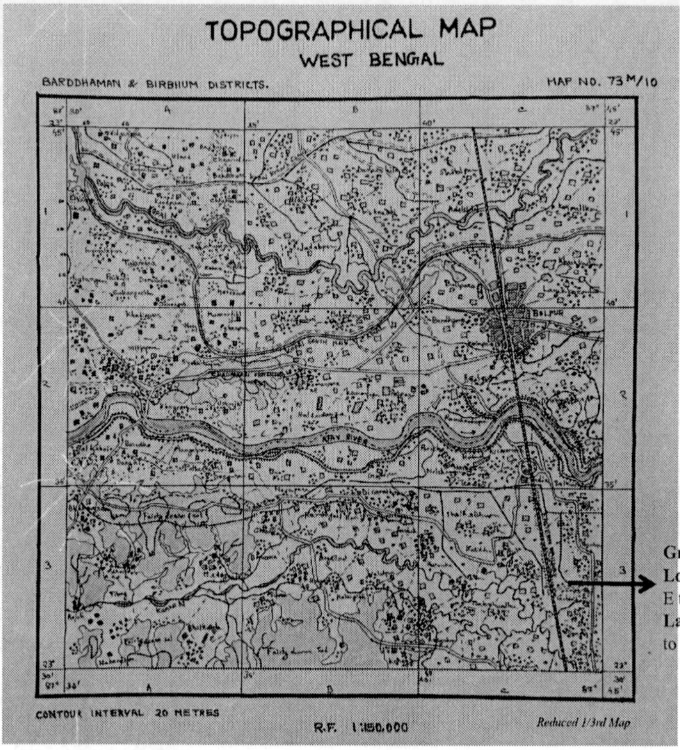

Grid No. C3
Longitude: 87°40′ E to 87 °45′ E
Latitude: 23°30′ N to 23°35′ N

Contour Interval: 20 Meters **R.F. 1:50,000**

Figure 4: *Grid Reference of the Topographical Map*

of dense forest may indicate that the area probably receives high rainfall. Cultivated land is likely to be found to a large extent in lowlands and river valleys, whereas, regions of comparatively high relief and steep slopes are likely to have forest or scrub jungle and they are not suitable for cultivation.

Settlements: The average distance between villages in different parts of the area signifies the settlement to be nucleated or dispersed. Settlements are likely to be more closely spaced in the valleys and lowlands than in the uplands, plateaus, and hilly areas. The site of settlements may be on the banks of rivers, near tank bunds, or near wells. These settlements are called *wet-point settlements* as they are near the sources of water supply. In such cases, one may infer that the region probably received low rainfall, and settlements seek sources of water for their location. In low-lying areas such as the flood plains of rivers or in the deltas, settlements are located on higher ground above the level of floods and may be located on river terraces, higher mounds, or near the foothills away from river courses. Such settlements are called *dry-point settlements*. Some settlements are located on hilltops or near forts. These are *strong-point* settlements.

In some areas, typical patterns of settlements may also be found. Linear patterns of settlements are found along roads, river banks, levee ridges, etc. Nucleated, square, or rectangular patterns of settlements may occur at the junction of roads. Dispersed settlements may be found in highly dissected forest regions, plateaus, coastal area and the areas vulnerable to floods.

There may be evidences of urban or suburban growth, nodal centres with transport routes and other amenities. It may be of square, rectangle, star-like, or circular shapes. The deserted village sites may be in forested regions or flood-prone sites.

Transport and Communication: The railway lines may be broad gauge or metre gauge; single track or double track and there may be cuttings, embankments, or tunnels along the railway line. Railway lines are not found in the low-lying zones of the land, rugged relief features, or in areas of sparse population.

The patterns of roads in the area may be metalled or non-metalled. Many settlements are linked by cart tracks or footpaths, bridges, or causeways across rivers which may have ferry services. The route pattern of roads and railway lines are related to the physical features of the area.

Occupation of the People: The occupations of the people may be inferred directly and indirectly from the evidence regarding land use and settlements. If the area has large tracts of cultivated land, the dominant occupation is likely to be agriculture which may be emphasised by irrigation from tanks, wells, or canals and cultivation is likely to be of the intensive type with wet crops like rice and sugar cane, being the common ones. If the area does not exhibit any evidence of any source of irrigation, cultivation is likely to depend on rainfall only. Places of quarries or mines are likely to provide occupation to the people. In the case of villages only along the coast, fishing is likely to be important. Large cities are likely to provide employment in industries, trade and services. These may be brick kilns, textile or rice mills, carpet factory, cement factory, etc. In the case of towns which are headquarters of districts and *taluks*, a large number of offices, administrative jobs, and other related services of the tertiary sector may be marked on the map which provides employment.

Transect Chart

Usually, *Transect Charts* are drawn to present the correlation between the physical and cultural elements present in a particular topographical area. An ideal section line is taken cutting through all the elements. This section line represents almost all the physical and non-physical (cultural) elements possibly (maximum extent) present in the map (may be rural or urban based).

Usually, it is within a frame of 5 to 6 tiers. The lowest tier marks the relief features (relief typicals present) with undulations in accordance with the cross-section taken from the topographical map. The second tier comprises the drainage network (both perennial, and non-perennial rivers), water bodies, and broken banks of rivers. The third tier depicts the natural (RF and PF) and planted vegetation with species-types. The fourth tier presents the transport (road, railways, water or ropeways or may be airways) and communication network (Post Office, Telegraph Office, Post and Telegraph Office,

Police Station). Minor transport routes like cart tracks and pack tracks in rural areas need to be considered. The top-most tier represents the settlement characteristics, types, and patterns. These may be around favourable sites of low relief, transport linkage, availability of water, cleared vegetation, agricultural land and other infrastructural amenities. The maps of the older edition had settlements (black in colour) whereas the newer editions have settlements (red in color).

These elements are arranged and designed in a frame so as to draw a correlation between all the above-mentioned elements and also to draw the inference about the climate, economy and, occupational pattern. Moreover, the problems and prospects of the region can also be identified. This process of identification would enable the planner to bring about regional development in the area by the initiation of several beneficial projects.

Precautions need to be taken so that the main river channel has no overlaps of vegetation, communication transport linkages and, settlement points in their respective tiers.

Conventional Signs or Symbols Used in Topographical Maps

The different geographical elements are described with different signs in topographical maps. So, to study the topographical sheets it is very important to have a concept about such conventional signs. These signs are developed and used by the Survey of India. Such signs have the following characteristics:

(a) Conventional signs are not drawn to the scale.
(b) The signs are easily understandable.
(c) In some special cases signs are selected in such a way that a separate explanation is not needed.
(d) Use of colour is very important like blue for water bodies, yellow for agricultural area, green for vegetation cover etc.

To interpret the topographical maps the following signs and symbols are to be studied:

1. (i) Metalled road, according to importance, with milestone, (ii) unmetalled road, according to importance, with milestone. 2. Cart-track 3. Pack-track 4. Footpath with bridge 5. Pontoon bridge 6. Ferry or Ford 7. Railway and roadway on river-bridge 8. (i) Roads in dry bed, (ii) Unlined Stream 9. Canal 10. Dam 11. Big stream—(i) perennial, (ii) sandy bed, (iii) waterline in sandy bed, (iv) riverbed with islands and rocks. 12. Tidal creek 13. Waterfalls with altitude 14. Rapid 15. Steep stream bank with height and broken bank 16. Submerged boulders 17. Shallow submerged bank 18. Swamp or Marsh 19. Reeds 20. Well—(i) lines, (ii) unlined. 21. Tube-well 22. Spring 23. Tank—(i) perennial, (ii) fluctuating, (iii) perennial with an embankment. 24. Excavated tank—(i) perennial, (ii) fluctuating, (iii) perennial with an embankment. 25. Embankment along roads 26. (i) Unmetalled road on an embankment along tank, (ii) unmetalled road beside embankment along tank 27. Broad-gauged railways—(i) double line, (ii) single line with stations, (iii) under construction. 28. Other railway- (i) double line, (ii) single line with stations (iii) under construction. 29. Tramways 30. Railway under tunnel 31. Railway station with siding 32. Telegraph line 33. Level crossing 34. Contours—(i) lines in shape, (ii) stony waste (iii) cliff. 35. Sand dunes—(i) flat, (ii) heaped, surveyed, (iii) transportational 36. Village—(i) permanent, (ii) relict. 37. Huts—(i) permanent,

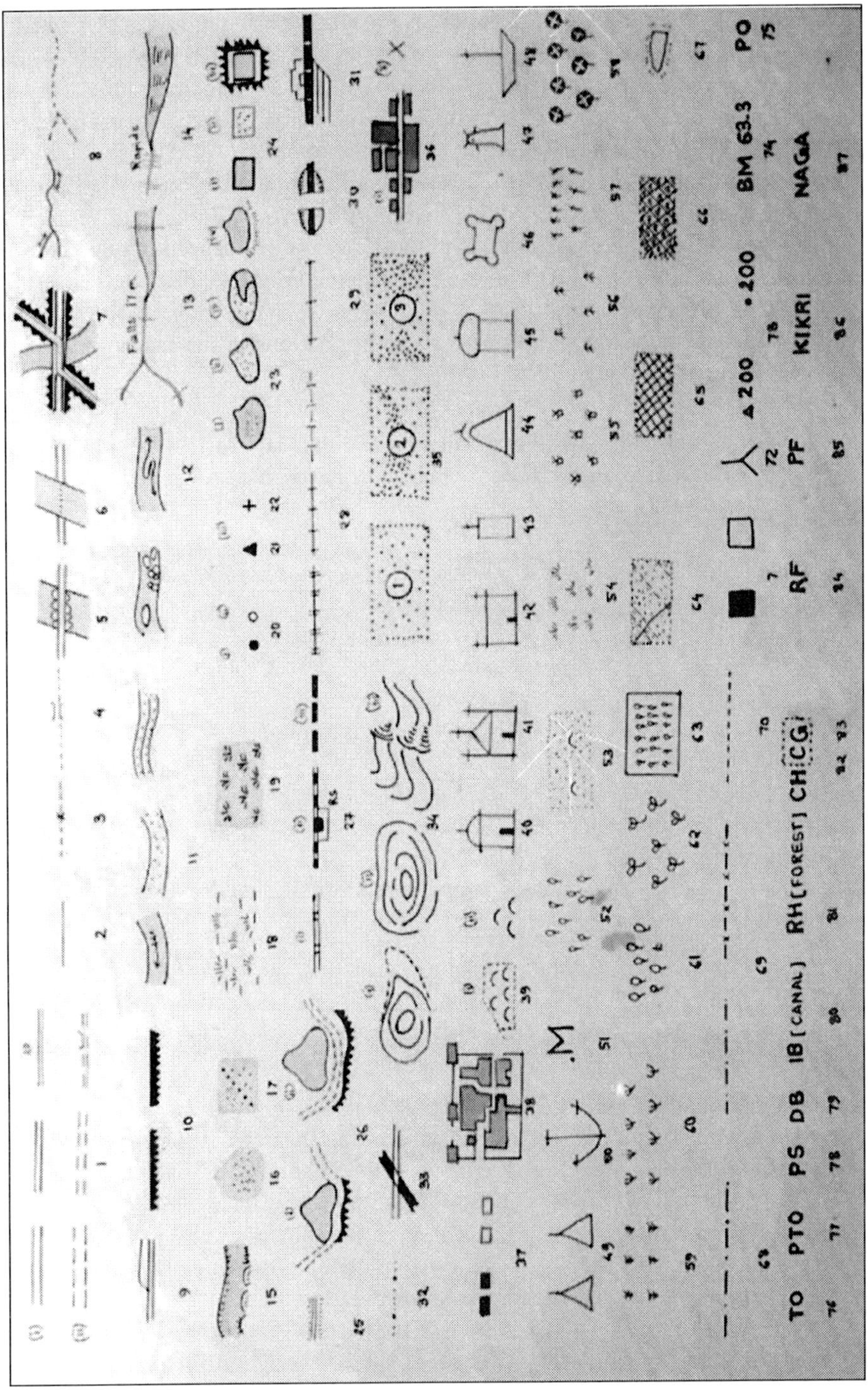

Figure 5: *Conventional Signs or Symbols Used in Topographical Maps*

(ii) temporary. 38. Village with walls 39. (i) Graveyard, (ii) Graves 40. Temple 41. Mosque 42. Idgah 43. Church 44. Pagoda 45. Tomb 46. Fort 47. Lighthouse 48. Lightship 49. Buoy—lighted, without light 50. Anchor 51. Mines 52. Trees 53. Shrubs 54. Grasses 55. Bet 56. Coniferous tree 57. Palmyra trees 58. Palms 59. Bamboos 60. Cactus 61. Other trees 62. Banana trees 63. Gardens 64. Tea garden 65. Betel 66. Vegetables 67. R.F. boundary 68. International boundary 69. State boundary 70. District boundary 71. Boundary pillar- (i) Surveyed, (ii) Unsurveyed. 72. Triangular point of village. 73. Heights surveyed 74. Benchmark- 63.3m 75. Post Office 76. Telegraph Office 77. Post and Telegraph Office 78. Police Station 79. Dak Bungalow 80. Inspection Bungalow 81. Rest house 82. Chimney 83. Circuit house 84. Reserved Forest 85. Protected Forest 86. KIKRI (Tribes) 87. Naga (Tribes).

A few specimens of topographical map interpretations of the plateau and plain regions have been discussed in the following section.

Topographical Map Interpretation (Map No.1)

(This map represents features of a plateau geographical area)

Introduction

1. Topographical sheet number : 73 $^{A}/_{7}$
2. State : Part of Bihar
3. Districts : Parts of Palamau and Ranchi
4. Extent : Latitude- 23°15′ N to 23°30′ N, Longitude-84°15′ E to 84°30′E.
5. Scale : 1 : 50,000 (2 cm to 1 km)
6. Contour Interval : 20m
7. Area : 703.8 sq.km
8. Surveyed in : 1979-80
9. Published in : 1982
10. Surveyor : Major General Girish Chandra Agarwal, Surveyor General of India

From the primary observation, it is revealed that the region has two distinct subdivisions- the northern portion represents a dissected plateau while the southern portion represents an erosional plain.

Relief

The relief of the map area is dominated by high plateau region which is eroded by the rivers, therefore making it a dissected plateau region. The general slope of the land is from south to north along the direction of river flow. The map area is sloping towards the river Koel which flows through the middle of the area. The highest peak is 1068m in the west-central part of the map (Grid A_2). Though the river valley has the lowest relief, yet it is 600m. The region can be broadly divided into three sectors based on physiography. They are discussed in the following section-

Above 900m (hilly region): The entire western part and east-central part of the map falls under this category with average height of 980m. Though this is a plateau region, many eroded hills are found here, like conical hill, flat top hill, low ridges etc. along with a number of widened 'v'-shaped valleys.

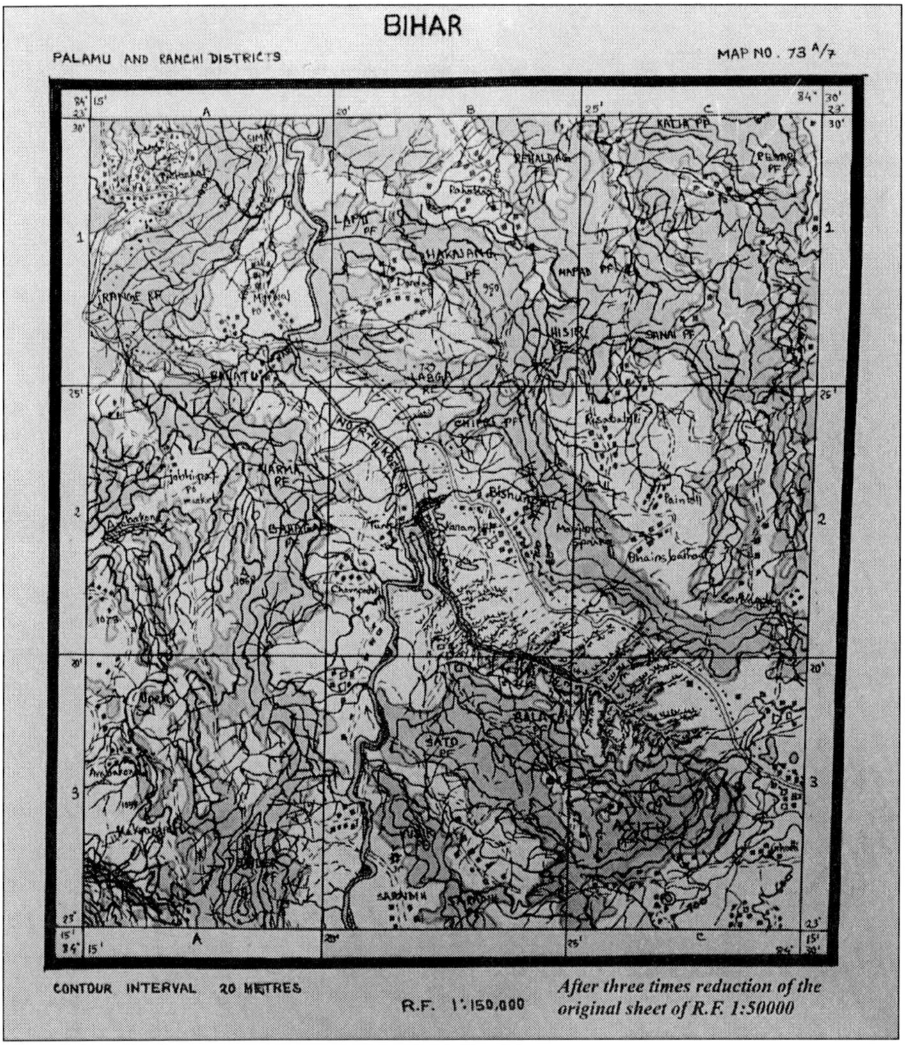

Figure 6: *Topographical Map of 73 $^{A}/_{7}$*

700-900m (higher plateau region): The region is mainly found in the west, east and central part of the area along with the middle part of the river valley. Though the average height is 900m, the region is sloping from the high hills towards the river Koel. In some places cliffs are noticed while in other parts gentle slopes along with Cols and Saddles are also found.

Below 700m (river valley region): This region has gentle slopes, but on the basis of height this can still be considered as a lower plateau region with average height of 650m. This region has a general slope from south to north.

Drainage

The main river of the area is Koel, flowing from south to north direction in the central part of the map. The river valley, though narrow, it is quite deep. It has divided the area

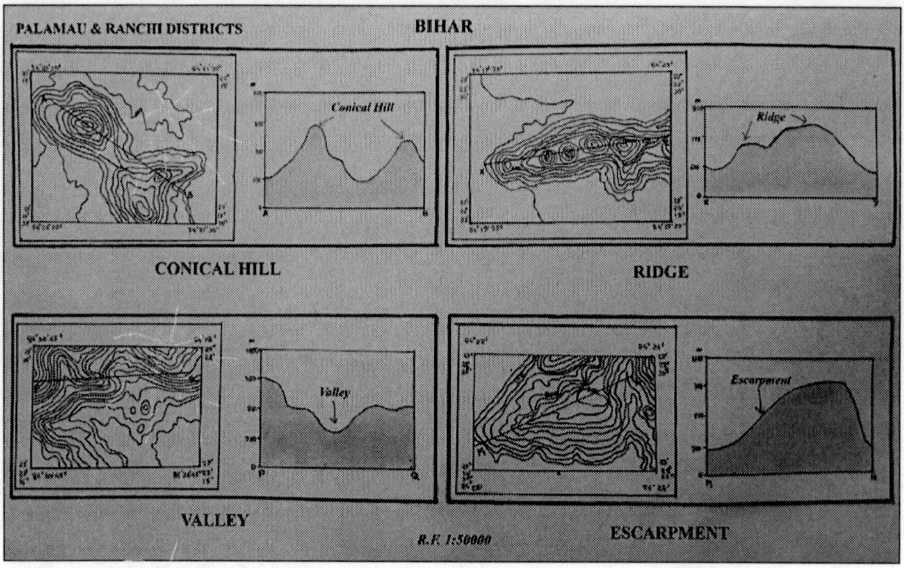

Figure 7: *Different Relief Features of the Topographical Map of 73* $^{A}/_{7}$

into two halves. Though perennial, huge amounts of sand and boulder are deposited in the river bed. The river is fed by numerous small streams on both sides. A big tributary Phuljhar river joins Koel in the central part of the area. Like Koel, Phuljhar also has a meandering course but it is non-perennial with the deep valley. Many drainage patterns are created by the small stream network throughout the area.

Drainage Patterns

Dendritic Drainage Pattern: It mainly occurs in those regions with uniform lithology. Here the alignment of tributaries and sub-tributaries are analogous to branching arrangement of the tree with a confluence angle typically acute. It is thus a well-developed drainage system with subsequent, consequent and obsequent streams.

The map area depicts a number of dendritic drainage patterns, all over it. A typical example of it is found near the Bishnupur settlement at 23°25′8′′ N to 23°26′6′′ N and 84°24′94′′ E to 82°24′4′′ E.

Parallel Drainage Pattern: It has developed due to alternate beds of hard and soft rocks. This is mostly found in areas having closely spaced counters, in high slope areas. In the map it is found at the extreme south- western corner near the Deoragang P.F. at 23°27′33′′ N to 23°27′57′′ N and 84°10′45′′ E to 84°20′ E.

Radial Drainage Pattern: It is mainly found in places where conical hill or dome-like structures are present from where the streams radiate out in different directions and is mainly controlled by relief. The most prominent example of radial drainage is at 23°27′33′′ N to 23°27′28′′ N and 84°22′9′′ E to 84°25′6′′ E.

Ring-like Pattern: It is mainly developed in places where hill and dome-like structures are present. The streams flow roundly of the hills and makes a ring-like, pattern. The most prominent example of ring- line drainage is found at 23°17′55′′ N to 23°19′ N and 84°15′ E to 84°17′ E.

Most of the rivers, here, form waterfalls. A large numbers of channel features like river-meander, badland topography and braided channels have also developed.

Natural Vegetations

About 80 per cent of the entire map area is covered by forest. Both protected and reserved forests are found here. The notable protected forests are Chorkakhar P.F., Barang P.F., Manata P.F., Porha P.F., Birmala P.F., Balato P.F., Sato P.F., Tahdar P.F., Saraideh P.F., Nindi P.F. etc. Parts of this Ranchi and Palamau districts also have reserved forests of which the most noteworthy are Rema R.F., Ranga R.F., Musir R.F. and the Katia R.F. Most of these forests are fairly dense with mainly the sal trees. There are also some open mixed jungles as well as some open scrubs or dense scrubs. However, plantations have been made in almost everywhere.

In some regions forests have been cleared for building up human settlements like parts of Menne P.F. Most of the forests are drained by a large number of tributaries of the North Koel river.

Settlements

The term 'settlement' refers to the establishment of inhabitation in a particular region. A number of settlements have been found in the Palamau and Ranchi districts of Bihar. As the area is quite rugged with thick forest cover, the settlements

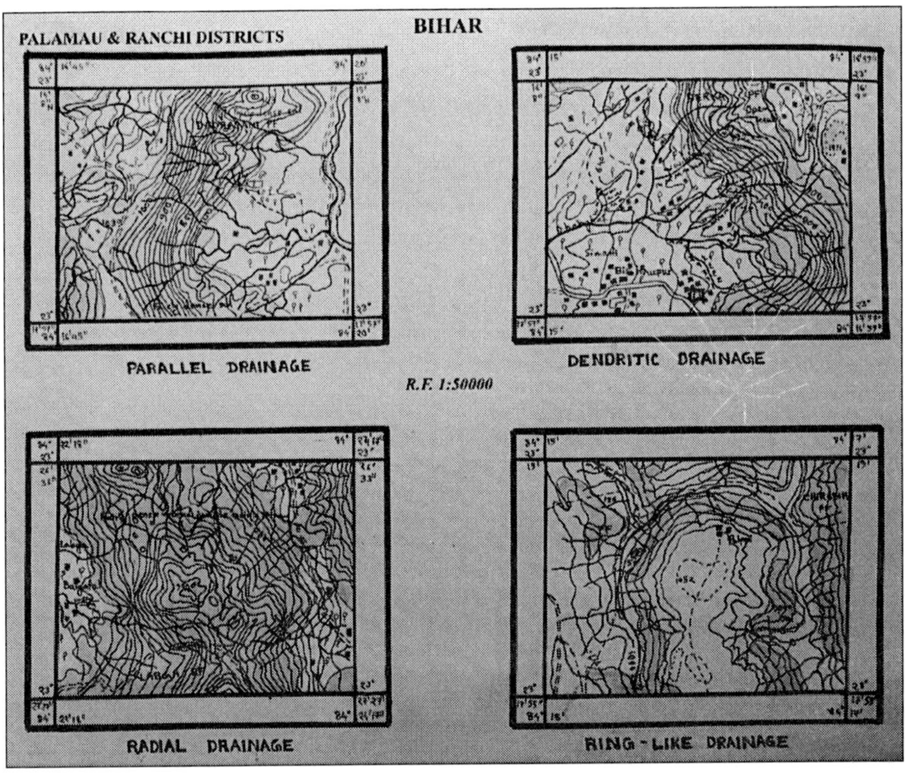

Figure 8: *Different Drainage Pattern of the Topographical Map of 73 $^{A}/_{7}$*

are not well developed here. Different types of settlements are found. These are mainly of three types:

Nucleated Settlements

Nucleated settlements are found at the junction of metalled roads and river. They develop in a region where many modern day civic amenities are present like post and telegraph office, electric sub-station, hospital, tanks. The communication network facilitates the growth of settlement.

Such a settlement is found in the area extending from 84°22′12′′ E to 84°23′57′′ E and 23°21′33′′ N to 23°22′28′′ N. The settlement is known as the Bishnupur settlement and has the amenities like overhead tank, dispensary, market, post office and telegraph office, police station and electric line and telephone line.

Linear Settlements

Linear settlements are usually developed along some important roadways or by the banks of the rivers. Such a settlement is found in the Palamau and ranchi districts of Bihar extending from 84°18′48′′ E to 84°19′20′′ E and at 23°26′51′′ N.

Isolated Settlements

Isolated settlements emerge as scattered hamlets which are set up surrounding to certain basic civic amenities like post office, market etc. Such a type of settlement is found in a latitudinal extension of 23°13′14′′ N to 23°30′ N and a longitudinal extension of 84°26′53′′ E to 84°30′ E. There are three settlements namely Jurni, Asurtali, Pirtaul which are located amidst the Birtau P.F.

Transport and Communication

Typically, the region does not possess a well-developed transport network. It is because there are extensive tracts of reserved and protected forests hindering the transport development. Mainly all kinds of roads *i.e.* both metalled and non-metalled are found in this area. The only broadest metalled road of the map stretches from north-western part *i.e.* Neterhat to south-eastern part of the map to Ranchi. The road covers a distance of about 100 km.

The area is also criss-crossed by a number of unmetalled road segments, which are usually motorable in dry season. It has been found that the metalled road extends from Neterhat, which is the focal point of traffic in this region to Ranchi. There are a lot of pack-tracks which serve as the connecting point between two villages. Hence transport development as a whole remained at a poor level except in some places like on levee, flood-plain margins of river North Koel and also on valley plains of some tributaries. The general patterns are –

(i) The valley of river North Koel is as such better served by roads.
(ii) The north-eastern part of the area has no means of communication except pack-track. Other parts enjoy more or less, better transport facilities.

Conclusion

The map area is not well-developed with proper infrastructural facilities due to the ruggedness of topography and extensive dense forest. The transect chart, a correlation technique, has been used to show the relationship between the physical and cultural features of a region. It has been found that in regions of high relief there is dense

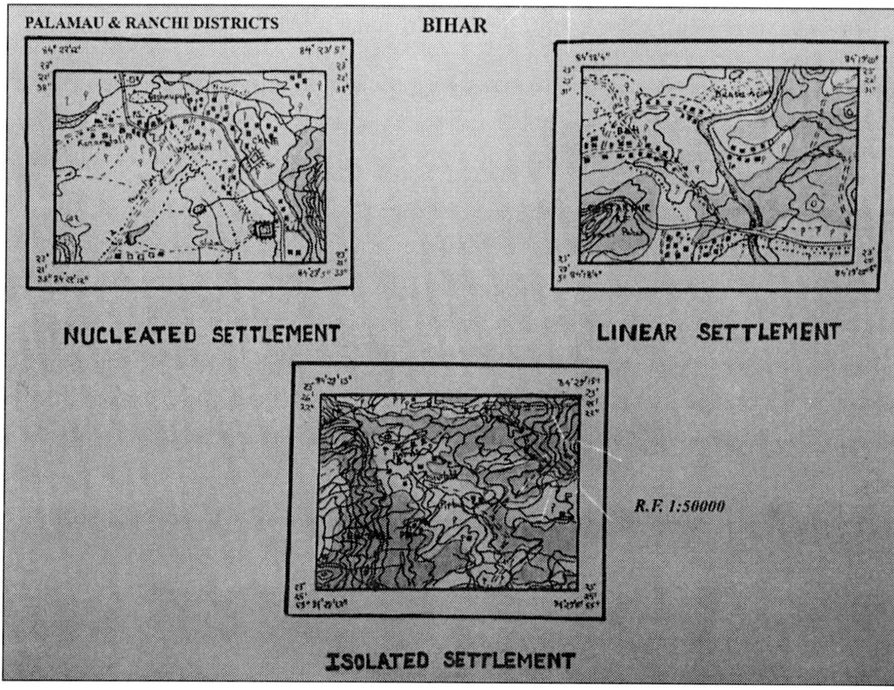

Figure 9: *Different Transport, Communication and Settlement Patterns of the Topographical Map of* $73\,^{A}/_{7}$

vegetation as the forests are not cleared to prepare agricultural fields in the regions. Moreover, these high relief zones have fewer roads and fewer settlements. The presence of comparatively non-resistant rock masses and the numerous drainage network have led to the denudation of the land surface and lowering of the gradient of the land. Only in these areas, the settlement with transport and communication system has developed. In the near future, this region too can develop with proper land-use planning.

Topographical Map Interpretation (Map No. 2)

(Only Relief and Drainage features have been discussed for this map)

Introduction

1. Topographical sheet number : $73^{B}/_{5}$
2. State : Bihar and Madhya Pradesh
3. Districts : Ranchi and Raigarh
4. Extent : Latitude- 22°45′ N to 23°0′ N, Longitude-84°15′ E to 84°30′E.
5. Scale : 1 : 50,000 (2 cm to 1 km)
6. Contour Interval : 20m
7. Area : 703.8 sq.km
8. Surveyed in : 1978-79
9. Published in : 1980
10. Surveyor : Kishori Lal Khoshla, Surveyor General of India

The general relief of the region represents dissected plateau topography. In the southern part, the relief is a high plateau region, where the highest contour is 784m. But the highest point of the region is 940 m which is observed in the eastern and north-eastern parts of the region. In the northern region, scattered high hills with a height above 700m are found in this region. The slope of the map is from the north to the south. In the western part, the highest height is 840m. There is a decline in slope from west to the middle of the map and from east to the middle of the map because of the

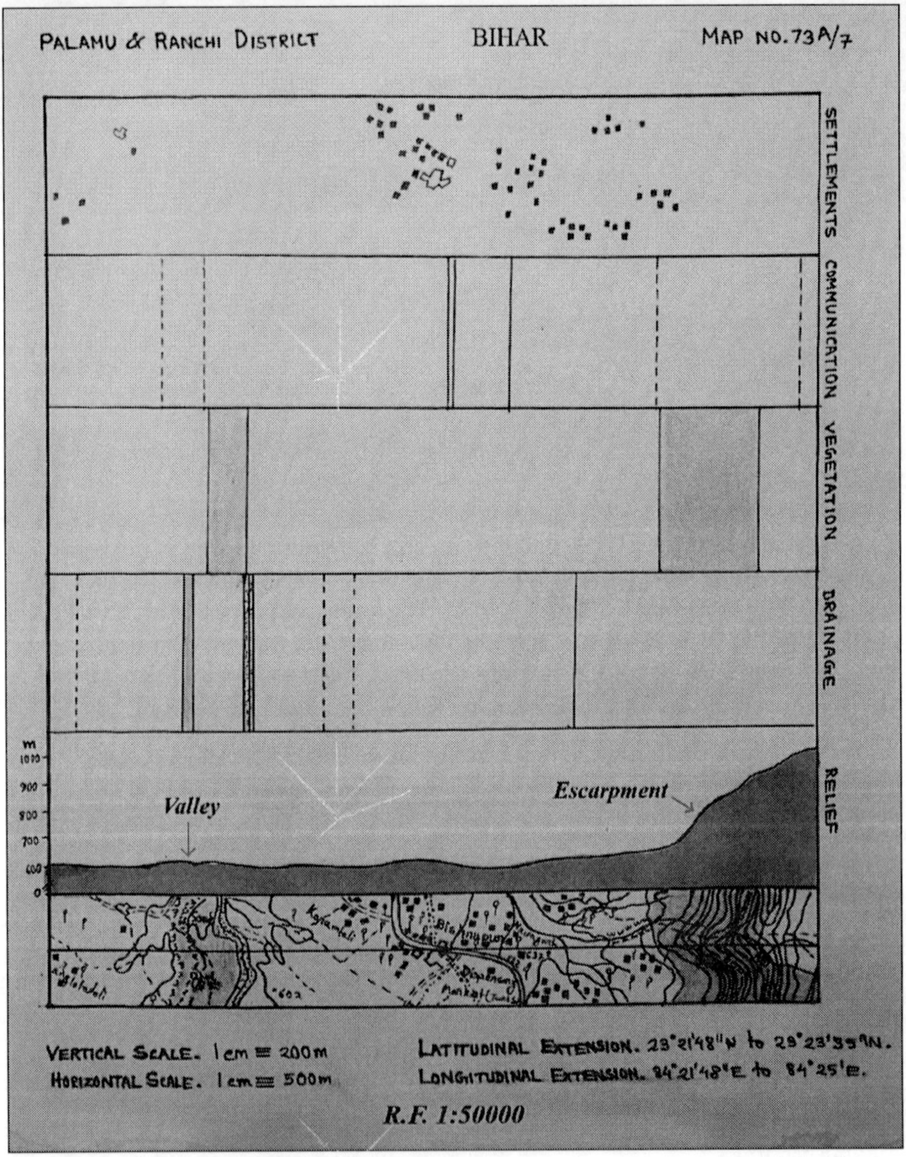

Figure 10: *Transect Chart depicting Interrelationship among Physical and Cultural Features of the Selected Area of the Topographical Map of 73 A/7*

river flowing through the middle of the region. But as a whole, the general slope is from north towards south. So it can be said the river Sankh dissected the region into two parts—eastern and western. In the north-eastern part, the highest relief is 893m and lowest contour is 740m. The lowest contour is found 440m in the southern region. The lowest elevation of the relief along the middle of the region has been observed because of the erosion of River Sankh. So many cliffs have been found in the southernmost part of hill-slopes. The dense drainage network made the region undulated in topography. The badland topography has resulted because of the erosion of numerous dip streams. A number of conical hills, dome-shaped hills, 'V'-shaped valleys, flat topped hills, ridges and saddles are observed. The stony waste and boulders are also typical features of this region. From all these features it can be said that this region of Ranchi and Raigarh districts is in the part of high plateau region. Some of the relief typical has been discussed in the following section.

Relief Features

Ridges and Saddle: The ridges are the elongated, sloping stretches of high-relief ground area. The middle portion of the ridges is seemed to be narrow and the slope beside the two sides of the narrow high regions is steep in nature. The contours of the ridges are elongated and spaced very close but some highly scattered contours are found in the middle, those mainly indicate the highest peaks of the ridges of resistant rock masses.

Saddles are dips between mountains and hills and along the erects of the ridges, ridge lines and spurs and are so named because they resemble a riding saddle viewed from the side. They can be wide or narrow, shallow or deep.

In the topographical map, the saddle and ridges are found in grid C3 with the extension of 22°53′ N to 52°52′ N and 84°27′30′′ E to 84°30′ E. This depicts simple ridges and saddle as it appears on a topographical map and in profile given. The saddles here are mainly formed because of the erosion of the rivers.

Conical Hills: A conical hill is a hill with a uniform slope and is represented by concentric contours spaced almost regularly. Highest contour value is observed in the middle of the circular contour and the lowest contour is observed in the outermost part of the circular contours. This is the hill that is round at the base and rises up to a sharp peak like a cone.

In the topographical map, the conical hill is found in the northern portion within the extension of 22°57′ N to 22°58′ N and 84°23′20′′ E to 84°25′9′′ E. Here, the highest elevation is 940m.

Sheet Rock and Rocky Knob: The sheet rock are the large area of rocks where the overlying soil layers have been eroded and removed due to mechanical weathering. In the topographical map a sheet rock is found in A3 grid. The highest relief in this portion is 740m.

Rocky knob is totally a barren rock and exposed as a rock exposure. The trails of the rocky knob vary from easy to strenuous.

Cliff: A few small areas of steep slopes have formed cliff like features with contours adjacent to each other. Such cliffs may be related to collapse of the lower part of the slope or to any other structural cause.

In the map, a cliff has been found within the extension from 84°27′ E to 84°28′30′′ E and 22°47′ N to 22°48′ N. The cliff has been formed because of erosion by the dip stream.

Isolated Hills: These hills are found at advanced stage of dissection. An isolated hill may be a monadnock too. Over such hill radial pattern of drainage network develop.

In the topographical map, isolated hills have been observed within the extension of 22°53′ N to 22°54′55′′ N and 84°23′53′′ E to 84°24′50′′ E. The lowest contour is 640m and highest contour value is 720m. This dissection has occurred by the distributaries of Sankh river in grid B2.

Valley: The valley is a low area between hills or mountains, often with a river running through it. The valley of the river may be of different shapes, such as 'I' shaped, 'V' shaped, or wide 'V' shaped.

In the topographical map a widened valley with 'V' shape has been observed. These widened 'V' shaped valleys are formed mainly for the erosion of the river in the youthful stage. The contour valleys are increased from the inner portion toward the outer portion of the valley. The highest contour is 600m and the lower contour is 580m in the grid C3.

Flat-Topped Hills: A flat-topped hill is an elevated area of the land. A flat top and the sides that have usually steep cliffs are known as butte, mesa. The mesa has a top wider than its height and they are all formed by erosion and weathering of horizontally layered rocks. In the given map, the highest contour value of flat-topped hill is 784m situated in the B3 grid extending from 22°46′30′′ N to 22°47′ N and 84°21′30′′ E to 84°22′30′′ E. Buttes, though absent in the map have greater height and top is less wide.

Dome: The irregular contour, and slightly spaced, rounded in shape are the domes. These are not very high in relief. They are also found in an isolated and scattered manner in the given map.

Drainage Patterns

The drainage pattern means the form of the drainage systems and the spatial arrangement of streams in a particular locality or region. Many drainage patterns are created by small streams network throughout the area.

Parallel Drainage Pattern: Parallel drainage pattern comprises numerous rivers which are parallel to each other and follow the regional slope. This pattern is mostly found in the area having closely spaced contours in high slope areas. In the map it is found in the northern part extending, from 22°58'30" N to 22°59'30" N and 84°17'30" E to 84°19' E.

Radial Pattern: Radial drainage pattern also known as centrifugal pattern is formed by the streams which diverge from a central higher point in all directions. It is obvious that dome structures, volcanic cones, conical hills etc. are favourable for the development of the radial pattern. In the map, a radial drainage pattern has been found on a conical hill in the northern part.

Annular Pattern: Annular pattern also known as circular pattern which is formed when the distributaries or the tributaries of the master consequent streams are developed in the form of a circle. The most prominent example of annular pattern is found in the south east corner of toposheet. Such pattern is developed over a mature and dissected dome mountain characterized by series of alternate bands of hard rock and soft rock beds.

Rectangular Pattern: Rectangular pattern shows some degree of resemblance drainage patterns join their master streams almost at right angle. The rectangular pattern is characterized by widely spaced tributaries. This is found in the map area extending 24°51'30" N to 22 °55' N and 84°25' E to 84°26' E.

Pinnate Pattern: Pinnate pattern is developed in a narrow valley flanked by steep ranges. The tributaries originating from the steep sides of the hills and join the master

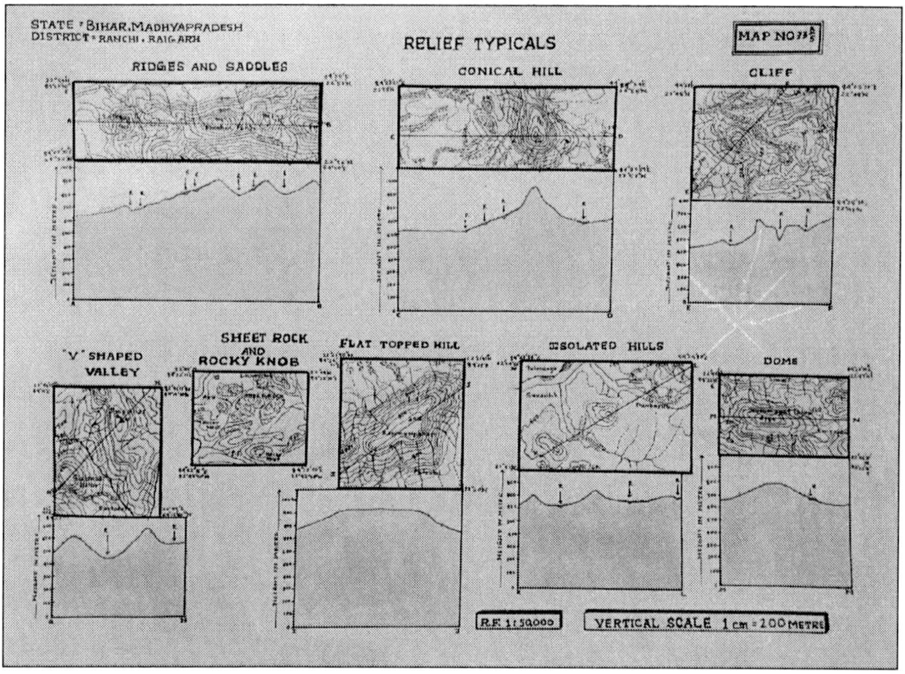

Figure 11: *Different Relief Features of the Topographical Map of 73 $^{B}/_{5}$*

stream in acute angle. This pattern resembles the veins of a leaf. In the southern part a pinnate drainage pattern have found.

Barbed Drainage Pattern: Barbed drainage pattern is developed in a region where the surface in underlying with the hard rocks. This is a rare kind of drainage pattern is formed when the distributaries are flow in opposite direction to their master streams. The tributaries join their master streams in a hook-shaped bend. This type of drainage pattern have found within 22°41' N to 22°50' N and 84°25'40" E to 84°28' E.

Herring Bone Pattern: Herringbone drainage pattern also known as rib-pattern, is developed in mountainous areas broad valleys are flanked by parallel ridges having steep hill side slopes. The longitudinal consequent streams as master streams are developed in the longitudinal parallel valleys while tributaries as lateral consequent after originating hill slopes. Just this type of drainage pattern found in the region extending 84°27' E to 84°28' E and 22°54' N to 22°54'50" N.

Centripetal Pattern: Centripetal or inland drainage pattern is opposite to the radial drainage pattern because it is characterized by the streams which converge at a point which is generally a depression or a basin. This pattern is formed by a series of streams which after emerging from surrounding uplands converge into the basin in the topographical sheet extending 22°53' N to 22°54'40" N to 84°29'40" E to 84°53' E.

Drainage Characteristics

The main river of the area is Sankh river – flowing from north to the southern of the area. This river dissected the whole region into two equal halves. This river is perennial in nature, huge amount of sand and boulder is deposited in the river bed and is fed

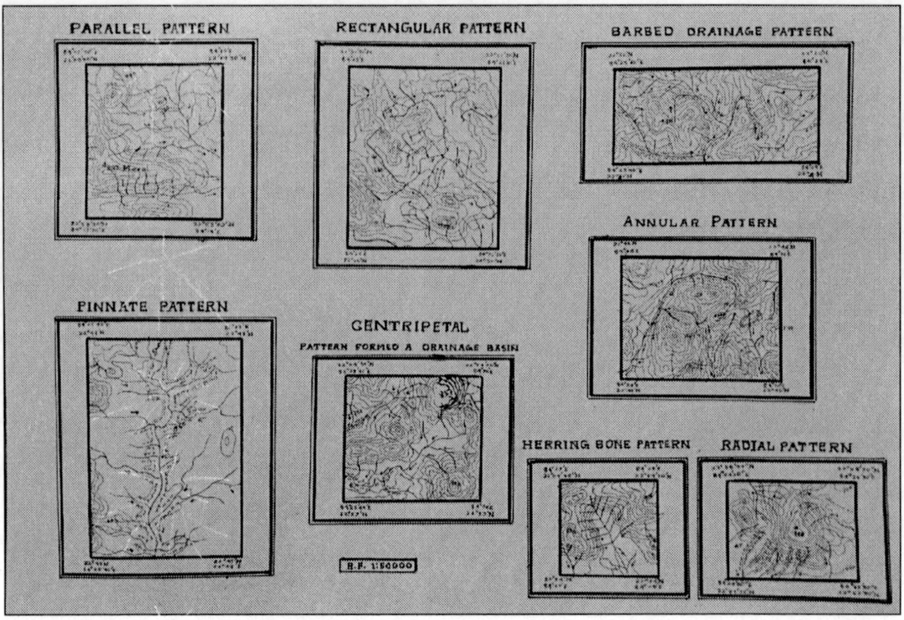

Figure 12: *Different Drainage Pattern of the Topographical Map of* $73\ ^{B}/_{5}$

by numerous small streams on the both sides. A big tributary Bala Nadi joins the Sankh river at the northern part of the area. River Sankh is meandering in nature. The distributaries which join with Sankh in its left side are Milmili Nala, Silphi Nadi and the distributaries which join with Sankh in the right sides are Bahri Nala, Bala Nadi etc. The river is fordable during November to June. In this lower course of the river several riverine features are ideally depicted in the region. They have been discussed in the following section.

Point Bar: A point bar is a depositional landform attached to the bank. In a meandering channel it is crescent shaped. Point bars are formed along the linear banks of the channel in the river Sankh.

Mid Channel Bar: The mid-channel bar has been observed in Sankh river, where the river bifurcates itself into two main parts. These mid-channel bars sometimes lead towards the formation of a braided channel.

Meander Bend: The meandering channel is characterized by pools (deeps) and riffles. In the river Sankh, a meander bar has been observed in loops within the extension of 22°24' N to 22° 30' N and 84°22' E to 84°23'40" E.

Straight Channel: "Truly straight channels scarcely exist outside the laboratory, except where a headwater stream is firmly held on the line of fault" – stated by G.D. Dury, 1969. In the natural system straight channel may be possible up to stretches of 100 meters or so. The straight reaches are generally characterized by riffles at crossovers. In the topographical sheet of Ranchi district the river Sankh is passing through a part of plateau region which formed granite- gneiss and some highly metamorphosed soft rock. So, the southern part of Sankh river became straight channel because of the geology of the area.

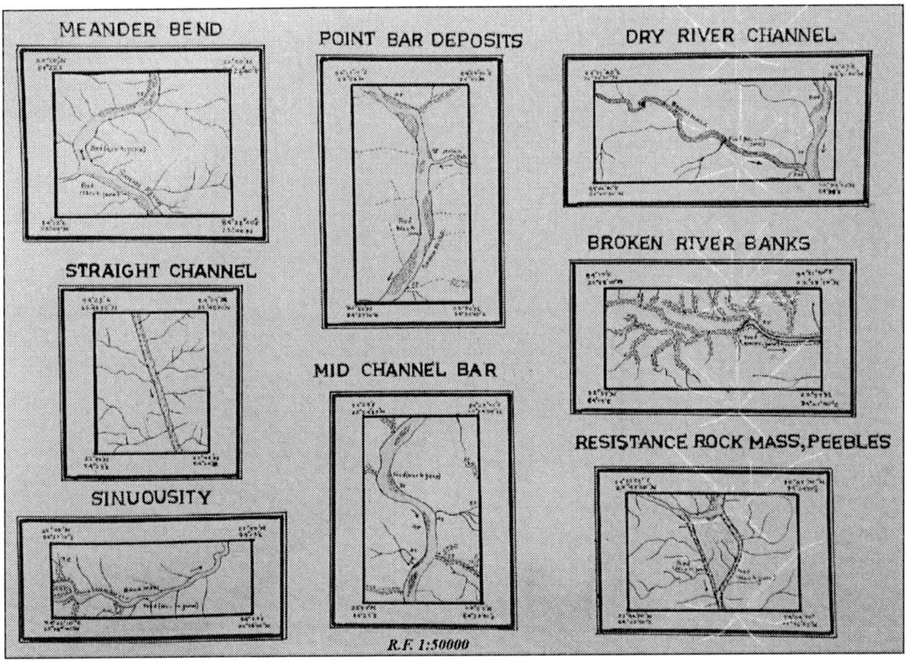

Figure 13: *Different Drainage Characteristics of the Topographical Map of 73 $^{B}/_{5}$*

Sinuosity: The shape of longitudinal segments of river channel is considered in terms of sinuosity which refers to departure of actual channel course from expected theoretical straight line. A channel having the sinuosity of 1.05 to 1.5 is defined as the sinuous channel according to Leopold and Wolman (1957). Sinuosity is the ratio of the river length (distance travelled it one floated down the river) divided by the length of straight river flood (bird's flight distance from source to mouth).

Topographical Map Interpretation (Map No. 3)

(This map represents features of a plain geographical area)

Introduction:

1.	Topographical sheet number	:	73$^{M}/_{10}$
2.	State	:	Part of West Bengal
3.	Districts	:	Parts of Birbhum and Barddhaman
4.	Extent	:	Latitude—23°30′ N to 23°45′ N, Longitude—87°30′ E to 87°45′ E.
5.	Scale	:	1 : 50,000 (2 cm to 1 km)
6.	Contour Interval	:	20m
7.	Area	:	703.8 sq.km
8.	Surveyed in	:	1970-71
9.	Published in	:	1972
10.	Surveyor	:	Dr. Hari Narain, Surveyor General of India

From the primary observation, it is found that the area covered by the map has a great variety of physical and cultural features, and it can be said that it is a plain area.

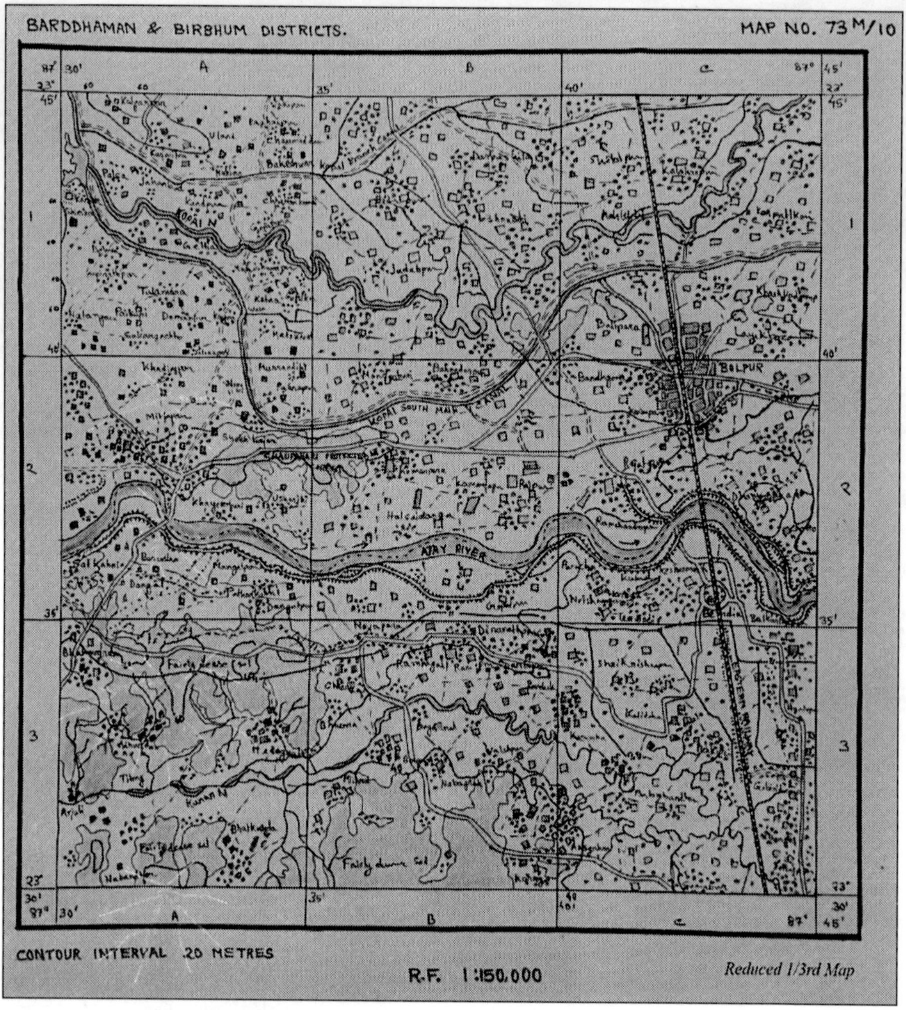

Figure 14: *Topographical Map of 73 $^{M}/_{10}$*

Relief: The general slope of this area is from west to east as is revealed by the value of contours and from the direction of river flow. The highest points are found in the east-southern portion and the lowest points are found in the west-southern portion. The general value of the contours ranges from 40m-100m. In the northern side, it can be seen that mainly the contours are varying between 40m- 60m. The south-eastern part of the map is consisting of the contour which is below 80m. The highest area in the map is on the northern side. By observing the map it can be said that it is mainly a depositional plain.

Drainage: An overall view of the topographical map shows that the region has one important perennial river and two non-perennial rivers. The main river is Ajay. It flows from western side to eastern side. It can be seen that the river is in its lower stage and have some depositional features on both sides of the banks.

There is the presence of another river named the Kopai Nadi. It flows from the north-western side to north-eastern side. It passes over Monohorpur. This river is perennial and a canal is also there named 'Kopai South Main Canal'.

Kunur Nadi is another river. It flows from south to western zone. Almost half of the year the river remains dry. In this area there are many tanks. In the northern side there is one reservoir near Kanchannagar. Kopai South Main Canal is joined with it.

Natural Vegetation: The vegetation cover in this area is almost absent. The vegetation is mostly restricted to trees planted by man. But some parts of the upper west-southern side have natural forest cover where some fairly dense sal forests have been found. However, there is a protected forest in Chaupahari in the western part of the region. Open scrubs are also found in a small portion of southern side near Warlshpur. There are some reserved forests near Santiniketan.

Settlements: The term 'settlement' refers to the establishment of inhabitation in a particular region. A number of settlements have been found in the Barddhaman and Birbhum districts of West Bengal because of the availability of the favourable factors for establishing settlement. Some settlement patterns have also been identified in this region. They are as follows:

Scattered Settlements: Scattered settlements have developed in the north-western side of the map extending from 23°39′30′′ N to 23°40′ N and from 87°31′ E to 87°32′30′′ E. The settlement like Sakargachho, Khadimpur, Jalanagar are present on unmetalled roads. These are localised near the points of available resources and amenities.

Linear Settlements: The places which are very much prominent for this type of settlement pattern is in the north-eastern side. The metalled road coming from Siuri and the Eastern Railway of Barddhaman Branch is responsible for this type of settlement. These have developed on both sides of the metalled road in C_1 grid between 23°43′30′′ N to 23°44′ N and 87°42′ E to 87°45′ E. This settlement named Sarbanandpur settlement enjoys the facilities of Kopai river, post office, dispensary and other communication facilities etc.

Dispersed Settlements: In the southern side of the map there are dispersed settlement. The houses and buildings are isolated away from each other. It includes Natunnagar, Deolbardh, Hedagarld with a latitudinal extent from 23°32′ N to 23°33′ N and longitudinal extent of 87°30′30′′ E to 87°32′ E. The availability of large tracts of agricultural land and the forested area has resulted in the dispersal of settlements.

Compact Settlements: This type of settlement is seen in the north-western side of the region. The name of the town is Bolpur. It is situated near Ajay river. Power line passes through the area. The site is marked with Post Office, Telegraph Office, Police Station, Visva-Bharati University, Hospital, and Market. This significant settlement is extending from 23°40′15′′ N to 23°41′30′′ N latitudes and from 87°40′ E to 87°42′30′′ E longitudes. There are also some reserved forests in this area.

In the western part of the map, another clustered settlement is found. The name of the places are Islamabad and Bhedia Nagar. They are also situated near Ajay river and Eastern Railway. The presence of PTO PWD Office, Police Station and Market (which opens every Monday and Friday) make the place important.

Transport and Communication: The transport and communication are through railway, roadway and motorway. The following discussion can give a brief idea about the modes of transport in the area.

- Railways: Eastern Railway is mainly found in this area. The railway line connects from north to southern side (to Barddhaman at a distance of 32 km) from Ahmadpur. There are four railway stations *i.e.* Prantik R.S., Bolpur R.S., Pickri and Dihal R.S. There are also some foot-bridges.
- Roadways: There is a national highway in the south-eastern side of the area. Two roads coming from Durgapur and Pivlganga meet with the highway and make a junction in Ilambazar and goes forward Kesarmath. Now from Bolpur, the road makes straight towards southern side. Regarding communication, a well-established network has been introduced through Telephone lines also.
- Waterways: The rivers are fordable during October to June which facilitates cheap connectivity.

Conclusion: The map is of a developing area of the Barddhaman and Birbhum districts. The transect chart shows the relationship between the different physical and cultural features. The relief varies from 45m to 60m. It has been drawn with a latitudinal and longitudinal extent of 23°36′30′′ N to 23°36′ N and 87°35′30′′ E to 87°37′30′′ E respectively. There is an important river *i.e.* Ajay and some non-perennial streams also. The Chaupahari Protected Forest is in this area. There are settlements like Gopalpur, Karnagar, Shichendrapur developed beside the railway and metalled road. These transport routes add special importance to the development of the area.

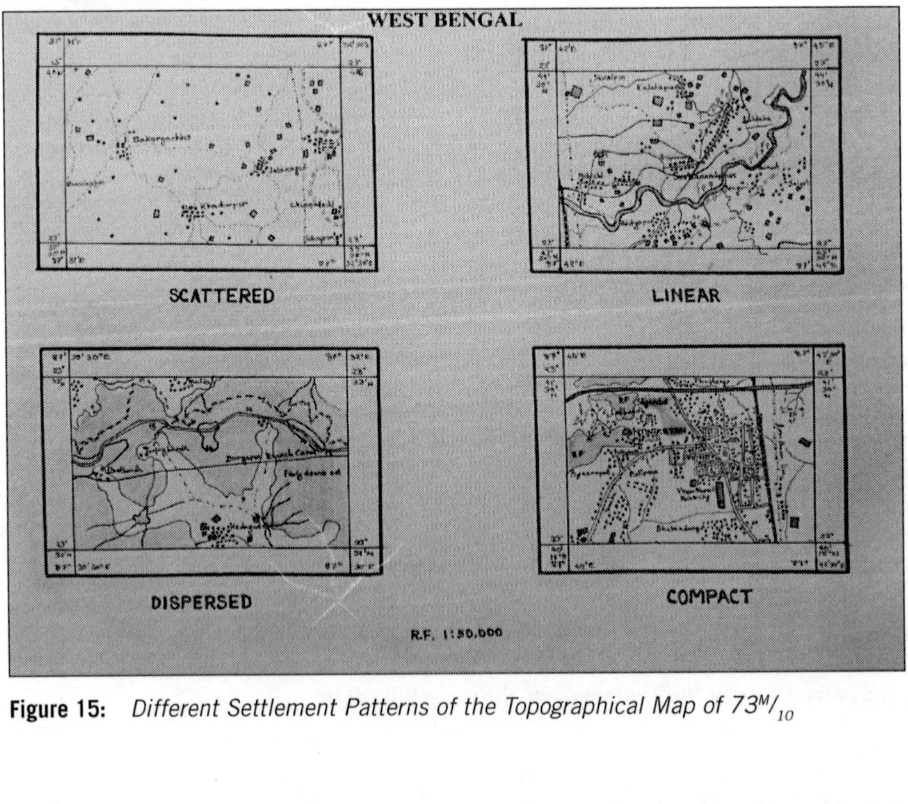

Figure 15: *Different Settlement Patterns of the Topographical Map of* $73^{M}/_{10}$

Due to the lack of fertile agricultural tracts and industrial establishments, the growth and prospect of settlements in this region are very feeble. The local inhabitants have mainly settled around hospitals, market places, post offices from where they can avail services and communicate with others. Man has constantly endeavored to make the region suitable for their living. Thus, the advancement of science and technology and the fruitful implementation in the betterment of practical life will definitely help to surpass the hindrance caused by lacking resources for the development of the region.

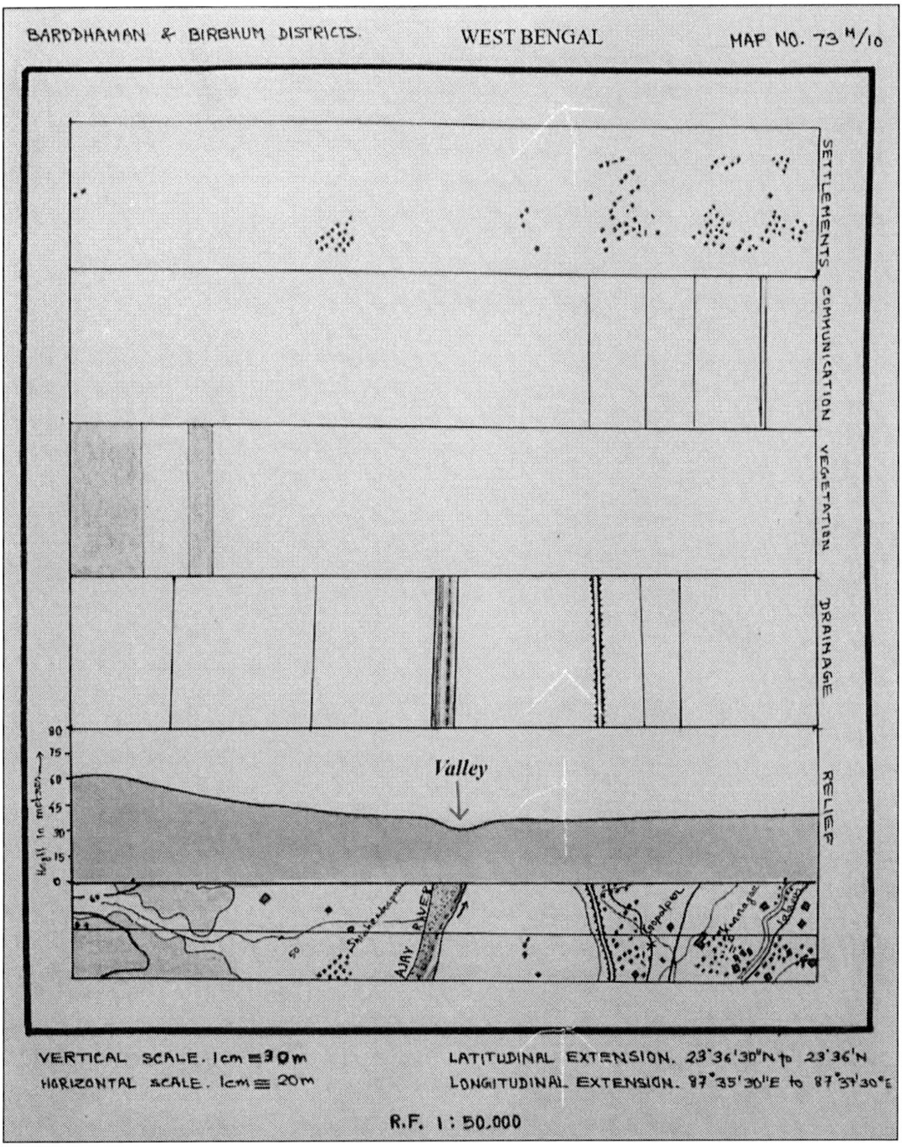

Figure 16: *Transect Chart depicting Interrelationship among Physical and Cultural Features of the Selected Area of the Topographical Map of 73 $^{M}/_{10}$*

Interpretation of Serial Profiles

Including Superimposed, Projected and Composite Profiles – A Specimen

Problem: To study the serial profiles of the area which has a latitudinal extent of 24° 20' 40" to 24° 25' N and longitudinal extent of 83° 50' to 83° 55' E.

Data base: The topographical sheet number is 63 P/5. The area lies between 24°15' to 24°30' N and 83°45'E to 84° E. The map is published in 1930. The magnetic variations

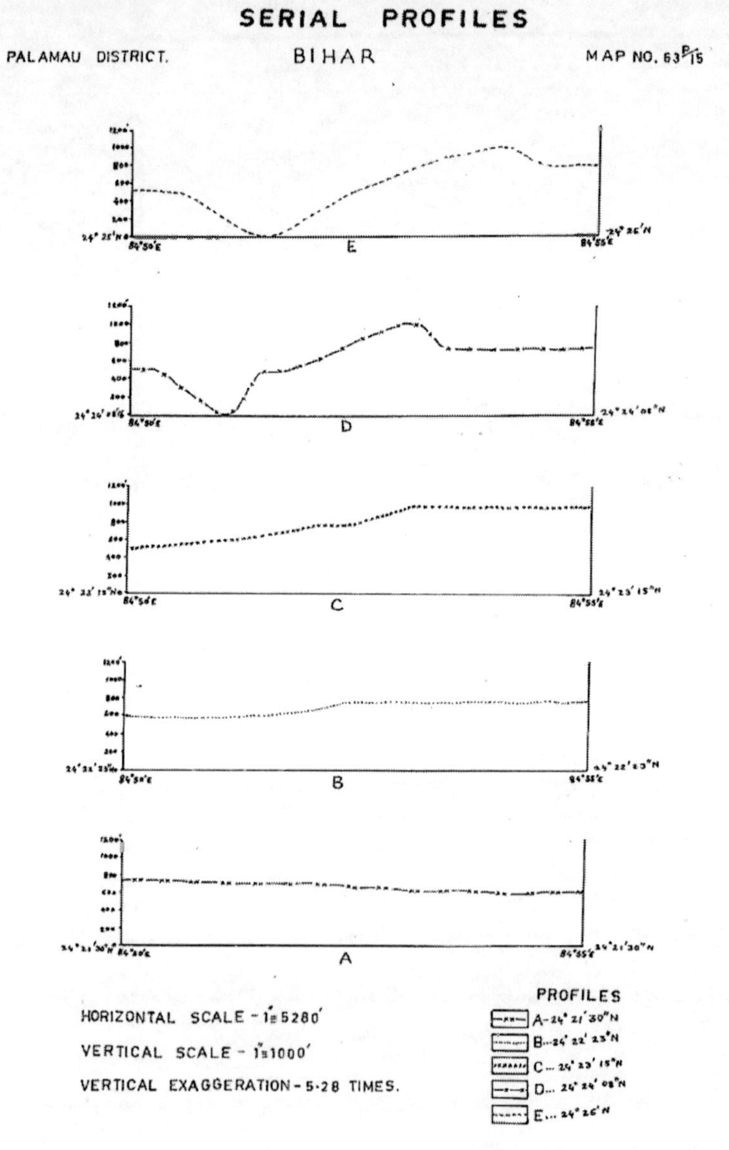

Figure 17.

from true North were about I' W in 1953. The contour interval is 50'. The region is situated in the Palamau district of Bihar. The scale of the map is 1 inch to 1 mile.

Objective: The main objectives of the drawing of serial profiles are:
(i) To show important geomorphic features of the area concerned
(ii) To compare and interpret the relief and
(iii) Finding out the probable controls of relief

Methodology

(i) Primarily, a typical block is taken for its depiction and detailed study.
(ii) Suitable cross-sections are chosen on the topographic map.

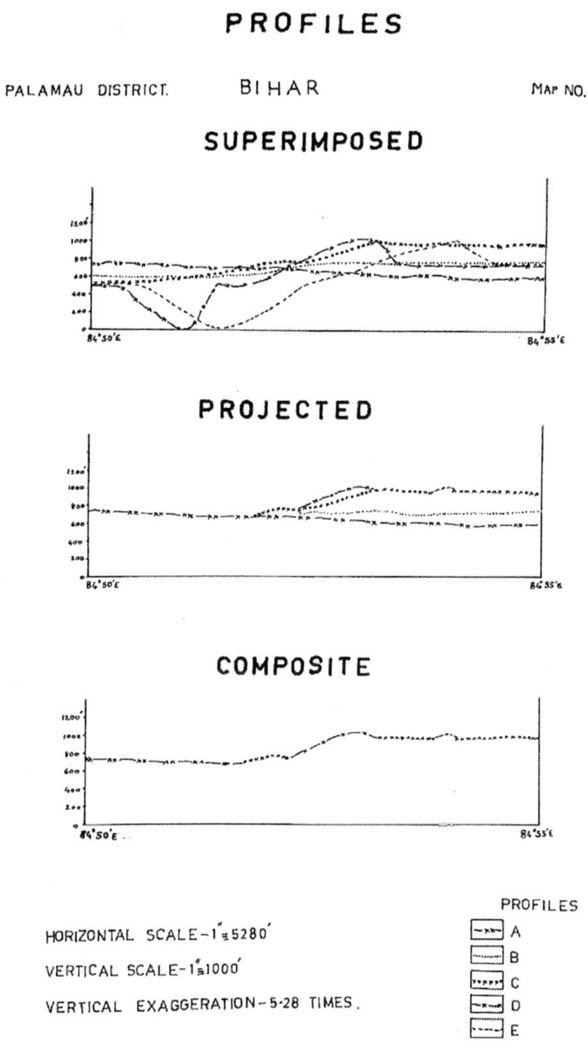

Figure 18.

(iii) Conversion of V.S. (1" ≡ 1000') and H.S. (1" ≡ 5280'). Thus, the Vertical Exaggeration (V.E.) becomes 5.28 times.
(iv) Five cross-sections have been selected latitudinally at definite intervals, having the same frame of longitudinal extension, and on the basis of these profiles – superimposed, projected and composite profiles are also drawn.

Interpretation: From these serial profiles ABCDE, it is evident that the topographic elevations of the plateau section are higher in the eastern part and lower in the western parts. This is due to the lithologic conditions and the geomorphic agent (fluvial action) curving on it. For better specificity names of fluvial channels and the sites may be relevantly mentioned.

In *superimposed* profiles, at the centre of the profiles D and E, it is more rugged ie. the evidence of the plateaux being dissected by the numerous non-perennial channels. While in the eastern portion, the profile exhibits a flat-topped resistant plateau section with one or two monadnocks. But in the western section, much of the relief of the plateau has been reduced due to the fluvial erosion of the main river present in the map and as a result, a relatively flat terrain has been formed whose elevation is about > 600'.

A *projected* profile from the south to the north has been drawn to compare the intervisibility. It exhibits only those features which are not obscured by higher intervening forms. It exhibits first an abrupt- increase of relief in the centre of the profile. On the western section, the land is relatively flat due to fluvial action. It gives a panoramic effect, with a foreground, middle ground, and distant skyline.

The *composite* profile shows only the ruggedness of the skyline. It is constructed to represent the surface as viewed in the horizontal plane of the summit levels from an infinite distance. So, from this profile, one can see that it is a highly dissected plateau area with isolated monadnocks, steep escarpments, and, gentle hill-slopes. But in the western section of the map, relief is comparatively low than in the eastern portion due to the lithologic conditions and the fluvial action on it.

11

Geological Map

A geological map shows the geographical distribution of different rock formations and strata at the ground surface. An appropriate Section Line exhibits the presence and thickness of all rock types in the area which led to the formation of the resultant topography. Its interpretation leads to an understanding of the lithology and the structure of the underlying rocks which help in the study of the evolution of landforms in a particular area.

In the following section, 2 Geological Maps have been discussed as specimens.

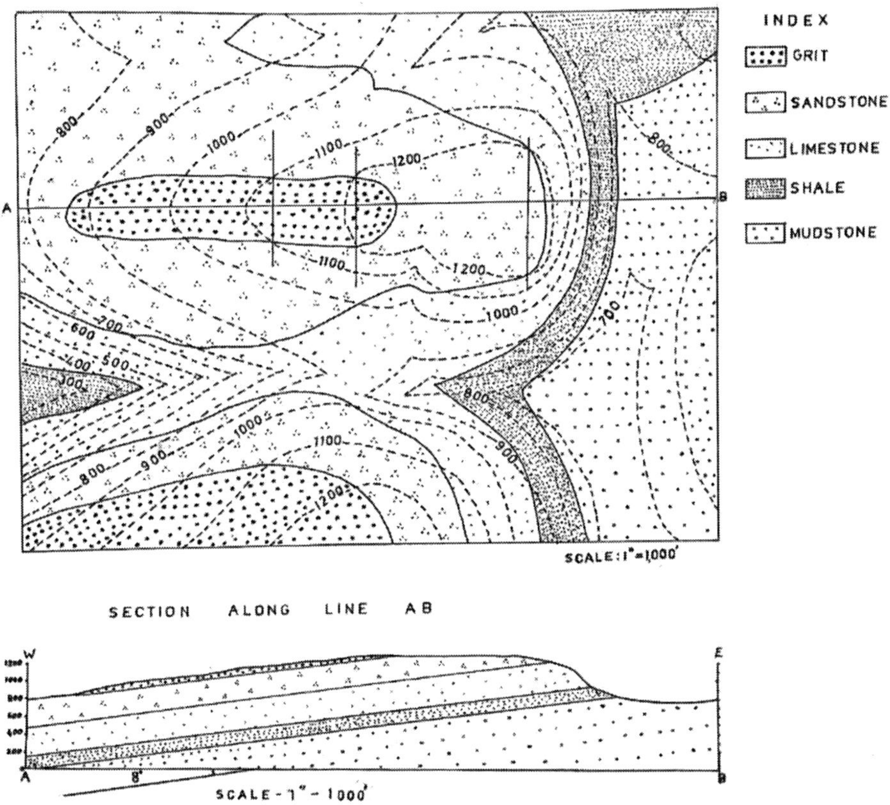

Geological Map No. 1

Introduction: The geological sections have been chosen along AB which runs west to east. It consists of a simple series and there is no unconformity.

(a) *Geological Descriptions: -*

Succession	*Vertical thickness*
Grit	100'
Sandstone	300'
Limestone	300'
Shale	100'
Mudstone	800'

(b) *Structure:* - The given section shows that it has a uniclinal structure which has a dip angle of about 8 degrees and the slope is from the east to the west.

(c) *Relation of rock group:* - It consists of sedimentary rocks like grit, sandstone, limestone, shale, and mudstone. The sequence of the beds is from the newest to the oldest which is grit, sandstone, limestone, shale, and mudstone.

(d) *Topography and its relation to structure:* - On the eastern side, there has been a valley due to the fluvial action of the soft rock like mudstone, whereas the limestone and sandstone being resistant have formed escarpments. Grit being softer has been exposed due to erosion.

(e) *Geological history:* - The beds have been laid down in order under marine conditions, and sub-ariel denudation has started forming the present topography.

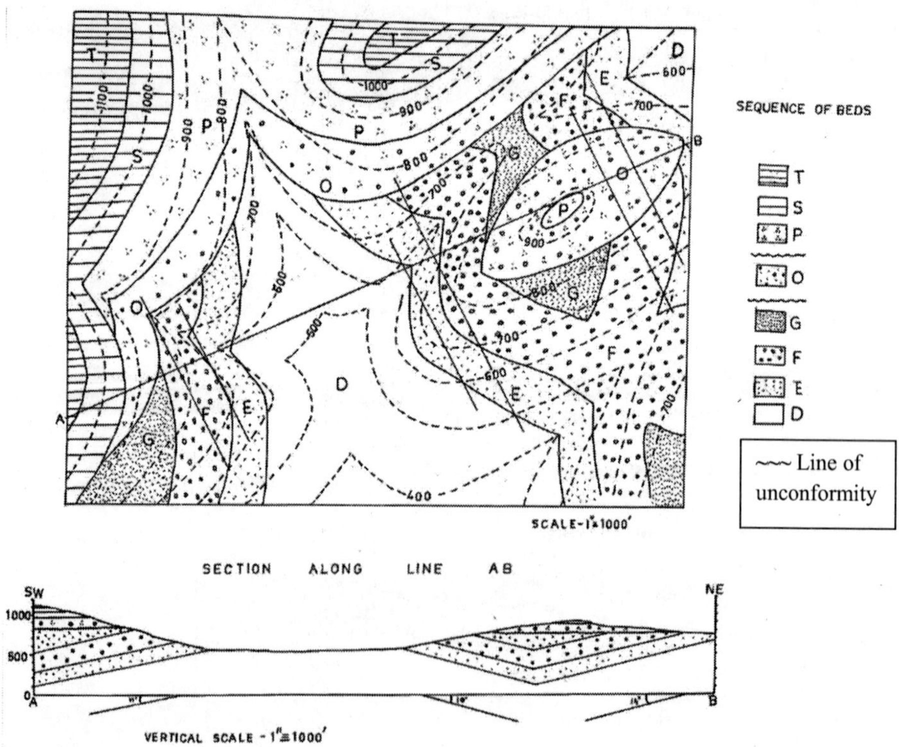

Geological Map No. 2

Introduction: The geological section has drawn from the southwest to the northeast along line AB. Due to folding; the dip of one bedding plane will vary from one limb of the fold to the next. The dip should, therefore, be determined for each limb of a fold.

(a) *Geological Descriptions: -*

Succession	*Vertical thickness*
T	undetermined
S	>50’
P	100’
~~	
O	
~~	
G	>500’
F	250’
E	200’
D	>700’

(b) *Structure: -* The beds show folded structure with an anticline in the central part of the section and a syncline on either side of it. The sedimentary beds represent a composite structure with the younger bed being horizontal and the older beds being folded. Within the older beds, there is one anticline in the mid-western part and one syncline in the eastern part. The western and eastern limbs of the anticlines dip at an angle of 11 degrees and 10 degrees respectively to the west and east respectively. The eastern limb of the syncline dips at an angle of 14 degrees towards the west.

(c) *Relation to rock groups: -* There are three series of rock beds that are separated by two lines of unconformities. T, S, and P comprise of the younger series; O is the middle rock layer and the rock beds of G, F, E, and D are the older series.

(d) *Topography in relation to structure: -* The region is characterized by ridge and valley topography. The ridge is located in the southwestern part and attains a height of over 900’. The valley has an average elevation of less than 600’. From the valley, the land rises gradually both to the east and west. In the western part of the land rises to a height of over 1100’. The valley occupies the anticlines while the ridge is located on the syncline. The anticlinal valley and the synclinal ridge represent an inversion of topography. Thus, it is evident that the topographical features have been structurally adjusted and the region of folded structure has attained the last stages in the process of the cycle of erosion. The younger beds occupy the highest parts in the southwestern parts, while the older beds have been successively exposed in the central and western parts. This has been due to differential erosion by the different geomorphic agents.

(e) *Geological history: -* The beds of the older series were deposited under the sea under shallow and marine conditions in the order from D to G. The beds were uplifted followed by folding. Sub-ariel erosion started, and the land has been peneplained and again submerged under the sea. The beds of the younger series were deposited over the beds of the older series. The whole area was again uplifted. Sub-ariel erosion started forming the present topography.

12

Survey Formats

Concept and Definition

Fieldwork or survey is an important pedagogical part of learning geography. Each and every region on this earth is unique in its physical and cultural elements. Therefore, to experience the traits of the natural region and to enhance the comprehension of the learners, field surveys are indispensable. To study the patterns and spatial distributions their interactions and relationships, survey works are essential. The studies can be done with any geographical issues like physical elements, settlement, economic activities, or contemporary societal issues.

The survey can be done at a local level or at a distant land. Surveys may differ in accordance to the fund outlay, purpose, grade, or capability of the learners and permissible time.

At the Primary, Secondary, and Senior Secondary stages of education, survey works encompass both primary and secondary sources.

The teacher has a vital role to guide and supervise the students or learners. The learners need to communicate well for their purpose and bear a positive gesture. All precautions and care should be taken so that no unwanted events occur and all data and information acquired are stored properly.

Formats of Questionnaire for Geographical Survey

Land-Use Study

Land is the most important resource. The primary, secondary and tertiary sectors of economy can be developed.

The land-use study is a particular method of study in which how plots of land are utilized is examined. In doing so, visit the plots of land personally and take care to collect information about the land from land-owners and local people with regard to its utilization. This can be done by visiting and observing the places and thus verifying the collected information.

After the collection of all information regarding the plots of land and their respective uses, need to be proceeded for drawing a map for it. This land use map will show how the land of an area has been used. It revealed the physical, social, economic, and other cultural aspects of the area.

Necessity of the Study of Land Use

Land-use study is an important part of Geography:

1. It helps to build up an important comprehensive geographical knowledge of the local environmental area.
2. A land-use study offers a clear idea about an area and its utilisation in both rural and urban perspectives.

3. It also builds up knowledge about the utility of land and the limitations of its development and some ideas about maps and land surveys.
4. The information table which comes out from the local land-use study often paves the way for further development of the plan of an area or at least an idea of it, often makes aware of the environment.

Procedure of Land-Use Study

First Phase: Collection of mauza maps or ward maps

The first step in this work is to collect a map of the area which is intended to study and know the land-use pattern. (a) If the region is a rural one, one has to collect a map of it where the mauza is located. For this, one has to collect a mauza-map according to J.L No. (Jurisdiction List Number) from the Collector's Office. (b) If the region is an urban area, one has to collect the ward map from the municipality or corporation in which it is located.

Second Phase: Drawing of the map, making it suitable for land-use study

The mauza-maps or ward-maps which are collected cannot be used readily as these are very small. These maps are usually in the scale of 40 cm = 1.6 km = 1600 cm or 1cm = 40 metre. As the scale of the map is too small to work, enlargement along its respective *Dag Number* (Plot No.) to make it usable for the study is necessary.

Third Phase: Selection of symbols or signs for the representation of the information collected

The selection of symbols or signs for the identification of the use and utilization of plot land is an important base work in this study. The land is used for various purposes like agriculture, housing, building roads, railways, water tanks, and for other services. Some lands remain as fallows. To mark various types of land accurately, selection of symbol or sign is necessary, such as, (i) Cultivated land-C; (ii) Housing land-H; (iii) Constructed building-B; (iv) Kutcha building (mud-built)-KB; (v) Pucca building (brick or stone built-PB; (vi) Road-R; (vii) Water bodies-W; (viii) Industrial estates or factories-I; (ix) School-S; (x) Hospital and health centre-HS; (xi) Post Office-PO; (xii) Religious Centre-RC; (xiii) Fallows-F, etc. For the perfect use of symbols, these selections may be grouped categorically in some graded classes as given below:

1. *Cultivated land (C):* Rice field-(R), Wheat field-C(U), Sugarcane field-C(S), Jute field-C(J), Vegetable field-C(V), Mango garden-C(M), Flower garden-C(F).
2. *Housing (H):* Kutcha (mud built) House-H (K), Pucca (brick built) House-H (P), Two-storied Pucca House-H (P2), Two-storied Kutcha House-H (K2), Three-storied Pucca House-H (P3), Three- storied Kutcha House-H (K3), School Building-H (S) etc.
3. *Road (R):* Unmetaled Road-R (U), Metaled Road-R (M), Brick built Road-R (B) etc.
4. *Water bodies (W):* Pond-W (P), Tube well-W (T), Canal-W (C), River-W (R) etc.
5. *Market-place, Business centre and Shop (M):* Market area-M (a) Shop-M (s), Factories-M (f), Cloth shop-(c).
6. *Religious Centre (RL):* Temple-RL (T), Mosque- RL (NI), Church-RL (C) etc.

Fourth Phase: Collection of information by visiting and observing the plots personally

In this phase, student should visit the place personally and inspect the plots and see how these are utilized and what purpose do they serve. Following the ward number or dag number, students will visit the plots and collect information which they will write in the survey note book and insert the symbols on the map.

Fifth Phase: Drawing of the map for the area surveyed
In this phase, students will draw a land-use map from the information so collected. In drawing the map, different shades, colours and symbols are to be employed to show the land-use pattern. An index should be furnished on one corner of the map, giving an explanation of colours, shades, and symbols.

[Directions and instructions about the use of colour: There are some different directions for the use of colour in the map, such as yellow or green for agricultural land, red for houses, buildings, and settlements, grey for fallows, blue for water bodies, rivers, canals, ponds and lakes. Again for showing different crops, such as rice, wheat, jute, coffee, sugarcane, oilseeds etc. different symbols may be used]

One or another corner of the map, name of the area, district, mauza or ward no. should be given clearly. North-Line direction should be shown at the one corner by an arrow.

After completion of the drawing usually on the bottom, the student should write his or her name of the school, class, section, roll number, and date.

[The land-use study of the rural and urban areas shows the following distinctive features:

a) In rural areas, most of the lands are utilized for agriculture; there are few roads and lines of communication as well as settlements and habitable areas.
b) In urban areas, most of the lands are occupied by habitable areas, houses, buildings, markets, shops, business centres, factories, etc. Open land and fields are hardly seen there.]

A Household Survey of a Locality

[Family Size/ Age-Sex Structure/ Educational Background/ Occupation]

There is a socio-economic survey along with the study of land-use of a region; the study on economic and social aspects is essential to have comprehensive knowledge about a region which is intended to study for a complete geographical account of a particular area.

Procedure of Study

1. *First Phase of Work: Preparation of questionnaire*

It is the first and foremost task for this survey. A series of questions relating to the social and economic conditions of the people of the region (which is going to be surveyed), is to be prepared. Questions may be set on the following items [A specimen of this questionnaire is given in the following.]

a) Name of the family, address, age, occupation, educational qualification, type of work, number of family members employed, monthly income etc.
b) Condition of the house (Kutcha-Mud-built, Pucca-Brick built- One or two storied), arrangement of drinking water, mother tongue, religion and, other matters if any.

In the case of rural studies, rural conditions, such as the total land area in possession, cultivated land, land used for housing, land used in different crop production, name of crops cultivated, the amount of agricultural income, amount of agricultural expanses, profit and loss etc. Considering the nature of place and economic conditions, a few more questions may be set, for an accurate survey, such as,

a) In the case of a shop or shopkeeper, a few more questions may be interrogated on the following: (i) Name of the shop and shop-keeper (ii) Name of the items which the shop-keeper deals in (iii) Number of buyers and from where they come (iv) Income, loss and profit etc.
b) In the case of factories and manufacturing centres, questions may be asked like (i) Name of the firm (ii) Producing items (iii) Raw materials used (iv) Volume of raw materials required (v) Source of these raw materials (vi) Number of workers employed (vii) Amount of investment (viii) Income, profit, loss etc.
c) In case of any problem or issue in the local or global community, the following questions may be arranged are like: (i) Name of the problem, (ii) Cause of the problem, (iii) Present sources of the problem, (iv) Arrangement to combat or handle the situation by local community or state level or central level government or NGOs, (v) Medical or financial assistance (if any) sources, (vi) Symptoms (acuteness or chronic) (vii) Mitigation or problem-solving techniques by different sections of the society, (viii) Any particular organisation involved, or their contribution, (ix) Mention the role of education in this field.

Surveyor should make comment on the findings of his or her survey and studies and should write everything and state whether the statements made by the local people are true or not. It may appear to the surveyor that the statements made by local people are not matching with me data or information collected by the surveyor. In this case, the surveyor's comment may help to evaluate the proper conditions.

After doing these, the surveyor should write his own name, name of the institution, class, section, roll number, data etc.

2. *Second Phase of Work: Door-to-door visits and collecting information from various sources*

In this case, the surveyor should proceed with the field study with his or her questionnaire and should visit door-to-door with his or her questionnaire and collect information from various sources. In doing so, he or she would make a friendly gesture and behave accordingly.

3. *Third Phase of Work: Setting up the information so collected in a graded pattern*

It is the most important phase of work. The surveyor should write down the collected information in a graded pattern and the population mentioning male or female, the standard of education mentioning the number of literate or illiterate persons and the incomes of the surveyed people.

4. *Fourth Phase of Work: Writing a description of the study*

In this phase, the surveyor should write the study report in the form of a description with the help of the findings in his or her study, mentioning the following information: (a) location of the area (village, mauza, P.S, district, J.L. No. etc.), (b) objective of the study, (c) feature of the socio-economic condition of the region, (d) general description relating to the mode of living of the inhabitants, environmental influences on them, nature of adaptation of the people to the geographical environment etc.

The study of land-use and socio-economic conditions presents some aspects of the comprehensive idea about the region. It helps in the study of Geography and enables one to realise the relationship of man to the local environment.

Suggested Topics for the Survey

1. Visits to concerned polluted sites, interviewing inhabitants, and preparation of reports of the environmental problems (causes, how affecting flora and fauna, and their solutions.
2. Visits to industries and studies on alternative strategies of environmental management.
3. To study the contribution of NGOs in improving a particular issue in a local area.

Questionnaire on Socio-Economic Condition

Date: ____________ Serial No.: ____________

Name of the Surveyor: ____________

Name of the School: ____________

Class: ________ Section: ________ Roll No.: ________

1. Name of the head of the family:			13. Head of the family: Tiller ☐ Worker ☐
2. Age:			14. Land-holder: Yes ☐ No ☐
3. Number of the family members:			15. Qty. of land:
Male	Adults ☐	Minors ☐	Where:
Female	Adults ☐	Minors ☐	Description:
4. Mother-Tongue:			16. Name of crops cultivated:
5. Religion:			17. Qty. of production:
6. Schedule Caste or Tribe: Yes ☐ No ☐			18. Income from land:
7. Number of educated member(s):			19. Income of the others:
8. Highly educated:			a)
Doctor:			b)
Engineer:			20. Income from other sources:
Other:			a)
9. Birthplace of the head of the family:			b)
District:			21. Occupation of the other members:
State:			a)
10. Causes of migration:			b)
a) Job:			22. Unemployed, if any, in the family:
b) Education:			How many:
c) Other:			Female:
11. Monthly family income:			Male:
12. Occupational qualification of the head:			23. House (Kutcha/Pucca):
a) Service:			Single/Two Storied:
b) Place :			Owner/Tenant:
c) Business:			24. Roof made of:
d) Type of business:			(Thatched/Tally/Tin):

25. Is there any electricity/internet facility (as relevant):	32. Source of water:
26. Means of lighting:	Municipal water:
27. Mode of conveyance:	Tank:
Train/Bus:	Others:
Cycle/Car:	33. Distance of market or shop:
Foot/Any other:	34. Any other market nearby:
28. How do children go:	35. Inconveniences of that area, if any:
29. How far is their school:	a)
30. Classes studying:	b)
31. Highest degree attained in family:	c)

[*N.B.* Relevant questions may also be attached in accordance with the purpose of the study]

Remarks of the Surveyor:
Signature of the Surveyor:
Name of the School:
Class:
Section:
Roll No.:

____
Signature of the Geography Teacher
Date:

Rural Land-Use Survey

Date of Survey: **Time:**

Mauza: ____________ J.L. No.: ____________ Police Station: ________ District: ________
Location: ______________________________ Areal Extent: ______________________________
Relief: ______________________________ Soil: ______________________________
Drainage/River: ________________ Canal: ______________ Pond: ____________
Total Inhabitants: ______________________ Density of Persons per sq. km. __________
Male: ______________________________ Female: ______________________________
Literate: ______________________________ Illiterate: ______________________________
Number(s) of Plots/Plot No. (As Required): ______________________________

Identifying Household Plots:
Settlement: ____________ Pucca: ______________ Kutcha: ______________
School: ____________ Primary: ______________ Madhyamik/School Final: ________
Post Office: ____________ Health Centre: __________ Bank: ______________
Clubs/ Organizations: ____ Panchayat Office: ________ Any Other Office: __________
Shops: ______________ Fair/Market: ____________
Temples: ______________ Mosque: ______________

Identifying Agricultural Plots:
Rice Cultivation: ________ Jute Cultivation: ________
Vegetable Cultivation: ____ Other Crops: __________

Identify Other Land-use:
Playground: ____________ Pond: ________________Garden Barren Land: __________
Use of Fertilizer: ________ Production of Crops: ____

Occupation of the Inhabitants:

Agriculture: __________ Business: __________ Service Inherited/Family: __________
Handicrafts: __________ Small Scale: __________ Animal Husbandry: __________
Poultry: __________ Transport Activities: ______

Remarks of the Surveyor:
Signature of the Surveyor:
Name of the School:
Class:
Section:
Roll No.:

____
Signature of the Geography Teacher
Date:

Urban Land-Use Survey

Date of Survey: **Time:**

Name of City: __________ Police Station: __________
Location: __________ Name/No. of Ward: __________
Areal Extent: __________ Population/No. of Inhabitants: __________
Literate: __________ Illiterate: __________
High-Income Family: ______ Middle-Income Family: ____ Low-Income Family: ______

Residence:

Temporary/ Unstable Huts: ___ Pucca House: __________
Park: __________ Playground: __________
Temples: __________ Mosque: __________ Church: __________
Clubs/Organisations: ______ Library: __________
Theatre Halls: __________ Cinema Hall: __________
Offices: __________ Shops/Shopping Mall: ______ Market: __________
Large-Scale Industries: ______ Small-Scale Industries: _____ Cottage Industries: ______

Total Schools:

Primary Schools: __________ Secondary Schools: ______ Colleges: __________
Universities: __________

Transportation Mode:

Metalled Roads/ National Highways: __________
Bus: __________ Trams: __________ Minibus: __________
Auto-Rickshaw: __________ Rickshaws: __________ Waterways: __________
Private Vehicles: __________ Cycle: __________ Motor-Cycle: __________
Cars: __________ Metro: __________ Rail-Station: __________
Airport: __________
Supply of Drinking Water: ___
Cemetery: __________ Graveyard: __________
Occupation of the Inhabitants: __________
Service: __________ Business: __________ Transport Activities: ______
Inherited Occupation: ______ Handicrafts: __________
Large/ Middle/ Small Scale Industries: __________

Remarks of the Surveyor:
Signature of the Surveyor:
Name of the School:
Class:
Section:
Roll No.:

____
Signature of the Geography Teacher
Date:

Survey of Agricultural Farm

Date of Survey: **Time:**

Name of the Teacher/Group Leader: ______
Total No. of the Students in Survey: ______
Name and Address of the Agricultural Plot Owner: ______
Area Extent of Land: ______
Soil Type and Characteristics: ______
Name of Crops during Survey and Type: ______
Sowing/Harvesting Time Period: ______
Cropping Techniques: ______
Type of Irrigation Applied: ______
Time of Applying Fertilizer (for Particular Crop): ______
Name of Fertilizers/ Insecticides: ______
Ripening Time of Particular Crop(s): ______
Estimated Production of Crop(s): ______
Production Cost Incurred by the Farmer/ Owner of the Plot: ______
Net Profit/Loss: ______
Other Producing States/Areas of the Particular Crop: ______
Comparison of the Crop in Relation to State/District Production: ______
Uses of the Crop: ______

[*N.B.* Relevant questions may also be attached in accordance with the purpose of the study]

Remarks of the Surveyor:
Name of the Student/Surveyor:
Class:
Section:
Roll No.:
Name and Address of the School:

______
Signature of the Geography Teacher
Date:

Survey on Exhibition / Market

Date of Survey: **Time:**

Name of the Teacher/Group Leader: ______
Total No. of the Students in Survey: ______
Name and Location: ______
Post Office: ______ District: ______
Occurrence of the Event (Time of the Year): ______
Duration of the Event: ______
Name of the Organiser: ______
Areal Extent of the Event: ______
Special Features/Points around the Event Ground: ______
Transport Linkage: ______
Rail Routes: ______ Road Routes: ______ Water Routes: ______
Name of the Participating Companies: ______
Products/Commodities Sold: ______
Special Incentives/Items Prevent: ______
No. of Seller: ______
No. of Customers (Buyers): ______

Advantages of the Event: ____________________
Defects of the Event: ____________________
Improvements of the Event: ____________________
Remarks (if any): ____________________

[*N.B.* Relevant questions may also be attached in accordance with the purpose of the study]

Remarks of the Surveyor:
Name of the Student/ Surveyor:
Class:
Section:
Roll No.:
Name and Address of the School:

____
Signature of the Geography Teacher
Date:

Survey of an Industrial Plant

Date of Survey: **Time:**

Name of the Teacher/Group Leader: ____________________
Total No. of the Students in Survey: ____________________
Name and Address of the Industry: ____________________
Type/Name of the Ownership: ____________________
Products (Processed/Finished): ____________________
Sources of Capital to the Plant: ____________________
Amount of Capital Involved: ____________________
Fuel/Power Supply: ____________________
Raw Materials Used: ____________________
Sources of Raw Material to the Plant: ____________________
No. of Labourers: ____________________
Cost of Production of Product(s): ____________________
Utility of the Finished Products: ____________________
Certified/Non-certified Products: ____________________
Marketing Network of the Products: ____________________
Financial Outlay of the Finished Products: ____________________
Net Profit/ Loss: ____________________

[*N.B.* Relevant questions may also be attached in accordance to the purpose of the study]

Remarks of the Surveyor:
Name of the Student/ Surveyor:
Class:
Section:
Roll No.:
Name and Address of the School:

____
Signature of the Geography Teacher
Date:

13

Importance of Studying with Relevant Typicals in Geography for Secondary Classes at Schools

Typicals are items or things that have distinctive qualities/traits and represent a particular group or category/class. Their signs, symbols, pictures, or diagrammatical representations or affixing of real-life objects (in very miniature forms) may be presented to depict the 'typical' nature. As these items/objects have the ability to exhibit similar/related essential characteristics pertaining to geographical concerns, they may help the learners of different grades to familiarize themselves and get sensitization about the various environmental issues. These *activity-based tasks* are related to collecting information or pictures/typicals, appropriate to the concerned geographical topic of study from different sources, generate curiosity and facilitate inquiry-based learning, add interest and also enable the understanding of the geographical concepts.

It is worth mentioning that these typicals (specimens) can be prepared appropriately and arranged by the subject expertise for an inclusive classroom in order to enhance and enrich the learning process.

Suggested Typicals in Geography:

1. Common fibres (*given as examples*) - Given as specimen
2. Digital assets at a glance (*given as examples*) - Given as specimen
3. Soil types & texture (*related to climatic and vegetation influences*)
4. Shapes and colours of plant leaves (*related to climate*)
5. Identification of minerals or rocks (*real-life miniature forms*)
6. Costumes and festivals (related to climate - *statewise*)
7. Flow, fund, and alternate sources of resources (energy)
8. House types (related to climate)
9. Items of Disaster Management Kits, Road Safety symbols, First-aid kit, COVID-19 Medical Kit
10. Species related to endangered/extinct/aerial/arboreal/deep sea
11. Modern Surveying types of equipment
12. Management of bio-degradable & non- bio-degradable wastes.

Different Disciplinary Perspectives on the Study of a Consumer Product – (*Activity Based On Thinking Skills*)

Problem: To trace any consumer product from its raw form and to present the study from different disciplinary perspectives.

In this case, the major disciplines taken for study are Geography, Biology, Physical Science, Economics, History, Politics, Environmental Education, Mathematics/ Statistics, and Art, and the consumer product selected for discussion is Petroleum – The Liquid Gold

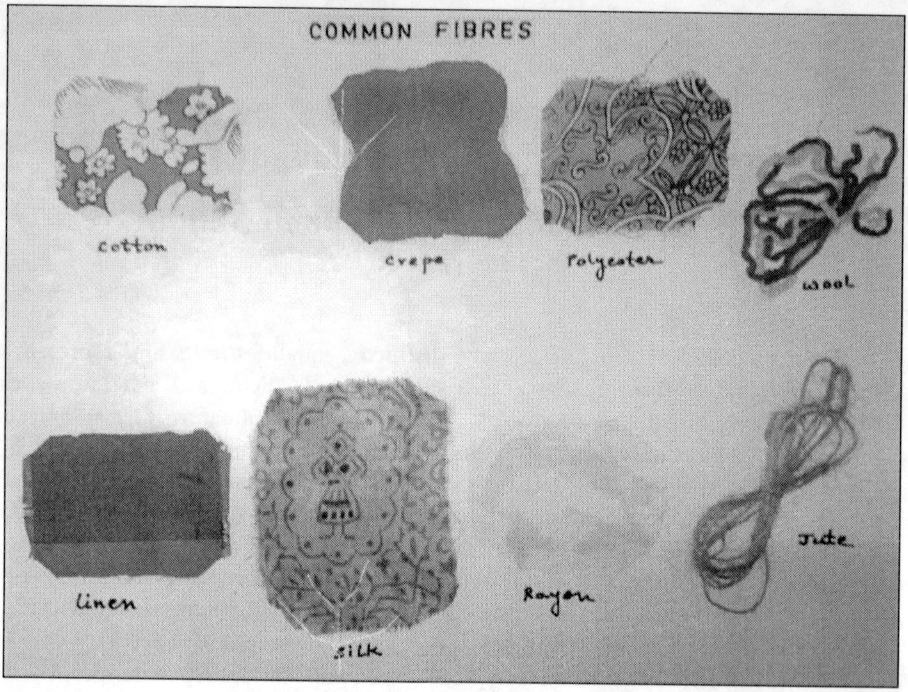

Figure 1: *Common Fibres*

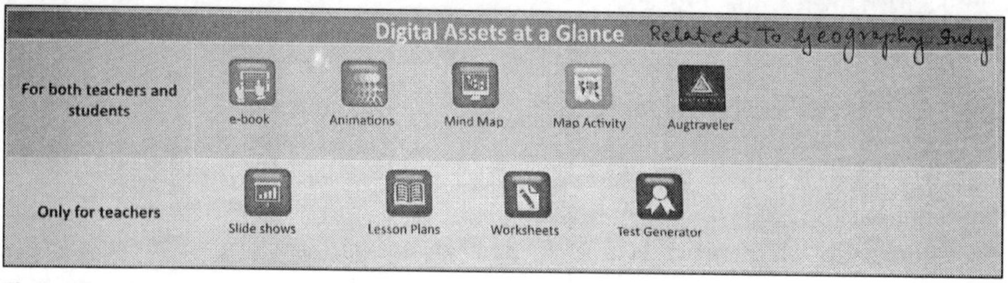

Figure 2: *Digital Assets at A Glance*

Geography

1. Properties and Nature
2. Oil prospecting and drilling
3. Oil transportation and storage
4. Distribution/ of oil belts with/without maps.
5. Problems and presence of oil/petroleum industries
6. Uses of oil

Biology

1. Origin/occurrence/formation of oil
2. Discoveries of new oil fields

Physical Science (Physics and Chemistry)

1. Oil refining process
2. Research and development for commercial production of by-products
3. Scientific and infrastructural facilities for the discovery of new oil fields

Economics

1. Factors of the location of oil refineries
2. Location of petrochemical industries
3. Economic consequences of the large oil fields in sparsely populated areas
4. Demand and supply chain of old and new oil fields
5. Concept of optimum utilization and market strategies
6. Fund allocation for flow resources/ non-renewable sources of energy
7. Government intervention in the context of the energy crisis

History

1. History of oil extraction from the earliest time till modem technological facilities
2. Increasing role of petroleum as a domestic and industrial fuel with the declining importance of coal
3. Formation and organization of OPEC

Politics

1. OPEC and energy crisis
2. Future of oil industries
3. Oil prices and investments in global markets
4. Exporting and importing countries—issues related to crude/refined oil

Environmental Education

1. Oil spill/oil pollution- causes, description, mitigation, and future preventive measures
2. Legal measures- Acts/Rules for environmental safety of oil transaction
3. Environmental protection in respect of the discovery of new fields

Mathematics/Statistics

1. Production/Trend of demand and supply of oil/petroleum products in the world
2. Consumption pattern and prices (comparison) with other fuels
3. Cost-Benefit analysis (comparison with alternate fuels)

Art

1. Drawing of occurrence of oil in rock layers (permeable and non-permeable)
2. Depiction with appropriate symbols of distribution of oil fields
3. Graphical representations of the export/ import or price trends of oil in different countries
4. Timelines showing the production of by-products of petroleum with time

Great Geographers

Importance of Study of Great Geographers' Contribution

It is noteworthy to study the contributions of the Great Geographers who have given varied dimensional thought and have rediscovered the subject from different perspectives. This scientific intervention provides a blueprint for the future of the discipline, recommending how to strengthen its thought-provoking domains so as to study natural and cultural relationships. Through the curriculum of Geography teaching, the "locality-sense" can be gradually developed to the horizon of global thinking. To open-up innovative avenues of research and build sub-disciplinary ideas and adoption of contemporary learning tactics an ideal geographer/ teacher can lead a learner virtually all over the globe through streamlined appropriate strategies.

ERATOSTHENES
c. 276 - c. 195 BC
He invented the system of Latitude and Longitude; calculated the circumference of the earth and tilt of the earth's axis; and was the first to use the word 'Geography'

CARL O' SAUER
1889-1975
He developed the idea of 'Cultural Landscapes' in his book, THE MORPHOLOGY OF LANDSCAPE (1925)

CARL RITTER
1779-1859
He developed the idea of 'Unity in Diversity', and declared geography as 'ERDKUNDE' i.e. an earth science

HALFORD MACKINDER
1861-1947
He was the founding father of 'Modern Geo-politics' and 'Geo-strategy'. He formulated the concept of 'Geographical pivot of History' or 'Heartland Theory' (1904)

ALEXANDER VON HUMBOLDT
1769-1859
He was a great traveller and wrote KOSMOS (1845) and laid the foundation of 'Bio-geography'

RICHARD HARTSHORNE
1899-1992
An ardent advocate of regional approach in geography, he wrote the famous book PERSPECTIVE ON THE NATURE OF GEOGRAPHY (1959)

DAVID HARVEY
b. 1935
Advocate of the 'Right to the City', he is the world's most cited geographer and social thinker. His famous books include EXPLANATION IN GEOGRAPHY (1969) and SOCIAL JUSTICE AND THE CITY (1973)

ELLSWORTH HUNTINGTON
1876-1907
Postulated the idea of 'Climatic Determinism' in his famous book CIVILIZATION AND CLIMATE (1915)

FRIEDRICH RATZEL
1844 - 1904
He propounded the concept of 'Social Darwinism' and wrote two seminal books, ANTHROPOGEOGRAPHIE (1882-91) and POLITISCHE GEOGRAPHIE (1897)

SIVAPRASAD CHATTERJEE
1903-1989
He founded the Department of Geography, Calcutta University (1941), the National Atlas & Thematic Mapping Organization (1956) and was awarded with PADAMBHUSAN

VIDAL DE LA BLACHE
1845-1918
He was known as the 'Founder of Human Geography', as he pleaded for the notion of 'Possibilism'

WILLIAM MORRIS DAVIS
1850-1934
He postulated the concept of 'Cycle of Erosion' in development of landforms. He was the most influential geomorphologist of his time, and he founded the Association of American Geographers (1904)

14

Statistical Representations

Meaning of Statistics

Statistics is a very familiar word nowadays. The word 'statistics' is associated with the term indicating quantitative information. The term Statistics was coined by the German scholar Gottfried Achenwall around the middle of the eighteenth Century. In the modern context, the word includes all branches of human activity concerned with numerical data. Statistics is used as a plural noun and also as a singular noun. It has two different meanings, in the plural sense, the word implies a set of numerical figures usually obtained by measurement or counting connected with any field of human experiences. In a singular sense, Statistics refers to the subject of scientific activity which deals with the theories and methods of collection, presentation, analysis, and interpretation of numerical data.

Functions of Statistics

- *To simplify complex data*- The fundamental function of the statistical method is to reveal the significance of complex data by means of various diagrams, different averages, etc.
- *To compare*- Statistics facilitates the comparison of data both chronologically and geographically.
- *Formulation and testing of hypothesis*- Statistical methods have their use in formulating and examining hypotheses.
- *Prediction*- Statistical techniques are helpful in forecasting future trends in different fields of activity and thus in relation to industry, business, developmental plans, and policies, they can be formulated well in advance of their time of execution.
- *Testing the laws of other Sciences*- Statistics is very useful for examining the facts of all sciences whether Natural or Social Sciences.
- *To conduct a survey*- Statistical methods are used in making the efficient design of experiments and surveys and to interpret the relevant information for better social development.
- *To help the government*- The important principles of Statistics have become necessary for the administration of a country.

Importance of Statistics

Statistical tools and techniques are found to be applicable in different fields of Business, Commerce, and Management for academic pursuits and also in scientific interventions.

Limitations of Statistics

It is applicable only to quantitative data and mainly related to the aggregate of facts and is not concerned with individual items. Secondly, statistical laws are true only on average and they may not hold in a particular situation. Thirdly, in a particular case, statistical methods, if not applied in the proper perspective of the collected data, may lead to false conclusions. Fourthly, though they are useful for prediction and are affected by a multiplicity of causes, exactness, cannot be guaranteed, and no estimates or forecasts for statistical decisions can be taken to be exact. They are all subject to a certain amount of inaccuracy whose limits may be stated in terms of probability. Lastly, statistical data must be uniform in the sense and need to be handled with care by experts or by learners and care should be taken so that it is not misused in any research pursuits.

Data

*Data m*eans a body of information collected in numerical form and that need to be arranged in a tabular form called *Data Matrix*. The unclassified data from which frequency distribution may be done may have an unsystematic arrangement of R*aw Data* and irrespective of any order. An *Array* is defined by an arrangement of data in which all items are placed sequentially in order of magnitude. This array helps one to know at a glance, the range of values, the nature of concentration, and the degree of continuity of the data and their distribution pattern.

Collection of Data

The collection of data is the primary task at the outset of any statistical activity which can be collected through scheduled questionnaires and pilot studies.

Primary Data and Secondary Data

It may be classified as Primary and Secondary, depending on the sources of the data.

1. Primary – The statistical data which are gathered directly from the field of investigation for the desired purpose is called primary data and these are obviously original in nature and have the reliability of data.
2. Secondary – Data collected by some agency and compiled from that source by the enquirer for his/her use, are called secondary data. These data are original and cheaper to collect and involve less manpower and time.

Variables

The items about which information has been collected are usually referred to as individuals. In the data matrix, each column gives the value of a specific property or characteristic that varies from one individual to another. These are called variables. The variables that can take on all sorts of fractional values are called continuous variables, for example, height, length, etc. In discrete variables, the values are all simple whole numbers so that no score can fall in the gaps between each pair of successive whole numbers.

Variables may also be classified as discrete or discontinuous and continuous.

1. Discrete or Discontinuous variable - A quantitative character that can take certain isolated values only in its range of variation is called a discrete variable. Example: Size of families of some locality, number of students in different colleges.
2. Continuous variable - A variable that can assume any value within its range of variation is termed a continuous variable. Example: Weights of individuals or the marks of candidates.

In the case of discrete variables frequency distribution can be represented by perpendiculars (or columns) having heights equal to the frequencies of the variable values which are drawn at the corresponding points (indicating variables values on the horizontal axis). The diagram so formed is called a column diagram or a frequency diagram.

Class

In organizing and summarizing that statistical data, a frequency distribution is normally constructed. In this construction process, the whole range of the data is divided into some groups defined by an interval, and the number of observations falling in each group is stated precisely. The groups are called *class intervals* or simply *classes* and the observations are called *frequencies.* The classes are defined by limits or boundaries on two ends. When one end of the class is not specified, it is called an *open-end class.* Normally in distribution, all classes are of the same size but in some cases, classes of *unequal* size may also be constructed, particularly for *highly dispersed* data. The number of observations included in a class is termed as absolute frequency or *frequency of the class.* The *class limits* are the two end-points of a class interval used for tally marking the given values. The *class boundaries* are the two real endpoints of a class interval. The upper-class boundary coincides with the lower-class boundary of the next class. The class boundaries are used for forming the frequency distribution of a continuous variable.

These are defined as the most extreme values that would ever be included in a class. In fact, the class boundaries are the real class limits. The lower extreme value is called the lower class boundary while the upper extreme value is called the upper-class boundary. For all practical purposes of graphical representation of data, class boundaries are used. The working formula is:

Lower Class Boundary = [Upper Class Boundary – (d/2)] and

Upper Class Boundary = [Upper Class Boundary – (d/2)]

where,
d = Difference between the two successive class limits.

Class Width: The width or size of a class is the difference between the lower and the upper class. It is expressed as,

Class Width (i) = (Upper-Class Boundary – Lower Class Boundary)

Class Mark: The Class Mark or mid-value of a class refers to the value that lies exactly in the middle of a class. It is used as the representative value of a class for the purpose of calculation of descriptive measures, The common formula is:

Class Mark (x) = ½ [Lower Class Limit + Upper Class Limit]
or, = ½ [Lower Class Boundary + Upper Class Boundary]

Class Frequency: The number of observations contained in a class is known as its class frequency. The sum of all class frequencies in a distribution is called total frequency. Classes with zero frequencies are empty classes. Therefore,

$$\text{Total Frequency, (N)} = \sum_{i=1}^{n} fi$$

Where,
n = Number of Class and
f_i = Frequency of the ith class

Frequency Distribution

Frequency distribution is a statistical table that shows the values of the variables and the corresponding number of observations or frequencies. Frequency distribution may be of two types – *simple and grouped.* In a *simple frequency distribution,* values of the variables are arranged individually in order of magnitude. In a *grouped frequency distribution,* values of the variables are arranged in groups in order of magnitude. Frequency distribution is generally used to:

i. Condense and summarize a huge body of data
ii. Visualize the range, average, spread and shape of the distribution and
iii. Approximate the nature of probability distribution in the population.

Relative Frequency: Relative frequency is simply the class frequency (f_i) expressed as the proportion of the total frequency (N) of a distribution. The sum of all relative frequencies in distribution is equal to unity. In expression form:

$$\text{Relative Frequency, } Rf_i = \left(\frac{fi}{N}\right)$$

and,
$$\sum_{i=1}^{n} R.fi = 1$$

where, n = Number of Class

Frequency Density: The Frequency Density of a class is defined as its frequency per unit width. It indicated the concentration of frequency in a class. The common expression is:

$$Frequency\ Density\ (fi) = \frac{frequency\ (f)}{class\ width\ (i)}$$

Frequency distribution may also be represented in terms of *relative frequencies or cumulative frequencies.* The *cumulative frequencies* are found by successively adding the class frequencies, where the additions begin from the top (that is the lowest class) or bottom (that is the highest class). Depending on – *less-than type or more-than type cumulative frequencies* are obtained.

Histogram

It is an appropriate diagram for representing the frequency distribution of a continuous variable. It consists of rectangles, each of which has breadth equal or proportional to the size of the concerned class intervals and height equal or proportional to the corresponding frequency. In a histogram, consecutive rectangles have a common side and for this, the class intervals are made overlapping in all cases.

A histogram displays the frequencies of different class intervals. As such it becomes easier to locate the class intervals having the greatest frequency or the least frequency and also helps to find the *mode* of distribution.

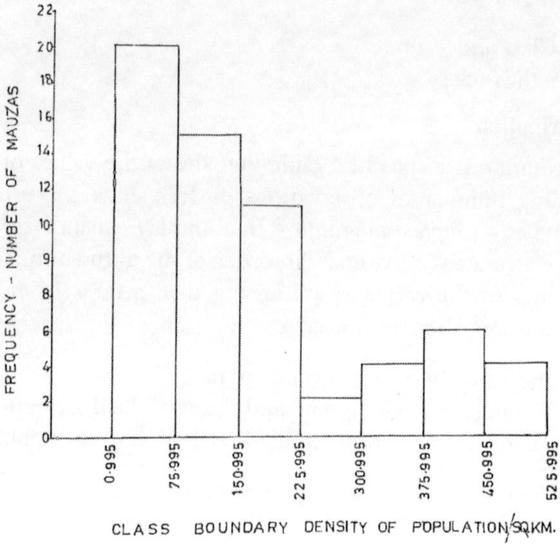

Histogram Showing Distribution of Mauzawise Population Density-P.S. Kalimpong Dist. Darjeeling, West Bengal - 1981

Frequency Polygon

It is a suitable diagrammatic mode of exhibiting the frequency of a distribution of a discrete variable. The variable values are located on the horizontal axis and the frequencies on the vertical axis in a rectangular system of coordinates. The available data are plotted as points on the plane where the abscissa and the ordinate for a point respectively indicate the variable values and the corresponding frequency.

Finally, when the points are successively joined by line segments the frequency polygon is obtained.

Therefore a *frequency polygon* is a polygon whose vertices are at the middle points of the tops of rectangles forming the histogram of the frequency distribution. These middle points correspond to the class marks of the corresponding class intervals.

There are two methods of constructing a frequency polygon for classified data:

(a) With the help of a histogram
(b) With the help of class marks

Interpretation - From the diagram, it is seen that the class boundary, 0.995 – 75.995 has the highest frequency of 22, whereas the lowest frequency is 2 lying within the class boundary of 225.995 – 300.995. The frequency polygon curve is double-humped, the first hump lies within the first-class interval, and the second hump in the second-last-class interval.

Ogive (Cumulative – frequency Curve)
Grouped data are also reprinted by a curve called *ogive or cumulative frequency curve.* In the case of less- than type cumulative frequencies, they are plotted against the upper-class boundaries as different points which are joined successively by line segments to

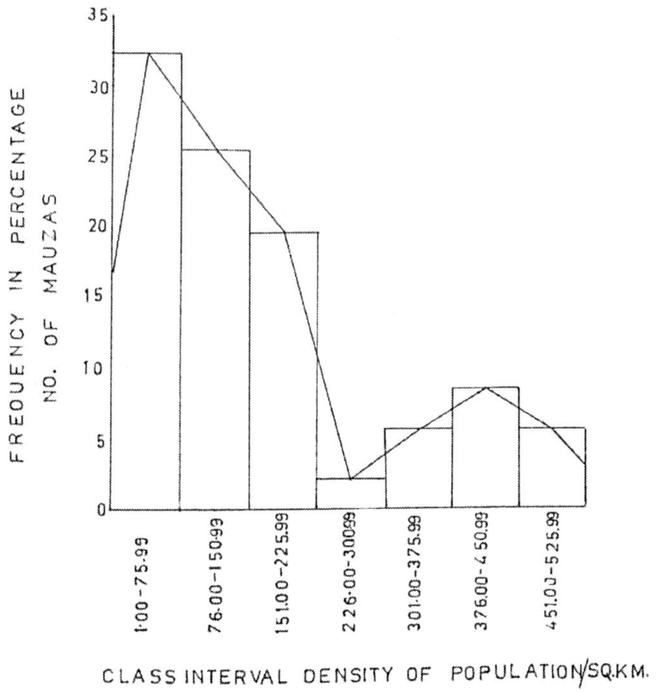

Histogram & Frequency Polygon Showing Distribution of Mauzawise Population Density-P.S. Kalimpong Dist. Darjeeling, West Bengal - 1981

get less-than-type ogive. Again, for a more- than type ogive cumulative frequencies of more-than-type correspond to lower class boundaries. Obviously, the cumulative frequency of the less-than type is zero for the lower boundary for the lowest class and it and the cumulative frequency of greater than type /more-than type are zero for the upper boundary of the highest class which has to be included in the drawing diagram.

An ogive for a frequency distribution displays cumulative frequencies of different class intervals and helps to find the median, the upper and lower quartiles of a distribution.

Shape of the Frequency Curve

The shape of a frequency curve is an important characteristic of a distribution. It is normally measured from two aspects – symmetry of the frequency curve (*skewness*) and peakedness of the frequency curve (*kurtosis*).

Skewness – the term skewness is used to denote the extent of asymmetry in the distribution any asymmetric distribution is a skewed distribution basically it measures the extent by which the bulk of the values in distribution is concentrated to one side or the other of the mean. A perfectly symmetrical normal distribution has, therefore, zero skewness. In any distribution, if Mean>Median>Mode, it is positively skewed whereas if Mean<Median<Mode, the distribution is negatively skewed. It has been shown in the diagram below.

Kurtosis – Kurtosis measures the extent to which values are concentrated in one part of the frequency distribution. Hence it refers to the degree of peakedness of the frequency curve. The two distributions may have the same average, spread and symmetry but they can represent different degrees of peakedness. For a normal distribution that is neither very peaked nor flat-topped, kurtosis is 3. This is called a *meso-kurtic* distribution. When kurtosis is less than 3 the frequency curve is flat-topped and the distribution is called *platy-kurtic*. When the frequency curve is highly peaked and the value of the kurtosis is more than 3, the distribution is called *lepto-kurtic*.

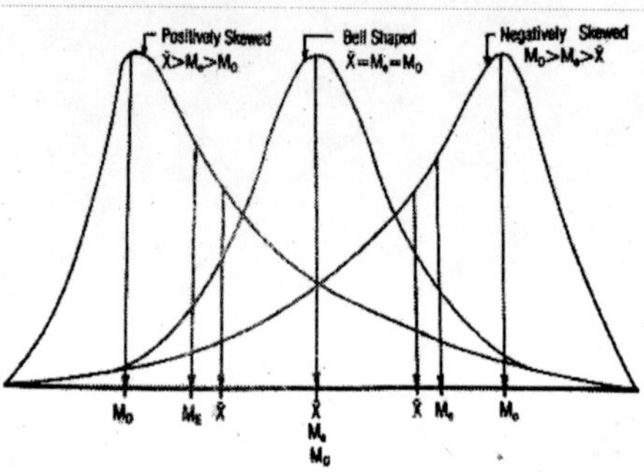

Shapes of the Frequency Curve

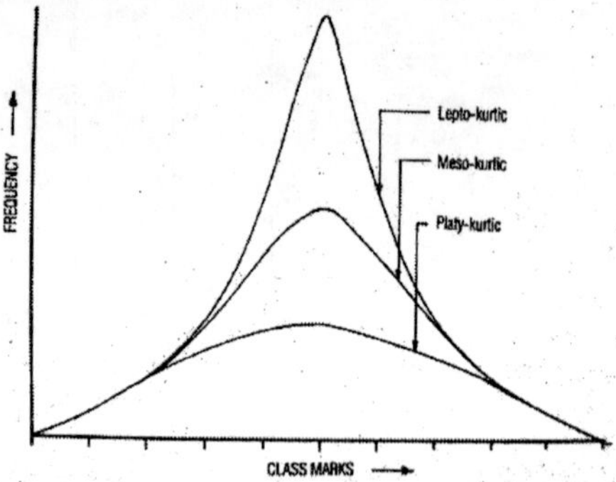

Peakedness of Frequency Distribution

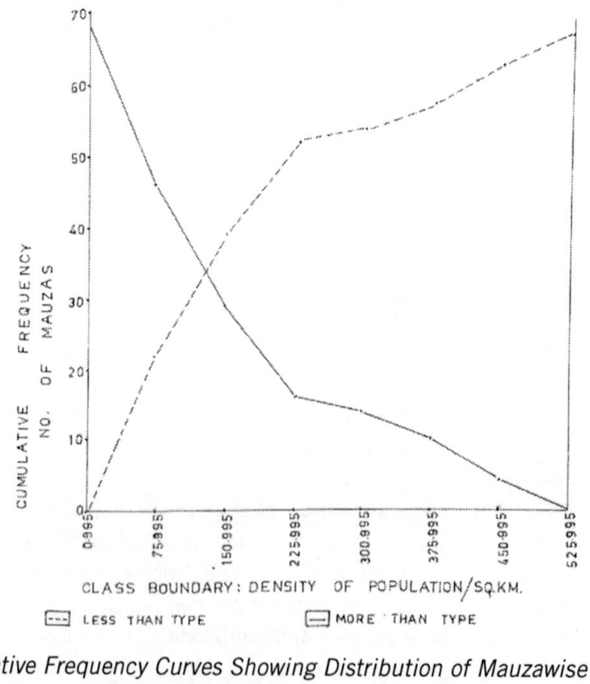

Ogives Cumulative Frequency Curves Showing Distribution of Mauzawise Population Density-P.S. Kalimpong, Dist. Darjeeling, West Bengal - 1981

Data for Histogram, Frequency Polygon and Ogives Cumulative Frequency Curves of Mauzawise Population Density – P.S. Kalimpong, Dist. Darjeeling – 1981

Class Interval Pop/Sq. Km	*Class Boundary*	*Frequency*	*Cumulative Frequency (less than)*	*Cumulative Frequency (more than)*	*% of Frequency*	% C_f (L)*	% C_f (M)*
1 – 75.99	0.995 – 75.995	22	22	68	33.25	32.35	100.00
76 – 150.99	75.995 – 150.995	17	39	46	25.00	57.35	67.65
151 – 225.99	150.995 – 225.995	13	52	29	19.13	76.48	42.65
226 – 300.99	225.995 – 300.995	2	54	16	2.94	79.42	23.52
301 – 375.99	300.995 – 375.995	4	58	14	5.88	85.30	20.58
376 – 450.99	375.995 – 450.995	6	64	10	8.22	94.12	14.70
451 – 525.99	450.995 – 525.995	4	68	4	5.88	100.00	5.88

*L – Less than type, *M – More than type

Interpretation – Both the cumulative frequency curves (less-than and more-than) show a gradual stopping tendency. But, near the value of 25.995, the sloping nature of the curve changes, i.e. from a gentle-slope to a more gradual sloping curve.

Measures of Central Tendency

In order to make a comparative study of Frequency Distributions, arithmetic number are needed. Depending on the nature of the distribution, such arithmetic numbers are calculated in different ways. Such number are called arithmetic measures of data. These numbers indicate how the data are distributed around the central part of the distribution. So, these arithmetic numbers are known as *measures of central tendency* which are as follows: (1) Mean or arithmetic mean (2) Median and quartiles and (3) Mode.

Mean

A. Mean of Ungrouped (Raw) Data

The mean (or arithmetic mean) of n observations (variates) $x_1, x_2, x_3, \ldots, x_n$ is given by

Symbolically, $A = \frac{\sum xi}{n}$. In words, mean = $\frac{\textit{sum of the variates}}{\textit{total number of variates}}$.

Note $\Sigma x_i = nA$, *i.e.*, sum of the variates = mean × number of variates

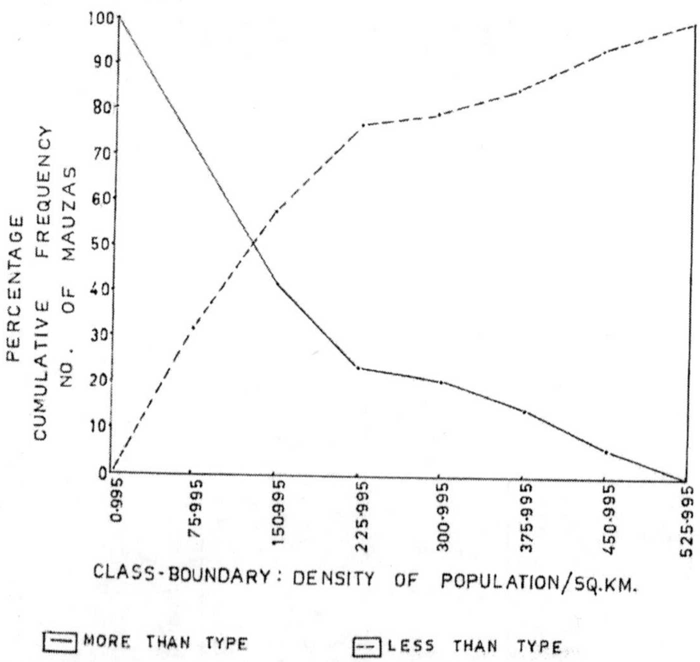

Ogives Relative Cumulative Frequency Curves Showing Distribution of Mauzawise Population Density-P.S. Kalimpong, Dist. Darjeeling, West Bengal - 1981

B. Mean of Grouped (Arrayed) Data

If the values of the variables (called the observations or variates) be $x_1, x_2, x_3, \ldots, x_n$ and their corresponding frequencies are $f_1, f_2, f_3, \ldots\ldots, f_n$ then the mean of the data is given by

Symbolically, $A = \dfrac{\sum xifi}{\sum fi}$

In words (by direct method),

$$\text{mean} = \frac{\textit{sum of the products of the variates and their corresponding frequencies}}{\textit{total frequency}}$$

Median

Like mean median is another measure of central tendency of a distribution. If the observations (variates) are arranged in ascending or descending order then the variate in the middle is called the *median* of the distribution. Thus the median divides the ordered observation in two equal halves.

i) If *n is odd* then the $\dfrac{n+1}{2}th$ variate is the median

ii) If *n is even* then the mean of the $\dfrac{n}{2}th$ and $\left(\dfrac{n}{2}+1\right)th$ variates in the median, i.e.,

$$\text{median} = \frac{\frac{n}{2}th\,variate + \left(\frac{n}{2}+1\right)th\,variate}{2}$$

$$\text{median} = L + \frac{\frac{N}{2} - F}{f} \times i$$

where,

L = Lower boundary of the median class

f = Frequency of the media class

i = Interval of the median class

N = Total frequency

F = Cumulative frequency upto the median class

Quartiles

The three variates which divides the data of a distribution in four equal parts (quarters) are called quartiles. The median is the second quartile.

If the data are arranged in ascending or descending order then the variate lying at the middle between the lowest variate and the median is the called the *lower quartile* (or the *first quartile*), and it is denoted by Q_1.

The variates of a data are real numbers (usually integers) so they are scattered over a part of the number line. A researcher would like to know the nature of the scattering of the variates. The arithmetic numbers associates with distributions to show the nature of scattering are known as *measures of dispersion*:

(a) Range (b) Interquartile Range

(a) Range: The difference of the greatest variate and the smallest variate in a distribution is called the range of the distribution.

(b) Interquartile Range: The interquartile range of a distribution is $Q_3 - Q_1$, where Q_1 = lower quartile and Q_3 = upper quartile.

$\frac{1}{2}(Q_3 - Q_1)$ is known as *semi-interquartile range.*

Median of Classified Data

The median of a distribution in a continuous classified form can be calculated either by formula or by the ogive of the distribution.

Median class

If in a continuous distribution the total frequency be N then the class interval whose cumulative frequency is just greater than $\frac{N}{2}$ (or equal $\frac{N}{2}$) is called the *median class.* In other words, median class is the class interval in which the median lies.

Percentiles

Percentiles are the points which divide the set of observations into one hundred equal parts. These points are denoted by P1,P2,P3,...,P_{99}, and are called the first, second, third... ninety ninth percentile. The percentiles are calculated for a very large number of observations like workers in factories and the populations in provinces or countries.

Deciles

Deciles are similar to Percentiles (sounds like decimal and percentile together), as they split the data into 10% groups: The 1st decile is the 10th percentile (the value that divides the data so 10% is below it), the 2nd decile is the 20th percentile (the value that divides the data so 20% is below it).

A decile rank arranges the data in order from lowest to highest and is done on a scale of one to ten where each successive number corresponds to an increase of 10 percentage points.

Mode

Mode (or modal value) is another measure of central tendency. The *mode* of a distribution is the value represented by the point about which the variates tend to be most heavily concentrated. In general, the mode is the variate which occurs more frequently, i.e., the variate having the highest frequency. Mode is usually denoted by M_0.

Mode of Classified Data

The mode of a distribution in a continuous classified form can be calculated either by formula or by the histogram of the distribution.

Modal Class

The class interval having the maximum frequency is called the *modal class* and its class mark is called the *crude mode.*

$$\text{Mode} = L + \frac{(f_m - f_p)}{(f_m - f_p) + (f_m - f_n)} \times \text{i}$$

where,

f_m = Modal class frequency

f_p = Frequency of the class just preceding the modal class

f_n = Frequency of the class just next or following the modal class

L = Lower boundary of the modal class
i = Modal class interval

Comparison between Mean Median Mode

The mean and the mode are rigidly defined. The median is also rigidly defined, except when there is an even number of observations. In finding each of these measures, all the observations are taken into consideration. However, only the mean is directly based on all the values and its value changes even if a single observation is altered. On the other hand, median and mode may remain unchanged even after the alteration of several observations. The significance of all the measures is quite easily comprehensible. In general, the labour involved in the computation of all the three measures is almost the same. But in practice, the exact determination of mode of a continuous variable is impossible, because practically an ideal distribution (or the frequency curve) is never obtained. The mean has several properties and by virtue of them it can be readily manipulated in varied situations. But the median and mode does not possess such desirable properties. Among the three measures, the mean is generally found to be least affected by sampling fluctuations. However, in this respect, the median or mode may be better than mean in some specific situations.

Thus it is evident that, in general, mean is the best measure of central tendency. But there are situations where mean cannot be used or should not be used. In the case of a grouped frequency distribution, if one or both of the terminal classes are open, mean is indeterminate. Again, if there be a few extreme values markedly different from majority of the values, mean should not be used as an average. In such situations, median or mode would be the appropriate measure.

Quartile Deviation

The Quartile Deviation is defined as the semi-interquartile range. It is dependent on the upper and lower quartile values and does not concern the variability of the largest 25% or the smallest 25% of the observations. However, it takes into account 50% of the observations lying between the first and third quartiles.

$$\text{Quartile Deviation (QD)} = \frac{Q_3 - Q_1}{2}$$

Standard Deviation

The variance is not normally used as a descriptive measure, instead the square root of the various is taken. It is called standard deviation (σ) which is defined as the *root-mean-squared-deviation from the mean*. It is superior to variance in that variance may become a large number, even greater than the individual values but in the standard deviation it is square-rooted so it becomes more comprehensive and manageable.

$$\text{Standard Deviation } (\sigma) = \sqrt{\sum_{i=1}^{n} \frac{(x_i - \overline{x})^2}{N}}$$

Scatter Diagram

The simplest way of a diagrammatic representation of bi-variant data is by using a scatter diagram or dot diagram taking two perpendicular axes of coordinates, one for X

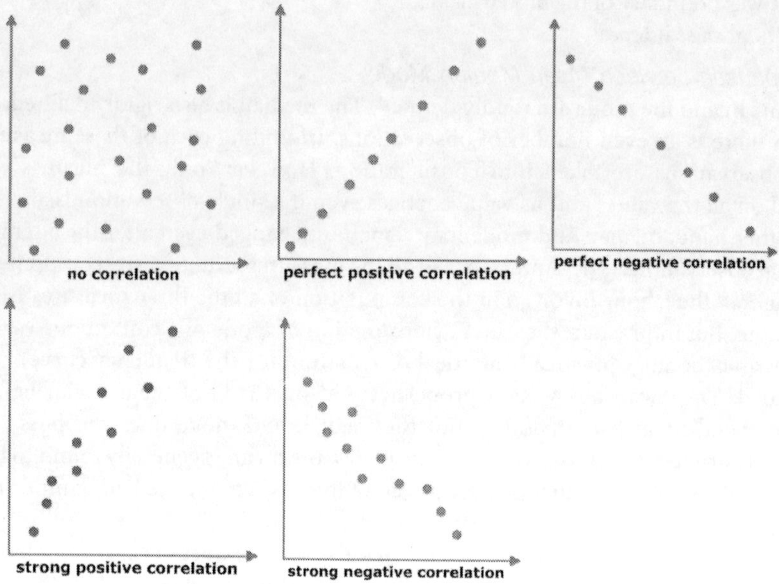

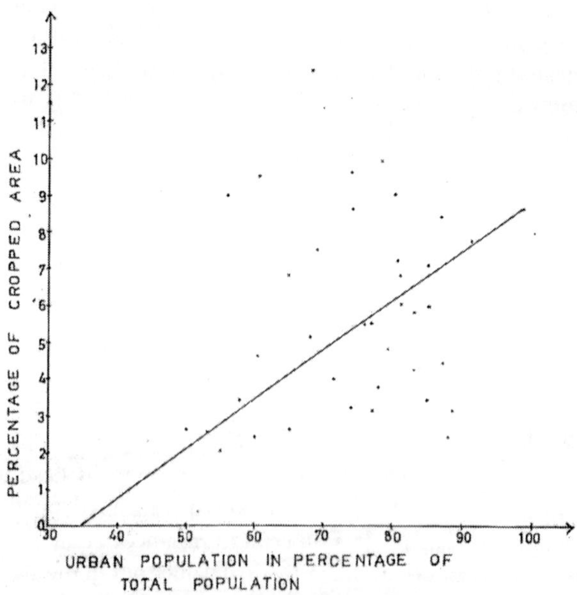

Scatter Diagram Showing Co-relation Between Cropped Area in (%) & Urban Population in (%) of Total Population in Districts of West Bengal

and the other for Y. Each pair of values is plotted as a point on graph paper. The whole set of points higher constitutes a scatter diagram. This method is, of course, not suitable when the number of individuals is very large. In such cases, one may use some three-dimensional analogue of histogram or frequency polygon, constructed on the basis of a bi-variant frequency table.

If the pattern of points or dots on the scatter diagrams shows a linear path diagonally across the graph paper from the bottom left-hand corner to the top right correlation will be *positive*. In other words, the association between the variables is direct indicating thereby that higher values of the other variables and low values are associated with low values.

On the other hand, if the pattern of dots is such as to indicate a straight line from the upper left-hand corner to the bottom right correlation is *negative* i.e., the association is direct, with high values of one variable being associated with low values of the other.

When the dots do not indicate any straight-line tendency but show a concentration around a curved line and correlation is small or will be *zero* or show *no correlation.*

When the dots lie exactly on a straight line, the *correlation is perfect or strong.* The correlation coefficient being +1 or -1 according to the shape of the straight line is positive or negative.

Coefficient of Variation

In order to compare the spread of two or more distributions, the relative measures of variability are used, it is independent of the units of observations and is defined as

$$\text{Coefficient of Variance, CV} = \frac{\sigma}{\bar{x}} \times 100\%$$

Coefficient of Variation is a relative method of measurement of dispersion, absolute measure of dispersion includes range of mean deviation, quartile deviation and standard deviation.

Standard Deviation method is the most important since it takes into account of all the values.

The formula of Coefficient of Variability is $CV = S.D. / \bar{x} \times 100$

Whereas, $\bar{x}$ = mean value

S. D. = Standard Deviation

$= \sqrt{\Sigma(x - \bar{x})^2 / N}$

Data and Calculations For Coefficient of Variability of Rainfall In Burdwan, Suri and Bankura (1950-1979)

	Burdwan			*Suri*			*Bankura*		
Year	x	$(x-\bar{x})$	$(x-\bar{x})^2$	x	$(x-\bar{x})$	$(x-\bar{x})^2$	x	$(x-\bar{x})$	$(x-\bar{x})^2$
1950	48.11	−12.86	165.38	46.43	−6.62	43.82	54.89	−0.99	0.98
1951	46.07	−14.9	222.01	62.05	9	81.00	43.13	−12.75	162.56
1952	57.56	−3.41	11.63	50.63	−2.42	5.86	46.03	−9.85	97.02
1953	44.71	−16.26	264.39	60.52	7.47	55.80	52.4	−3.48	12.11
1954	88.88	27.91	778.97	63.13	10.08	101.61	48.57	−7.31	53.44
1955	78.3	17.33	300.33	57.13	4.08	16.65	61.13	5.25	27.56
1956	44.32	−16.65	277.22	34.84	−18.21	331.60	54.67	−1.21	1.46
1957	65.37	4.4	19.36	60.24	7.19	51.70	50.07	−5.81	33.76
1958	66.15	5.18	26.83	70.31	17.26	297.91	69.02	13.14	172.66
1959	51.14	−9.83	96.63	60.64	7.59	57.61	48.08	−7.8	60.84
1960	47.04	−13.93	194.04	53.17	0.12	0.01	57.5	1.62	2.62
1961	51.33	−9.64	92.93	33.73	−19.32	373.26	33.37	−22.51	506.70
1962	71.99	11.02	121.44	56.6	3.55	12.60	56.6	0.72	0.52

	Burdwan			*Suri*			*Bankura*		
Year	*x*	$(x-\bar{x})$	$(x-\bar{x})^2$	*x*	$(x-\bar{x})$	$(x-\bar{x})^2$	*x*	$(x-\bar{x})$	$(x-\bar{x})^2$
1963	55.68	−5.29	27.98	48.48	−4.57	20.88	56.77	0.89	0.79
1964	54.89	−6.08	36.97	46.44	−6.61	43.69	44.71	−11.17	124.77
1965	77.61	16.64	276.89	48.48	−4.57	20.88	56.71	0.83	0.69
1966	66.35	5.38	28.94	46.44	−6.61	43.69	73.49	17.61	310.11
1967	68.25	7.28	53.00	48.03	−5.02	25.20	44.99	−10.89	118.59
1968	56.53	−4.44	19.71	50.6	−2.45	6.00	83.09	27.21	740.38
1969	53.57	−7.4	54.76	91.36	38.31	1467.66	54.67	−1.21	1.46
1970	51.78	−9.19	84.46	50.08	−2.97	8.82	50.03	−5.85	34.22
1971	75.36	14.39	207.07	47.78	−5.27	27.77	64.02	8.14	66.26
1972	68.72	7.75	60.06	46.81	−6.24	38.94	42.08	−13.8	190.44
1973	59.5	−1.47	2.16	48.96	−4.09	16.73	57.08	1.2	1.44
1974	51.31	−9.66	93.32	40.46	−12.59	158.51	62.3	6.42	41.22
1975	70.63	9.66	93.32	57.52	4.47	19.98	49.36	−6.52	42.51
1976	50.68	−10.29	105.88	52.5	−0.55	0.30	80.08	24.2	585.64
1977	63	2.03	4.12	50.09	−2.96	8.76	61.58	5.7	32.49
1978	69.01	8.04	64.64	52.32	−0.73	0.53	46.95	−8.93	79.74
1979	75.36	14.39	207.07	55.8	2.75	7.56	70.68	14.8	219.04

Burdwan	Suri	Bankura
$\Sigma x = 1829.2$ $\Sigma(x-\bar{x})^2 = 3991.52$	$\Sigma x = 1591.57$ $\Sigma(x-\bar{x})^2 = 3345.35$	$\Sigma x = 1674.05$ $\Sigma(x-\bar{x})^2 = 3722.04$
$\bar{x} = 60.973$	$\bar{x} = 53.052$	$\bar{x} = 55.802$
Variance = $\Sigma(x-\bar{x})^2/n = 133.05$	Variance = $\Sigma(x-\bar{x})^2/n = 111.51$	Variance = $\Sigma(x-\bar{x})^2/n = 124.07$
SD = sqrt (Variance) = 11.53	SD = sqrt (Variance) = 10.56	SD = sqrt (Variance) = 11.14
CV = (SD/$\bar{x}$)*100 = 18.91%	CV = (SD/$\bar{x}$)*100 = 19.90%	CV = (SD/$\bar{x}$)*100 = 19.96%

Here, coefficient of variability of rainfall 1950-79, in three districts, is represented by isopleths called equivariables. Isopleths may be used by plotting the places with similar deviations from their average weather conditions. For instance, mean annual figure for any given station does not affect the reliability of rainfall and of those stations. The exact degree of variability is difficult to calculate and various statistical formulae have been derived to represent it. The form of a coefficient of variability is one of such devices. It may be calculated and plotted and isopleths may be drawn. The coefficient of variability is expressed in percentage (%). Higher the coefficient of variability, the nature of distribution is more irregular and uniform distribution is least significant. Coefficient of variability depends on the nature of extreme normal distribution of rainfall and the variability is correlated.

Problem: - To draw the Coefficient of Variability of rainfall of West Bengal from the following given data.

Interpretation

The map of West Bengal shows the Coefficient of Variability of Rainfall from 1950 to 1979 in different districts for about 30 years. This shows that where the rainfall is high, the variability is low, as it can be seen that low variability is found in Medinipur (13.75%), Darjeeling (16.33%), Hooghly (16.77%) and Burdwan (18.91%). High variability is found near Balurghat (33.50%), Cooch Behar (26.23%), Howrah (21.13%). In Cooch Behar (26.23%), there is little rainfall but in Balurghat and Howrah due to absence of vegetation and city/urban influence with smoke and dust, rain bearing clouds have drifted apart. Due to the formation of heat islands in those urbanized areas, especially in Howrah and adjoining areas, the formation of clouds is inhibited. But in areas like

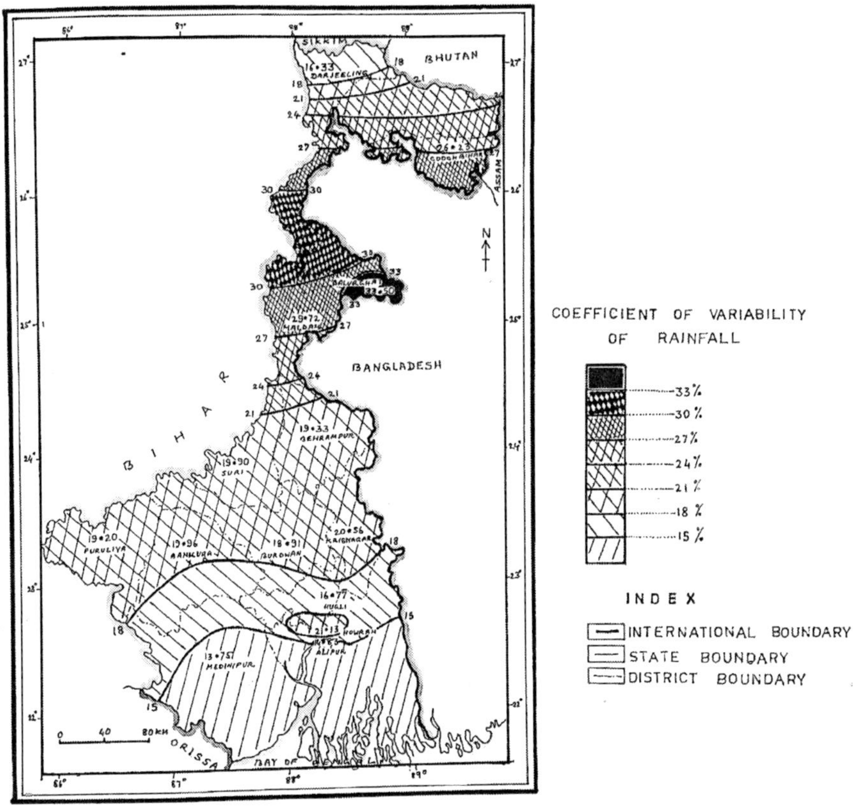

Coefficient of Variability of Rainfall in West Bengal 1950–79

Suri (19.90%), Bankura (19.96%), Berhampur (19.33%) and Purulia (19.20%) i.e., the Western plateau and Rarh region of West Bengal the coefficient of variability of rainfall though moderate but has an increasing trend. The amount of rainfall is decreasing, primarily, due to its geographical conditions accentuated by the high rate of urbanisation and industrialisation which simply impede rain formation. The existence of natural vegetation which is the prime factor of causing rain are being continuously cleared for human settlements and so there is marked variation in the decrease of rainfall.

PRACTICUM

15

Unit Planning for School Geography

Introduction

Unit planning serves as a roadmap for educators, guiding them in navigating the diverse terrain of geographical concepts, skills, and inquiries. By meticulously crafting each unit through the processes of Preparation, Presentation, Association, Generalisation, and Practical Application, educators create cohesive learning experiences that engage learners and foster deep understanding and critical thinking skills. In general, unit planning in geography education has evolved and kept to the updates and demands of society.

Understanding the Curriculum

In secondary and senior secondary geography education, curriculum standards emphasize geographic literacy, spatial reasoning, and critical thinking skills. For example, learning objectives may include understanding processes shaping Earth's landforms and analyzing patterns of population distribution and migration. Integration with other school disciplines enhances learners' holistic understanding of geographic concepts.

Analyzing Learner Needs

Before designing units, educators assess learners' prior knowledge, learning styles, and abilities. For example, a pre-test on weather and climate comprehension gauges learners' understanding, while accommodating diverse learning styles through visual aids and hands-on activities ensures inclusivity. Strategies for varying proficiency levels, such as differentiated instruction, promote equitable learning experiences.

Setting Clear Learning Goals

Clear and measurable learning objectives ensure alignment with curriculum standards and guide instruction effectively. For instance, in a unit on migration, learners analyze push and pull factors influencing migration patterns, aligning with specific standards related to population dynamics and human geography. Content-based and skill-based goals promote a balanced approach to instruction, fostering comprehensive understanding and critical thinking skills.

Selecting Relevant Content

Relevant content selection involves identifying core concepts, themes, and topics that resonate with learners' experiences and interests. For example, exploring urbanization encompasses core concepts such as population growth and land-use patterns, integrating contemporary issues like climate change's impact on urban environments. Interdisciplinary connections with subjects like history and economics enrich learners' understanding of geographical phenomena.

Designing Engaging Learning Activities

Engaging learning activities cater to diverse learning needs through hands-on experiences, fieldwork, technology integration, and group work. For instance, hands-on activities like creating topographic maps immerse learners in practical applications of geographical concepts. Fieldwork, such as conducting surveys or exploring geographical features, connects classroom learning with real-world experiences.

Assessing Learner Learning

Assessment strategies include formative and summative methods that cater to diverse strengths and preferences. Formative assessments, like quizzes and concept maps, provide ongoing feedback to guide instruction. Summative assessments, such as projects and presentations, evaluate mastery of content and skills. Authentic assessment tasks, mirroring real-world applications, enhance relevance and engagement. Diagnostic tests followed by remedial techniques enable a learner to perform and thereby appear for placement tests for higher knowledge acquisition.

Differentiating Instruction

Differentiated instruction addresses diverse learning needs through scaffolding, flexible grouping, differentiated assignments, and assistive technologies. For example, scaffolding techniques gradually remove support as learners master concepts, ensuring inclusivity and promoting independence. Flexible grouping allows educators to tailor instruction to learners' abilities and interests, fostering collaboration and peer learning.

Incorporating Reflection and Feedback

Reflection and feedback promote metacognitive skills and guide learner progress towards deeper understanding. Opportunities for reflection, like journaling and think-pair-share discussions, encourage self-awareness and ownership of learning. Timely and constructive feedback, aligned with learning objectives, guides improvement and fosters continuous learning. Encouraging self-assessment and goal setting empowers learners to track progress and take ownership of their learning journey.

In the context of unit planning, discussing both merits and demerits provides a comprehensive understanding. It recognizes the strengths of unit planning and acknowledges potential drawbacks ensuring a balanced perspective for educators.

Merits of Unit Planning

Alignment: Unit planning ensures all geography activities match specific learning goals, like analyzing weather patterns or understanding cultural landscapes.

Coherence: Structured unit plans in geography allow learners to understand and progress logically and ensure better retention say, studying landforms to exploring the impact of human activity on ecosystems.

Demerits of Unit Planning

Time-consuming: Crafting detailed geography unit plans requires significant time investment, diverting educators from other vital tasks like creating interactive maps or conducting fieldwork.

Inflexibility: Rigid adherence to geography unit plans might hinder educators from seizing impromptu teaching moments, like discussing current events shaping global migration patterns.

Conclusion

Effective unit planning and instructional strategies are essential for enhancing geography education. It creates a meaningful learning experience that empowers learners to become informed global citizens. As geography education continues to evolve, educators must embrace continuous reflection and refinement of unit planning practices to meet the evolving needs of learners in society and the discipline.

References: List of recommended further readings, resources, and websites for further exploration of specific required content can be done for enhancing geography education.

FORMAT FOR UNIT PLANNING -- Lesson No.

Class:
Unit/Sub-unit:
Total No. of Period(s) needed for Unit/Sub-unit(s):
Duration of each Period:
Time Duration needed for Evaluation:
Time Duration needed for Remedial Teaching:

Sl. No.	*Topic-Unit / subunit & No. of Period(s)*	*Specifying Sub-concept(s) with specific learning objectives*	*Appropriate Methods of Teaching (as required for General and Special needs)*	*Teaching-Learning Material(s)/ resources*	*Strategies of Teaching-Learning Activities - Teacher/ Learners*	*Evaluation/ Assessment Strategies*

RATING SCALES FOR EVALUATING INTERNSHIP ACTIVITIES

Rating scale for Evaluation of Activity: Preparation of Unit Plan

Teaching Subject: **Class/Section:**
Name of the Candidate: **Roll/ Regn No./Enrolment No.:**
Name of the school:

Rating is to be done according to the following criteria:

Criteria	**Rating**				
Unit plan has covered the required number of sub-units	1	2	3	4	5
Stated the learning objectives of the unit	1	2	3	4	5
Sequenced the contents as per the learning objectives of the unit	1	2	3	4	5
Suitable methods and media have been suggested	1	2	3	4	5
Suitable evaluation methods have been suggested	1	2	3	4	5
Prepared appropriate unit-end questions for evaluation	1	2	3	4	5
Reflections included in the Report	1	2	3	4	5
Quality of Presentation in the Report	1	2	3	4	5

Note: Rating Score: 1: Very Poor, 2: Poor, 3: Average, 4: Good, 5: Very Good

- Any specific material used or adapted for the learners with special needs..............
- Remarks /suggestations ....

Overall rating: **Signature of the Evaluator with Date**

16

Lesson Planning /Learning Design: Geography

Introduction

It is primarily important to mention that the progressive term of 'Lesson Plan' is taken as 'Learning Design'. Effective classroom teaching highly depends on the proper design of a lesson. Prior knowledge of the subject matter and mode of delivery helps a teacher to do justice to the students. Experienced teachers may not write learning designs in detail as they are quite efficient with the steps, but beginner teachers are always advised to be careful in preparing a detailed learning design for their success as a teacher in the class.

A learning design is a plan of action. It includes the working philosophy of the teacher, his knowledge of philosophy, his information about and understanding of his pupils, his comprehension of the objectives of education, his knowledge of the materials to be taught, and his ability to utilize effective methods or procedures. In the words of Bossing (cf. Sharma and Chandra, 2003: p-506), 'Lesson plan is the title given to the statement of the achievements to be realized and the specific means by which these are to be attained as a result of the activities engaged in, during the period'.

Learning design is a brief outline of the main points of the lesson to be covered by the teacher in a specified period for the realization of specified objectives. It indicates clearly what already has been done, what the pupils are to do, how the pupils are to be engaged in various activities, and what activities are to be pursued. It is a clear and precise statement of the aims and purposes of the lesson and the various techniques and devices used by the teacher. Daily learning design involves defining the objectives, selecting and arranging subject matter, and determining the methods or procedure.

In preparing a learning design a teacher has to consider the following points: i. Broader objectives of the subject ii. Instructional objectives of lesson iii. Organization of subject matter to be covered in the lesson for achieving the stipulated objectives iv. The decision about the way of presenting the subject matter, teaching strategies and tactics, classroom interaction, and management v. Appropriate provision for evaluation and feedback.

Importance of Learning Design

A good learning design can perform a number of functions as follows:

1. Learning design helps a teacher to define his aims and objectives more clearly. The teacher thinks of ways and means with which he can realize his aims most effectively.
2. It helps a teacher in the selection and organization of subject matter, materials, and activities, according to the abilities, aptitudes, and developmental levels of students.

3. Learning design helps a teacher in the selection of the most effective teaching procedure, which will lead to the modification of pupils' attitudes, habits, and information in a desirable direction.
4. Learning design helps a teacher to evaluate the outcomes of instructions. Evaluation is a very important part of the teaching-learning process. Evaluation is possible if definite aims and objectives are kept in mind.
5. Learning design helps a teacher to be systematic and prevents wastage. It saves him from haphazard teaching and repetitions are avoided.
6. Learning design establishes the proper connection between the different lessons and the units/subunits of study. Thus, it encourages continuity in the teaching process.
7. Learning design provides adequate lesson summaries and ensures a definite assignment for the class which helps in developing insight and understanding.
8. Learning design imparts confidence and self-reliance to the teacher. It enables the teacher to enter the classroom without anxiety and foresees the difficulties that are likely to arise and get prepared to deal.
9. It facilitates the teacher a series of plans well in advance to progressively modify children's attitudes, habits, and abilities towards desirable directions.

Principles of Learning Design in Geography

1. Principle of Writing the Learning design: A good learning design should be preferably written and should not remain at the oral or mental stage for the management of both theoretical and practical/project sections of Geography.
2. Principle of Statement of Objectives: A Geography learning design must contain both general and specific objectives which should be clearly stated.
3. Principle of Selection of Suitable Subject Matter: The Geography content of the lesson plan should be selected according to the needs, interests, abilities, and levels of the pupils (as specified in the syllabus).
4. Principle of Orderly Presentation: The subject matter should be presented in an orderly, organized, and effective manner with the appropriate geographical instructional aids and techniques. The effective presentation includes-
   a) Principles of Co-Relation: The subject matter which is following, need to be correlated with previous knowledge by the teacher. Moreover, it should be co-related with various other disciplines for best integration. It will make the lesson interesting and meaningful.
   b) Principle of Teaching Techniques: The plan should indicate the teaching techniques to be used by the teacher-how lesson is to be presented, what method is to be followed, what questions are to be asked, and what illustrations are to be used. It should clearly indicate the teaching-learning aids to be used by the teacher.
   c) Principle of Use of Chalk-Board: Learning design should clearly indicate chalkboard work done by the teacher. It should point out the diagrams and sketches which are to be drawn on the chalkboard.
   d) Principle of Child's Activity and Participation: Learning design should point out the child's activities and participation which are to be made in the classroom. Pupils should be encouraged to ask questions, solve problems and perform demonstrations/experiments.

5. Principle of Flexibility: Learning design should be planned so that lesson transaction is successful for all learners in an inclusive set-up.
6. Principle of Evaluation: Learning design should indicate evaluation activities. The evaluation exercise must consist of short answers and objective-type questions.
7. Principle of Assignment: Geography Learning design should include assignments based on contemporary issues for pupils.
8. Principle of Bibliography: Learning design should indicate selected references-printed/online.
9. Principle of Duration: The learning design should indicate the time within which the lesson is to be delivered. It must be balanced and well pervasive.
10. Principle of Age: The learning design should point out the target group i.e., age/grade of pupils. The lesson is to be planned for a particular stage of development and the mental level of the pupils.
11. Principle of Making Summary: The learning design should have an outline or summary of the entire lesson which may be also developed on the chalkboard with the help of the students.

Qualities of Geography Teacher for Preparation of Learning Design

The requirements of a geography teacher for the creation of effective learning designs can be outlined as follows:

- A teacher should hold an in-depth knowledge of the environmental conditions and also possess the managerial capacity of the students (individually and collectively) in an inclusive classroom
- Ability to make the students a thorough comprehension or consolidation to learn the lessons based on both the physical and psychological abilities of the students
- A mastery of the subject, motivating skill for the integration of co-scholastic activities along with the lesson transaction
- Efficiency in the appropriate handling of different instructional aids and techniques of teaching for flexible class management
- Ability to meet the specific needs of the diversified learners in the class in accordance with the topic

Steps to Prepare a Learning Design of Geography

There are different approaches to writing a learning design. H. C. Morrison and B. S. Bloom had given their own schemes on the basis of the unit approach and evaluation approach respectively. In India, the Regional College of Education, Mysore (RCEM) had also developed another model of the lesson plan. Herbart's model is, however, still the most popular one, and it is inclusive of the following five steps:

Preparation> Presentation> Comparison and Association> Generalization> Application Herbart's model was briefly described under the following heads: 1) *Information of identification* 2) *Instructional objectives 3) Previous knowledge 4) Announcement of the aim 5) Presentation with Teaching aids 6) Practice work/Homework 7) Evaluating a lesson*

A format to guide the students in constructing the learning design has been provided below:

Basic Format for Learning Design

Name of the School: Class: Duration: Date: Teacher's Name:	Subject: Lesson Unit: Sub-units: Today's Lesson:

Learning Goals/Objectives
[Goal- Statements based on Revised Bloom's Taxonomy]

Learner will be able to.... *Factual Knowledge* *Procedural Knowledge*	*Conceptual Knowledge* *Metacognition*
Remembering	←
Understanding/ Comprehending	←
Applying	←
Analyzing	←
Evaluating	←
Creating	←

Analyze Learners and Context
[General characteristics of the learners in terms of entry-level behaviour, to be ascertained by putting relevant questions]

To know the general characteristics and entry-level behavior of the learners, the following questions may be asked with respect to the present context.

Develop & Select Learning Materials
[Text Books and references; Teaching Learning Materials]
Learning Strategies

Learning Areas	Related Strategy

Design for Evaluation
[Test items/ work sheets/models/ experiments/ creative writings/ creative performances etc.]

Diagnosing the weaknesses for remedial class (if any)
[Teacher will diagnose the weaknesses after the evaluation]

N.B. Factual Knowledge may be described as the basic information about a particular subject or discipline that students must be acquainted with. *Conceptual Knowledge*-refers to the knowledge of, or understanding of concepts, principles, theories, models, classifications, etc. *Procedural Knowledge*-also known as imperative knowledge, is the knowledge exercised in the performance of some task. *Metacognition*- "thinking about thinking", "knowing about knowing", becoming "aware of one's awareness" and higher-order thinking skills.

Remedial Teaching

This is the most important and final phase of a lecture because the transaction of ideas and information between the teacher and the learner takes place to improve the learning skill and rectify a particular problem. The teacher explains and recollects the concepts and principles, and provides facts, data, figures, etc., to the learners very briefly. In order to explain the context matter, the teacher once again cites various examples, uses various communication aids, uses analogies and illustrations, or by writing them again on the blackboard, etc. The teacher, when required, adopts different non-verbal communication techniques such as gestures, postures, etc. to facilitate the teaching activity. During this phase, the teacher should be cautious of the lecture, otherwise, it becomes ineffective and a repeated one. In this concluding phase of the lecture, the teacher usually recollects and summarizes whatever has been covered during the course of the lecture and so one might also ask a few questions on the content matter, in order to evaluate the understanding of the lecture by the students. Through these questions, the teacher gets to know the learning difficulties of students and accordingly modifies the teaching. This reflective teaching identifies learners with diverse learning needs. The teacher also gives some home tasks to the students.

Application and Limitations of Learning Design in Classroom Situation

Effective classroom teaching highly depends on the proper planning of a lesson. Prior knowledge of the class, subject matter, and mode of delivery helps a teacher to do justice to the diversified learners. Experienced teachers may not write learning designs in detail as they are quite efficient with the steps involving a learning design, but beginner teachers are always advised to be careful in preparing a detailed learning design for their success as a teacher. However, learning designs also have certain limitations, as outlined below:

1. A learning design is actually a plan of action. But sometimes, it requires goals and objectives of teaching through proper selection of subject matter, teaching procedure, methods of evaluation, and other planning activities in accordance with the situation. This requires precision and efficiency on the part of the teacher in order to create a scope for self-evaluation.
2. A learning design is a well-defined track ensuring steady progress and a definite outcome of the teaching-learning process that needs to be abided by the subject-teacher. It should be formulated in a manner so that interest is generated and they are keen to assimilate the subject. However, often the real-world situation may not be conducive to executing a well-prepared learning design.
3. Learning design should be done in such a way that it facilitates the teacher to progressively modify children's attitudes, habits, and abilities towards desirable directions. Vital points, quotations, and 'telling phrases' on current geographic topics need to be readily used to make the class abreast of the present world situation.
4. Learning designs need to be well-knitted and interconnected so as to avoid repetition and monotony. Learning design needs an appreciable set, to meet the general and specific needs of the learners in an inclusive class.

5. Time management and the need for an inclusive situation should adhere to the flexibility that exists in the curriculum transaction. There should be no rigidity in handling different aids and techniques of teaching particularly in disseminating the subject matter in the class.
6. It is of absolute necessity that the teacher should possess an intimate knowledge of environmental conditions and have an academic influence on the students present in the class.
7. Questions framed should be relevant and applicable for all learners possessing individual differences.
8. Care to incorporate a list of further reading should be included as the essential part of the learning design.

Learning Design

Name of the School: Class: VIII **Time:** 40 minutes **Date:** **Name of the Teacher:**	**Subject:** Geography **Unit:** Oceania **Sub-units:** • Oceania expedition • Physical and political divisions * Murray-Darling basin * **Today's Lesson:** Murray-Darling basin

A. Learning Goals/Objectives (Based on Revised Bloom's Taxonomy)

Domain/Area	Goals/Objectives
1. Remembering	The learners will be able to identify and delineate the location of Murray-Darling basin. (FK) The learners will be able to state the extent and volume of Murray-Darling river. (FK)
2. Understanding	The learners will be able to interpret the significance of Murray-Darling river. (CK) The learners will be able to explain the flow and direction of Murray-Darling river. (PK)
3. Applying	The learners will be able to describe the agricultural practices and cattle-rearing activities and their developmental trends through time. (FK) The learners will be able to interpret the cause of the growth of temperate grasslands (Downs). (CK)
4. Analyzing	The learners will be able to analyse the economic activities prevalent in the area. (CK) The learners will be able to reason out the favourable causes of the various crops that are grown in various parts of the Murray-Darling basin. (CA)
5. Evaluating	The learners will be able to judge the contribution of Murray-Darling river which has facilitated the region to develop economically. (FK) The learners will be able to evaluate how Murray-Darling has affected in the growth of the population in the region. (FK)
6. Creating	The learners will be able to draw the flow of Murray-Darling in Oceania. (MK)

B. Analyzing Learners and Contexts

To ascertain the general characteristics of learners in terms of entry-level, the following questions are to be put up—

- Why is the term 'Oceania' used?
- What are the major rivers of Oceania?
- Which area/physical division of Australia comprises the Murray-Darling valley?
- Why is Murray-Darling a significant river system?
- Why are Downs famous?
- What economic activities are possible based on Oceania's physical characteristics?

C. Developing and Selecting Learning Materials

General Learning Aids:

1. A prescribed Geography textbook of the concerned Class VIII
2. Chalk-duster and blackboard (part of the classroom).

Specific Learning Aids: Achart prepared showing the location of Murray-Darling basin, and a flashcard showing the directional flow of Murray-Darling.

D. Learning Strategies

Learning Areas	*Related Strategies*
The location of Murray-Darling basin	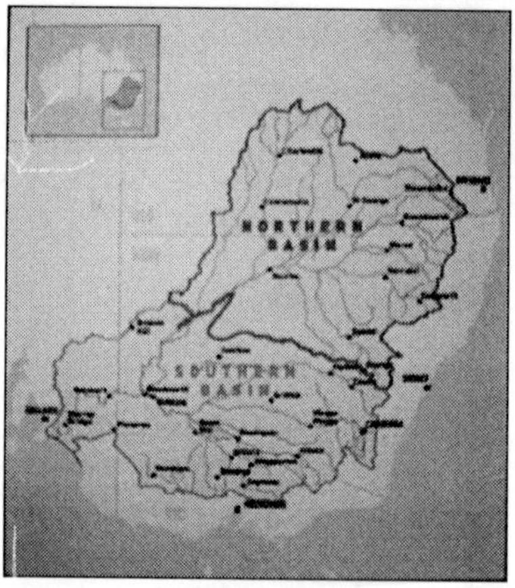 The location of the basin is pointed in the map of Australia using a chart through demonstration method.
The flow of Murray- Darling river	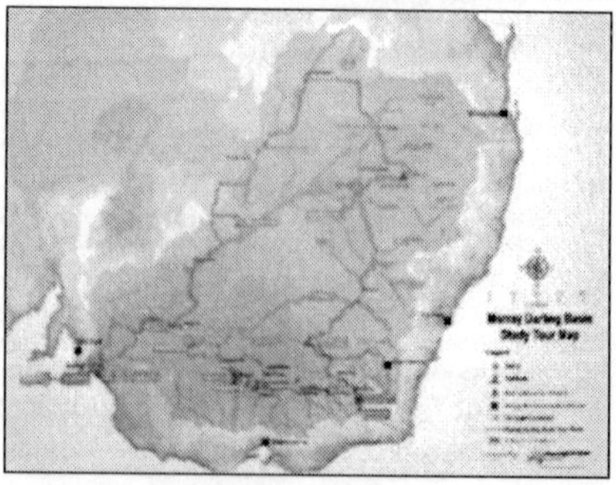 The flow is explained through lecture and demonstration methods using a flashcard.

Economic activities and areas under the economic activities	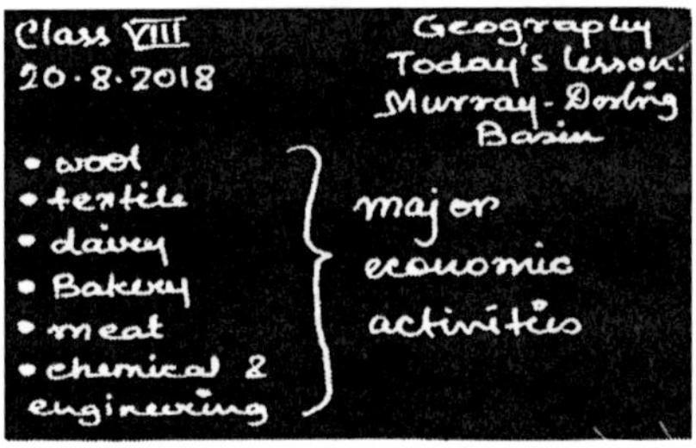 The name and area of economic activities are demonstrated using blackboard. The causes and reasons for the propagation of these activities are discussed.

E. Designing for Evaluation

In order to formulate the evaluation procedure, the teacher would design the formative and summative evaluation measures.

Formative Evaluation: To check the comprehensibility of the lesson concept, the teacher monitors the learning effectiveness by drawing attention and concentration, awakening motivation through goal-oriented behaviours, and promoting learner interaction through collaborative and socialization techniques.

Summative Evaluation: After the completion of the lesson, the teacher would conduct the performing tasks through the preparation of worksheets or experiments.

WORKSHEET (given in the class)

1. **Fill in the blanks:**
   (a) The grassland in Murray-Darling basin is _______.
   (b) _______ is called 'Silver City'.
   (c) The most important animal producing wool and fur is _______.
2. **State 'True' or 'False':**
   (a) The slope of Murray-Darling basin is towards west.
   (b) The Murray-Darling rivers have their confluence in Wentworth.
3. What type of climate prevails in Murray-Darling basin?
4. State the most populous area of Australia.

As a part of Homework:

A Project-based Learning activity on the *'Significance of the Murray-Darling basin in the lives of the people'* would be given by the geography teacher to the learners.

F. Diagnosing the Weakness for Remedial Class (if any)

In order to support/assist the learners with their learning difficulties, the teacher would formulate/ organise, the remedial class for them.

The aspect of climate on the basin environment was confused.	The role of climate on the basin and its relationship to be clarified through lecture and discussion methods and the location map on chart can also be used.

17

Pedagogical Analysis of Content: Geography

Definition and Purpose

Pedagogical analysis of content refers to breaking up the content of a particular topic of discussion into smaller parts, e.g. into units or sub-units for effective or scientific teaching. The teaching can therefore become meaningful, serving the intended purpose. A difference between content analysis and pedagogical analysis can be made, and this is important for a better understanding of the two. However, it is to be noted that the difference is solely dependent on the purpose or objective. In *content analysis* the objective is to break the whole subject matter into smaller parts, properly organized in sequential and meaningful order; whereas in *pedagogical analysis* similar partition of the subject matter is done following a suitable methodology for the purpose of scientific teaching. Effective teaching is defined as 'the ways and means provided to or utilized by a teacher for managing the task of teaching as smoothly and effectively as possible by involving his least efforts for the best possible outcomes' (Mangal and Mangal, 2008). Any course or syllabus comprises a number of chapters or units. Each unit consists of a quantity of knowledge that represents a particular theme. This body of knowledge has to be transacted in the class so that the objectives of teaching the unit are achieved successfully by the students. This necessitates that the unit/content is planned in a systematic manner to achieve the objectives of the teaching-learning process. The main objective of a unit plan is to manage student learning in that particular content. It represents the sequencing of learning activities of the content and the supporting learning materials that are required for the transaction of learning experiences. So, the content analysis serves the following purposes: 1) Smooth and orderly teaching 2) Maximum output with the least effort 3) Conceptual clarity among both the teachers and students 4) Effective management of time and skill and other necessary resources 5) Effective prior programming of teaching methods and selection of suitable teaching aids.

Operational Steps of Pedagogical Analysis and the Role of A Geography Teacher

In today's modern forms of teaching geography, both pre-service and in-service learners should be in a position to evidently: a) Acquire knowledge by themselves, understand issues, and be able to apply the knowledge and skills in life b) Build on their own efforts, gain knowledge, understanding, and experiences by working on activities and assignments supervised by the lecturers/teachers and engaging often in fieldwork activities and discovery expeditions in order to perfect their working skills for practical learning. c) The process of learning must model crucial aspects of critical thinking, reasoning, observations, and research. The learners should have ideas of what they want to learn and ought to establish facts and understanding of the topics coupled with a philosophical understanding of how the content will assist one in developing a deep understanding of issues learned is important. This will result in the learner developing action plans for handling issues in life.

The pedagogy involves interrelated discourse of four major elements of the teaching-learning process, namely- (1) Instructional objectives; (2) Content material and learning experiences; (3) Tools/ aids and techniques of teaching; and (4) Evaluation methods.

Content analysis is based on the interdependence of these four aspects of the teaching-learning process, and therefore it includes the following steps:

(a) ***Analysis of the content of the topic to be taught***
This requires a thorough and complete understanding of the content-material prescribed in the Geography syllabus of both secondary/senior secondary. So, a good textbook may be consulted to take note of the important points of the content.

(b) ***Setting the instructional objectives in behavioural terms***
After going through the teaching-learning process on a selected geographical topic, the students are required to show some definite types of behavioural changes in them. This can be identified with the help of certain questions/ tests based on various domains of educational objectives.

(c) ***Enlistment of the teaching aids and methods relevant to the contents of the topic***
The list of geographical teaching materials, methods, teaching aids, and activities should conform to the relevant conditions. Besides the textbook study, there can be demonstrations, exhibitions, or projections of the material; a teaching method may be selected from different alternatives for the benefit of inclusive learners.

(d) ***Selection of the suitable technique for proper evaluation***
The evaluation of the geographical learning outcome may be based on CCE/ competency-based or on written/ oral tests relevant to the theoretical/practical concept-formation. Other evaluation techniques like observation, rating scale, and interview may also be adopted.

Application and Limitations of Pedagogical Analysis in Classroom Situation

Pedagogical analysis of content refers to breaking up the content into units or sub-units for scientific and retentive teaching. This integrated teaching can become meaningful for serving the intended purpose. Although it has many rewarding applications, it has certain limitations too, as mentioned below:

1. Experience shows that class teaching would not be smooth and orderly with the help of pedagogical analysis only, because it obviously depends upon the particular classroom situation. In accordance with the demanding situation, the orientation of classroom teaching may be regulated. Therefore, strict adherence to the pre-determined pedagogical analysis of content is not feasible.
2. It is not always necessary that there would be maximum output with the pedagogical analysis because the handling of the classroom teaching requires appropriate mastery of the subject and conceptual clarity by the teacher concerned.
3. Effective prior programming of teaching methods and selection of suitable learning-aids demand efficiency on the part of the teacher, because it particularly depends upon the class size and the socio-economic background of the learners; and also, upon the infrastructural facilities available in the school.
4. Elaborate plans on teaching strategy, preparation of test items, table of specification, and worksheet does not necessarily lead to the best performances

in the class. The teacher may have the ability to prepare such items but may lack in-depth knowledge of the matter and its purpose. The trainee-teacher may work in a very mechanical manner and thereby the entire system may become futile.

5. Emphasis on the appropriate break-up and detailing of the sub-units obviously depends upon the class, the psychology of the learners, and the nature of the selected topic. This requires precision, understandability, and presentability on the part of the teacher.
6. Pedagogical analyses on a particular topic/unit/sub-unit and class are prepared beforehand on a very standard set of norms. But on the standard, the psychological capability of the students and the availability of the infrastructural facilities may vary from school to school. This may lead to the non-applicability of the entire procedure and everything might go in vain.
7. Too much time is spent preparing a well-planned pedagogy. And that's why the training procedure for the completion of the required number of pedagogies on a specified subject in such a short span of time is, to a great extent, limited.
8. Pedagogical analysis lays down a general concept of a teaching-learning programme for every student. But attention should be duly paid to serving all categories of students who may be gifted, average, and slow learners.
9. Unnecessary elaboration and exaggeration of the sub-units in the pedagogical analysis especially by inexperienced teachers may distort the spontaneous learning process in the classroom.

PEDAGOGICAL ANALYSIS

Breaking of Units into Sub-Units with no. of Periods

Class: X
Subject: Geography
Unit: Atmosphere

Sub-units:	*Number of Periods*
• Concept and Definition of Atmosphere	
• Composition of Atmosphere	1
• Structure of Atmosphere	3
• Heat, Temperature, and Global Warming	6
• Pressure Belts and Winds	7
• Humidity and Precipitation	4
• Major Climatic Regions of the World	3
• Remedial Class	1
Total Periods	25

Sub-unit Selected: Structure of Atmosphere

Previous Knowledge

1. The learners are aware of different elements of weather and climate.
2. Pupils have an idea of the concept of the atmosphere.
3. Learners know the varied components of the atmosphere.
4. Learners have cognizance about various proportions of gasses that exist in the atmosphere.

INSTRUCTIONAL OBJECTIVES IN BEHAVIOURAL TERMS

Domain/ Area	*Goals/Objectives*
1. **Remembering**	a. **Name** the hottest layer of the atmosphere. (FK) b. **State** the significance of the presence of the Ozone layer in the atmosphere. (CK) c. **Underline** the incorrect word of the statement given below: *The increase in temperature with an increase in height is known as the Normal Lapse Rate. (*PK) d. **Make facts- chart** on Thermosphere. (MK)
2. **Understanding**	a. **Interpret** the characteristic features of the Exosphere. (FK) b. **Distinguish** between the Homosphere and Heterosphere. (CK) c. **Discuss** the procedure of formation of auroras in the Ionosphere. (PK) d. **Find more information** on the causes behind Ozone Depletion. (MK)
3. **Applying**	a. **Investigate** the CFC generation sources from the household sector. (FK) b. **Illustrate** with an example the detrimental effects of depletion of the ozone layer. (CK) c. **Put together your knowledge** of the Thermosphere being responsible for enabling radio broadcasting and wireless communication. (PK) d. **Change** the following into the correct order of sequence from the surface of the earth: *Exosphere, Troposphere, Thermosphere, Stratosphere, and Mesosphere.* (MK)
4. **Analyzing**	a. **Explain** the reason for meteors' movements slowing down and burning up in the Mesosphere. (FK) b. **Compare and contrast** between the Troposphere and Stratosphere. (CK) c. **Analyze** the reason behind the thermosphere being the hottest layer in the atmosphere. (PK) d. **Investigate** the reason behind atmospheric pressure being lowest in the Exosphere. (MK)
5. **Evaluating**	a. **Discuss** the role of the Troposphere as a weather-making layer in the atmosphere. (FK) b. **Find the importance** of the presence of the ozone layer in the atmosphere. (CK) c. **Assess** the reason for the stratosphere providing ideal conditions for flying. (PK) d. **Validate** the need or importance of the presence of the Thermosphere and its positive impacts on human lives. (MK)
6. **Creating**	a. **Formulate** a chart showing different layers of the atmosphere with proper labeling. (FK) b. **Combine** the significance of each layer stating their importance in the earth's atmosphere in a write-up. (CK) c. **Create** a diagram of the depletion of the ozone layer and label it. (PK) d. **Prepare** a block diagram on the division of the atmospheric layer based on chemical composition and variation in temperature. (MK)

[Note: - FK- Factual Knowledge; CK- Conceptual Knowledge; PK- Procedural Knowledge; MK- Meta-cognitive Knowledge]

SUB-UNIT WISE CONCEPTS

Sub-Unit	*Broad Concepts*	*Brief Summary*
Structure of Atmosphere	**Concept 1:** Layers of Atmosphere according to the chemical composition	• The atmosphere is made up of a mixture of gasses that form a layer surrounding the earth. Water vapour and dust particles are other constituents of the atmosphere. • According to its chemical composition, the atmosphere can be divided into two layers, *i.e.* (a) The proportion of gasses varies greatly with the layers of height. Up to the height of 100 km, the composition of the atmospheric layer remains almost the same and it is popularly known as ***Homosphere.*** (b) Above this layer, the percentage of gasses varies and shows heterogeneous character; this upper layer nearly up to 10,000 km in height is known as the ***Heterosphere.***
	Concept 2: Layers of Atmosphere according to variation in temperature	• Based on temperature differences atmosphere can be divided into five concentric layers, *i.e.* • **The troposphere** is the lower-most layer of the atmosphere, nearest to the Earth. All daily weather changes take place in this layer (extends up to 14 km from the surface around the Equator and about 8 km around the poles.) • **The stratosphere** lies above the Troposphere (extends up to a height of 60 km) It provides ideal conditions for flying Jet planes and Commercial aircraft. • At the **Mesosphere,** the meteors that come towards the earth get burnt in this layer (which extends up to a height of 80 km above the earth's surface.) • In **Thermosphere,** ionization occurs in this layer, hence, this layer is also known as the Ionosphere (it extends from about 90 km to about 500-1000 km above our planet.) • **The exosphere**, the uppermost layer of the atmosphere (exceeding 700 km in height) merges into space gradually. • In the case of the Troposphere and Mesosphere, the Normal Lapse Rate of temperature operates where temperature decreases with increasing height but in the case of the other three layers i.e., Stratosphere, Thermosphere, and Exosphere, inversion of temperature is observed in which, with increasing height the temperature increases.

TEACHING - LEARNING STRATEGIES

The students will be able to understand the concepts of the chosen subunit by applying the following strategies:

Learning Areas	*Related Strategies*
Concept 1: Layers of the Atmosphere according to chemical composition.	• Through the Lecture and Demonstration methods, the layers of the atmosphere according to chemical composition will be introduced. • The trainee-teacher will draw a tree-chart to explain the concept of the Homosphere and Heterosphere and she/he will write down important features of these layers on the blackboard.

Learning Areas	*Related Strategies*

Subject: Geography ATMOSPHERE
Class: X
Date: 01/10/2020

Structure of the Atmosphere

Based on chemical composition:
a. Homosphere
b. Heterosphere

Based on temperature variation:
a. Troposphere
b. Stratosphere
c. Mesosphere
d. Thermosphere
e. Exosphere

Homosphere:
- The proportions of gases are constant.
- Extends up to a height of 100 km.

Heterosphere:
- The percentage of gasses varies.
- It shows heterogeneous character.
- Extends up to a height of 10,000 km

Black Board Work No. 1: Structure of Atmosphere

- The structure of the atmosphere based on chemical composition is taught using a Chart, namely *"Layers of Atmosphere Based on Chemical Composition"*

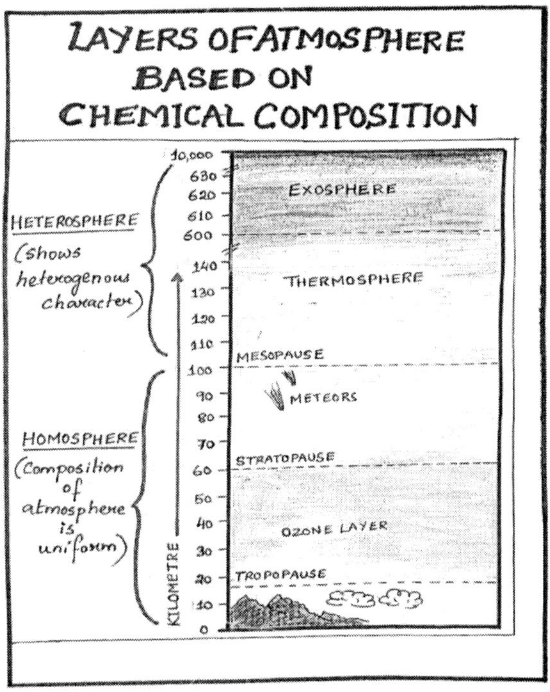

Figure No. 1: Chart Showing Layers of Atmosphere

Concept 2: Layers of Atmosphere according to temperature variation

- Demonstration, interactive, and lecture methods will be used to teach pupils about different layers of the atmosphere according to temperature variation.
- The trainee-teacher will teach the characteristics of different layers of the atmosphere, their altitudinal ranges, and temperature variation through the lecture method and relevant features of each layer will be written down on the blackboard.

Learning Areas	*Related Strategies*

Subject: Geography

Class: X

Date: 01/10/2020

ATMOSPHERE

Troposphere	Stratosphere	Mesosphere	Thermosphere	Exosphere
• Lowest Layer • Daily weather change • Normal Lapse Rate is observed	• Contains Ozone layer • Free from weather elements • Jet planes prefer to fly • Inversion of temperature	• Lies above Stratosphere • Normal Lapse Rate operates • Meteors burn up in this layer	• Temperature increases with increasing height • Enables radio broadcasting and wireless communication • Auroras are seen	• Uppermost layer • Gradually merges with space

Black Board Work No. 2: The Characteristics of Different Layers

- The trainee-teacher will demonstrate the structure of different layers of the atmosphere based on temperature variation with the help of a chart, namely *"The Levels of Atmosphere"*

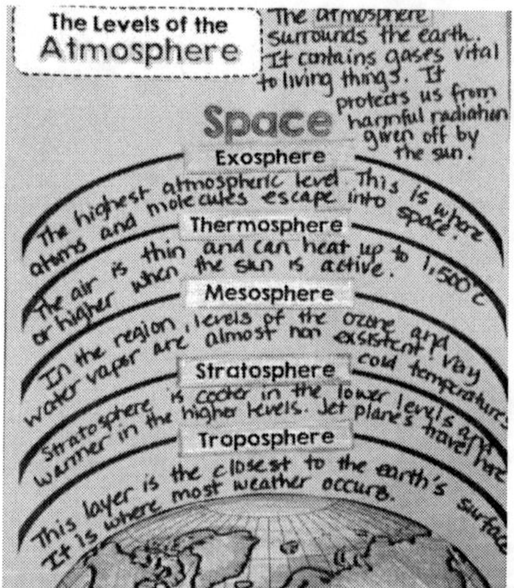

Figure No. 2: Chart Showing the Levels of the Atmosphere

- The trainee teacher will demonstrate some real-life pictures of Auroras in flash- card.

Figure No. 3: Auroras

Learning Areas	*Related Strategies*
	• The trainee teacher will show some short videos on the burning of meteors in the Mesosphere and Jet planes flying through the Stratosphere. [URL: https://www.youtube.com/watch?v=O4QrnC9vLs8 ; https://www.youtube.com/watch?v=pxuhyQLykao] • The trainee will also demonstrate that the troposphere is the densest layer by stacking books one above the other.

Figure No. 4: Stacking of Books

	• The trainee teacher will interact with the students by asking them some questions:
	  • When you visit the mountains, temperature increase or decrease?
	  • What will happen to us in the absence of the ozone layer?
	  • As student how can you contribute to preventing ozone depletion?
	  • In which layer do you think communication satellites are located?
	  • Why do you think Jet planes prefer to fly through Stratosphere?

USE OF TEACHING AIDS

The trainee-teacher will use the following aids to execute the lesson-

Type of Teaching Aid	*Specification of Teaching Aids*	*Broad Concept*	*Uses of Teaching Aids*
General Aids	Board/Council Prescribed Geography Textbook of Class 10	Detailed information about the composition of the atmosphere and its various layers	The book helps the students to connect the information in the textbook with the teacher's teaching
	Blackboard, Chalk, and Duster	• The Structure of atmosphere • The characteristics of different layers of the atmosphere	• Blackboard work is done to discuss the complex structure of the atmosphere • To make the students attentive the teacher writes down the important characteristics of each layer of atmosphere on the blackboard

Special Aids	Chart	• Layers of Atmosphere Based on Chemical Composition • The Levels of the Atmosphere	• The chart namely "Layers of Atmosphere Based on Chemical Composition" can help students to enhance their understanding of the chemical configuration of the atmosphere • The chart namely "The Levels of Atmosphere" helps to show the layers of the atmosphere based on temperature variation
	Real Images in Flash-card	Auroras	The teacher shows the real image of Auroras to demonstrate the distinctive characteristics of 'Ionosphere'
	Audio-Visual Aids	A short video is shown about the burning of meteors in the earth's Mesosphere and Jet planes flying through the Stratosphere	On this regard, to give exposure to the unique characteristics of the Stratosphere and Mesosphere two short videos can be shown by the teacher.

Probing Questions

i. Why thickness of the troposphere varies from the Equator to the poles?
Ans- The Troposphere is thicker over the equator than the poles because the equator is warmer. Heat differences on the planet's surface cause convection currents to flow from the Equator to the Poles. This implies that the warmer the weather, the thicker the troposphere.

ii. Why in the absence of an ozone layer the biosphere will turn into a blast furnace?
Ans- In the absence of an ozone layer all the UV rays of the sun will reach the earth's surface directly and consequently the temperature of the atmosphere will rise to such an extent that the biosphere will turn into a blast furnace.

Worksheet

A. Identify the correct answer:

i. The layer of the atmosphere where changes in daily weather take place is called: (A)
(a) Troposphere (b) Stratosphere (c) Exosphere (d) Ionosphere

ii. The upper limit of the Mesosphere is called: (R)
(a) Tropopause (b) Stratopause (c) Ozonosphere (d) Mesopause

iii. The densest layer in the atmosphere is: (U)
(a) Exosphere (b) Homosphere (c) Mesosphere (d) Troposphere

iv. Ozone protects mankind from: (E)
(a) UV rays (b) Infrared rays (c) X-rays (d) Gamma rays

v. The country that witnesses polar lights: (A)
(a) Spain (b) India (c) Norway (d) Mexico

B. Give a one-word answer:

i. The upper limit of the Troposphere that separates it from the Stratosphere. (An)
ii. Harmful rays that lead to skin cancer in human beings. (U)
iii. The Coldest layer of the atmosphere. (R)
iv. The continent over which a large ozone hole has developed. (A)
v. The layer that shows heterogeneous character. (R)

C. Fill in the blanks:

i. The decrease of temperature at the rate of 6.5°C/km is known as the __________. (R)
ii. The__________ layer absorbs the harmful rays of the sun. (R)
iii. Jet planes prefer to fly through __________ over Troposphere. (An)
iv. Communication satellites are located in __________ layer of atmosphere. (A)
v. Up to the height of __________ km, the composition of atmospheric gasses remains almost the same. (R)

D. Match column A with column B:

Column A	**Column B**
(i) Troposphere (R)	(a) Meteors burn up.
(ii) Northern polar lights (An)	(b) Aurora Borealis.
(iii) Mesosphere (U)	(c) Presence of Ozone layer.
(iv) Stratosphere (U)	(d) Enables radio broadcasting.
(v) Thermosphere (An)	(e) Lower-most layer of the atmosphere.

[Note: R- Remembering, U-Understanding, A- Applying, An-Analysing, E-Evaluating, C- Creating]

BLUE-PRINT FOR CRITERION REFERENCE TEST ITEMS

Subject: Geography
Class: X
Unit: Atmosphere

Full Marks: 30
Time: 40 minutes
Chosen Sub-unit: Structure of Atmosphere

	Cognitive Process Dimensions									
	R					*U*				
CONCEPT	*OT*	*VST*	*ST*	*LT*	*CT*	*OT*	*VST*	*ST*	*LT*	*CT*
CONCEPT 1	1(1)	1(1)	-	-	-	1(1)	-	1(2)	-	-
CONCEPT 2	3(1)	2(1)	-	-	-	1(1)	1(1)	-	-	-
TM	4	3	0	0	0	2	1	2	0	0
%	13.2%	10%	0%	0%	0%	6.7%	3.3%	6.7%	0%	0%

	Cognitive Process Dimensions									
	A					*An*				
CONCEPT	*OT*	*VST*	*ST*	*LT*	*CT*	*OT*	*VST*	*ST*	*LT*	*CT*
CONCEPT 1	-	-	-	-	-	-	-	-	-	-
CONCEPT 2	2(1)	1(1)	-	-	-	2(1)	-	1(2)	1(3)	-
TM	2	1	0	0	0	2	0	2	3	0
%	6.7%	3.3%	0%	0%	0%	6.7%	0%	6.7%	10%	0%

	Cognitive Process Dimensions											
	E					*C*						
CONCEPT	*OT*	*VST*	*ST*	*LT*	*CT*	*OT*	*VST*	*ST*	*LT*	*CT*	*FM*	*%*
CONCEPT 1	-	-	-	-	-	-	-	-	-	-	05	16.7%
CONCEPT 2	-	-	-	1(3)	-	-	-	-	-	1(5)	25	83.3%
TM	0	0	0	3	0	0	0	0	-	5	30	100%
%	0%	0%	0%	10%	0	0%	0%	0%	0%	16.7%	100%	100%

Key: Cognitive Process Dimension				*Key: Type of Questions*			
R	Remembering	An	Analyzing	OT	Objective type	LT	Long type
U	understanding	E	Evaluating	VST	Very short type	CT	Constructive type
A	Applying	C	Creating	ST	Short type	-	-

1(1) ⟶ 1 Question of 1 mark

TEST ITEMS

Date: 21/03/2024
Time: 40 minutes

Class: X
Full Marks=30

Instruction to the Examinees-

- Read the questions carefully before answering them.
- Extra time will be given for reading the questions.
- Do not copy the questions.
- Attempt all the questions.

I) Fill in the Blanks (OT) (5×1=5)

a) is the layer in the atmosphere that shows heterogeneous character. *(Remembering, concept 1)*

b) Up to the height of 100 km, the composition of atmospheric gasses remains almost the ____________ *(Understanding, concept 1)*

c) ____________ is the layer of the atmosphere where changes in daily weather takes place. *(Applying, concept 2)*

d) Jet planes prefer to fly through ________ over Troposphere. *(Analysing, concept 2)*

e) Ozone protects mankind from____________ rays. *(Remembering, concept 2)*

II) State True or False (OT) (5×1=5)

a) Thickness of the Troposphere remains constant around the Equator and near the poles. *(Understanding, concept 2)*

b) Exosphere is the lowermost layer of the atmosphere. *(Remembering, concept 2)*

c) Normal Lapse rate operates in Mesosphere. *(Remembering, concept 2)*

d) Temperature increases with increasing height in the Stratosphere due to the presence of the Ozone layer. *(Analysing, concept 2)*

e) Air-conditioners are major household sources of generating CFCs. *(Applying, concept 2)*

III) Answer the Following in One or Two Words (VST) (5×1=5)

a) State the extension of the Homosphere. *(Remembering, concept 1)*

b) Name the upper limit of the Troposphere. *(Remembering, concept 2)*

c) Name the continent over which a large ozone hole has developed. *(Applying, concept 2)*

d) Name the layer where meteors slow down and burn up. *(Understanding, concept 2)*

e) Give another name for the northern polar lights. *(Remembering, concept 2)*

IV) Write in One or Two Sentences (ST) (2×2=4)

a) Differentiate between Homosphere and Heterosphere? *(Understanding, concept 1)*

b) Give a reason for atmospheric pressure being lowest in Exosphere. *(Analysing, concept 2)*

V) Answer the Following in Brief (LT) (2×3=6)

a) Compare and contrast between the Troposphere and Stratosphere. *(Analysing, concept 2)*

b) Validate the need or importance of the presence of the Thermosphere in the atmosphere and its impact on human lives. *(Evaluating, concept 2)*

VI) Construct a Diagram on the Following (CT) (1×5=5)

a) Create a diagram showing the different layers of the atmosphere based on temperature variation and label it properly in accordance to their extension and significance. *(Creating, concept 2)*

ITEM-WISE ANALYSIS

Item No.	*Concept*	*Type of Question*	*Instructional Objectives* R	*U*	*A*	*AN*	*E*	*C*	*Marks Allotted*	*Approximate Difficult Level*	*Proposed Time*
I(a)	1	OT	✓						1	E	0.5
I(b)	1	OT		✓					1	M	1
I(c)	2	OT			✓				1	M	1
I(d)	2	OT				✓			1	M	1
I(e)	2	OT	✓						1	E	0.5
II(a)	2	OT		✓					1	M	1
II(b)	2	OT	✓						1	E	0.5
II(c)	2	OT	✓						1	M	1
II(d)	2	OT				✓			1	D	1.5
II(e)	2	OT			✓				1	D	1.5
III(a)	1	VST	✓						1	M	1.5
III(b)	2	VST	✓						1	E	0.5
III(c)	2	VST			✓				1	M	1
III(d)	2	VST		✓					1	M	1
III(e)	2	VST	✓						1	M	1
IV(a)	1	ST		✓					2	M	3.5
IV(b)	2	ST				✓			2	D	4
V(a)	2	LT				✓			3	M	5
V(b)	2	LT					✓		3	D	6
VI(a)	2	CT						✓	5	D	7
Total									30		40 mins

E- Easy, M- Medium, D-Difficult

Key: Cognitive Process Dimension				*Key: Difficulty Level*	
R	Remembering	AN	Analyzing	E	Easy
U	Understanding	E	Evaluating	M	Medium
A	Applying	C	Creating	D	Difficult

Distribution of Marks

Serial No. of Question	*Marks*
I	5
II	5
III	5
IV	4
V	6
VI	5
Total	**30**

GUIDELINES FOR COURSE OUTCOME (CO)

Semester	XX
Paper Code	XXX
Paper Title	XXX XXX
No. of Credits	X
Theory/Practicum	Theory
No of periods assigned	X teaching hours in accordance to the period duration & content
Course descriptions/	Type of paper- general/ Special/ Method
Outcome	Mention the 4/5 broad objectives of the content in the paper.

GUIDELINES FOR PROGRAMME OUTCOME (PO)

1. To foster an effective integration of diverse knowledge domains, aligning with the essential training requirements for prospective educators.
2. To ensure the curriculum remains contemporary, professional, and imbued with practical, skills-oriented learning experiences.
3. To equip trainee-teachers with the competencies necessary to address academic, societal, and national educational demands.
4. To cultivate a positive life outlook and nurture holistic personality development among trainee-teachers, fostering a commitment to community service.
5. To instill resilience and adaptability in trainee-teachers to confront the multifaceted challenges of the modern age.
6. To instill discipline and foster the cultivation of moral values among trainee-teachers.
7. To motivate and empower trainee-teachers to inspire and mentor future generations.
8. To incorporate opportunities for trainee-teachers to engage with innovative teaching methodologies and emerging educational technologies, preparing them for dynamic educational landscapes.

A Final Summarised Format of the Item Analysis

Name of the Institution

Data Sheet for Mapping Course and Programme Outcomes of the Question Paper

SESSION:			*PROGRAMME CODE:*		
COURSE CODE:			*SEMESTER:*		
DURATION:			*FULL MARKS:*		
Question Number	*Allotted Marks*	*Type of Question*	*Difficulty Level*	*Course Outcome (CO)*	*Programme Outcome (PO)*

Signature of Co-ordinator (with date) **Signature of Moderator (with date)**

*Type: MCQ/ Descriptive/ Objective/ Problem Solving/ Analytical

** Difficulty: Basic/ Moderate/ High

SCORING KEY

I. a) Heterosphere
   b) Same
   c) Troposphere
   d) Stratosphere
   e) UV

II. a) False
   b) False
   c) True
   d) True
   e) True

III. a) From surface up to 100 km
   b) Tropopause
   c) Antarctica
   d) Mesosphere
   e) Aurora Borealis

IV. a) The differences between Homosphere and Heterosphere are as follows:

Homosphere	*Heterosphere*
1. Extends from the surface up to a height of 100 km.	1. Extends from 100 to up to 10,000 km above the surface of the earth.
2. The composition and proportion of atmospheric gasses remain constant.	2. The percentage of gasses varies and shows heterogeneous character.

IV. b) Exosphere being the uppermost layer of the atmosphere has air that is extremely thin and rarefied. Hence, this makes the atmospheric pressure the lowest in this layer.

V. a) Troposphere and stratosphere layer of the atmosphere show some different as well as some similar characteristics:

Similarities:

i) In both Troposphere and Stratosphere atmospheric pressure decreases with increasing height.

Differences:

i) All dust particles and water vapour are present in the Troposphere, therefore all changes in daily weather conditions take place in this layer. On the other hand, absence of water vapour in the Stratosphere makes this layer free from weather disturbances.

ii) Temperature decreases with altitude in the Troposphere following the Normal Lapse rate. Whereas in the Stratosphere temperature increases with increasing height.

V. b) From the point of view of the present world scenario, the need or importance of the Thermosphere is immense in our lives:

- Electrically charged particles of ions and electrons are found here, so radio waves coming from the earth's surface are reflected by the earth, so it enables radio broadcasting and wireless communication across the world.
- Polar lights called the Aurora Borealis and Aurora Australis of the Northern and Southern hemispheres respectively are developed in this layer which acts as an important source of lighting during the months of darkness in the poles.

VI. a)

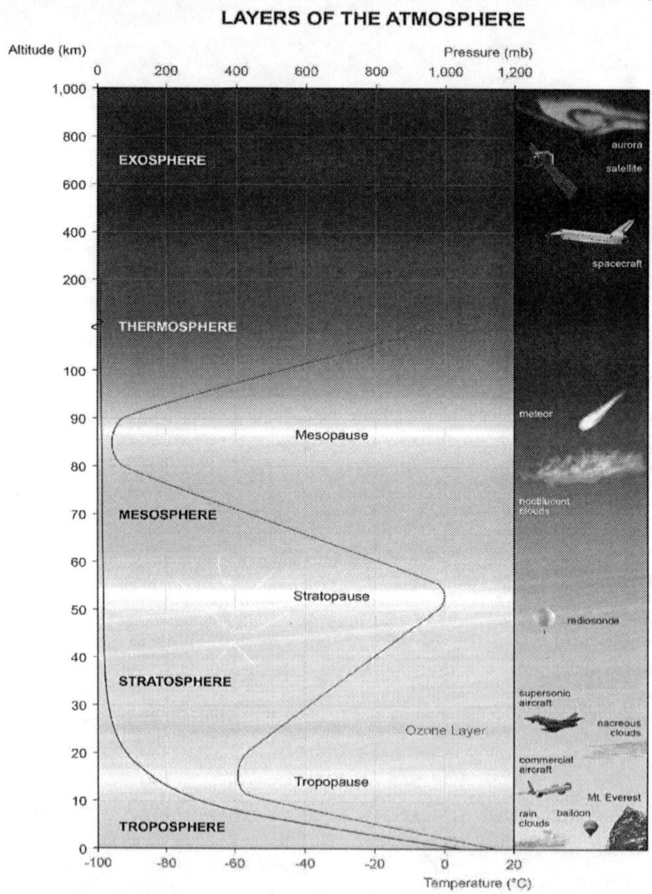

Diagram showing the Different Layers of the Atmosphere based on Temperature Variation

18

Definition of Achievement Tests: Geography

Introduction

The progress of a student can be measured on the basis of an achievement test, and a number of educational and vocational decisions on their performance can be taken by a teacher in a classroom environment. Since achievement tests help in the instructional development of an individual, it is necessary that a teacher be deft in various aspects of it.

An achievement test is defined as 'a systematic procedure for determining the amount a student has learned through instruction' (Gronlund, 1977). It is a test of performance to measure the knowledge and skill of a student on a particular content area compared with the performance of a general reference group of students on the same. However, the individual student's attainment is time-specific, and it may change under different circumstances. According to Thorndike and Hagen (1969), 'The type of ability test that describes what a person has learned to do is called an achievement test'.

Purpose of Achievement Tests

Achievement tests have practical utility for teachers and administrators of academic institutions. On the basis of it a teacher can find out the general range of capabilities of students in a class, and accordingly, one may select appropriate materials of instructions so that maximum instructional benefit can be provided to all individuals. The test is also helpful in the categorization of the students, working at their best capacity, as either weak or advanced, and necessary measures can be taken up separately to enhance the progress of both.

For the administrators, achievement test help in classifying school objectives; improving curriculum; preparing for measures to be taken separately for the weak and advanced students; providing incentives to the students based on their performances; and determining the overall educational attainment of an institution, etc. The management may take appropriate steps in light of the achievements for the betterment of it.

A test is defined as a standardized situation designed to elicit a sample of an individual's behaviour. Tests are classified as 1. Test of ability 2. Test of typical performance and 3. Educational achievement test.

Types of Evaluation or Tests

Achievement tests are practically a type of evaluation. Following are the different types of achievement tests (Mangal and Mangal, 2008):

1. ***Standardized tests (usually written with objective type of questions)***
   (a) General achievement (i) Test Battery (ii) Specialized
   (b) Diagnostic

2. ***Teacher-made informal tests***
   (a) Written paper test (i)Short answer types (ii) Very short answer types (iii) Objective types
   (b) Oral test
   (c) Practical test
   Tests are categorized as: (1) Formative, e.g., Unit test; (2) Summative, e.g., Half-yearly/Annual examinations (3) Diagnostic, e.g., Remedial test and (4) Prognostic, e.g., career counseling.

Essential Qualities of A Good Achievement Test

A good achievement test should be based on all difficulty levels and it should include a sufficient number of test items to cover all aspects of the topic. The questions should be unambiguous and properly framed to seek different knowledge and skill according to the suitable age groups. The test should be as objective as possible, and it should be standardized so that it can be compared to any other test of a similar type. The pupils should not feel bored or fatigued while going through the test. A good achievement test always encourages students to become creative and self-reliant. This also should be easy to administer and score.

Essential Points for Construction of Test Items

The following points are to be adhered to in developing balanced and objective-based questions:

1. Instructional objectives 2. Content analysis 3. Form of question 4. Language of questions
5. Directional words 6. Difficulty level 7. Marking schemes.

Stages of Planning for an Achievement Test in Geography

The following are the important steps normally adopted in planning for an achievement test:

(i) *Determination of the purpose* - There should be a specific purpose of an achievement test, e.g., to find out and categorize a group of students, say of Class VIII, into advanced learners, average learners, or slow learners according to their performance in the test.

(ii) *Selection of the target group* - The target group is selected on the basis of purpose and it should be clearly stated. Usually, the target groups may include either a class or a certain group of students of similar age.

(iii) *Selection of the topic* - The topic or sub-topic may be selected on the basis of availability of time, weightage of the test (i.e., marks assigned), and other resources affecting the test.

(iv) *Analysis of the content* - The content of the test should be chosen purposively from a suitable area of the subject of a particular class and then it is to be unitized and sub-unitized according to the specific conditions.

(v) *Development of the instructional objectives* - A group of cognitive psychologists, curriculum theorists and instructional researchers, and testing and assessment specialists published in 2001 a revision of Bloom's Taxonomy with the title *A Taxonomy for Teaching, Learning, and Assessment* to examine the educational objectives of the learners.

(vi) *Preparation of the table of specifications or blueprint* - This includes a detailed question-wise distribution of marks over specific objectives, topics, and forms of questions. Thus, all important factors like objectives to be tested, contents to be covered, types of questions to be asked and total time to be given should be kept in mind while preparing this table of specifications.

Table of Specifications Based On Criterion-Referenced Test (Crt)

Main topic / Unit :	*Essay Type*						*Short Answer Type*						*Very Short Answer Type*						*Objective Type*						*Total marks*	*% of marks*
Sub-topics/ Sub-units:	R	U	A	A	E	C	R	U	A	A	E	C	R	U	A	A	E	C	R	U	A	A	E	C		

Note: R = Remembering, U = Understanding, A = Applying, A = Analyzing, E = Evaluating, C = Creating

(vii) *Selection of item format on the basis of the table of specifications* - This can be on the basis of either weightage to objectives or weightage to forms of questions or sometimes on the basis of weightage to contents. This table of specifications should give self-confidence about the justice of the test. However, usually, it is done according to the table of specifications on the basis of weightage to forms of questions. This is shown in the following table.

(viii) *Development of the test items (i.e., question paper) as per the table of specifications* - Questions to be constructed as per ***(vi)*** (Table of specifications) and ***(vii)*** (Item format). However, the following points may help in preparing the test format (e.g., a question paper) :

(a) Long and short items should be kept in separate sections with specific instructions against each of them.

(b) The test items should be arranged in order of ascending difficulty level; that means it should proceed from easy to more difficult ones.

(c) For the objective items the multiple-choice format is the most suitable one because of its reliability and validity. Other types may also be chosen, but they should be similar to prevent loss of time in reading the instructions.

(d) It is better to frame a few extra items so that at the final review the unsuitable ones can be excluded.

(ix) *Preparation of the answer key* - Appropriate answer key should be prepared to compare the main points expected from the students. This is also necessary for proper evaluation so that a group of examiners can evaluate following the same set of the answer key, which will ultimately minimize any kind of biasness and personal error of the evaluators. The answer key may seek to incorporate the following points:

(a) Proper division of marks according to sections and sub-sections especially for the essay-type questions

(b) The points or steps expected in the answer

Item Format on the Basis of the Table of Specifications (TOS)

Class: Full Marks:
Subject: Topic: Time:

Unit	*Remembering*				*Understanding*				*Applying*				*Analyzing*				*Evaluating*				*Creating*				*% of Marks*
Sub-unit	*ET*	*SAT*	*VSAT*	*OT*	*ET*	*SAT*	*VSAT*	*OT*	*ET*	*SAT*	*VSAT*	*OT*	*ET*	*SAT*	*VSAT*	*OT*	*ET*	*SAT*	*VSAT*	*OT*	*ET*	*SAT*	*VSAT*	*OT*	
Marks assigned																									
Total marks																									
% of marks																									

Note: ET = Essay Type, SAT = Short Answer Type, VSAT= Very Short Answer Type, OT = Objective Type

(c) Number of lines or words expected for the short and very short type questions
(d) The weight given to the steps and important points to be covered
(e) Requirement of explanations and illustrations and their weightage
(f) Weightage for diagrams, tables, sketches, etc.

(x) *Administration of the test on the target group already selected*

This stage includes advance announcement of the test with regard to date, time, and venue; and other instructions related to the accessories, e.g., admit card, drawing and writing instruments, charts, models, etc. to be brought into the institution by the students. Sometimes, prior intimation is also given regarding the type of test to be conducted. However, personal problems- either physical or social, or psychological, of the pupils should not flout the test in any manner. Clear instructions should also be given to start or end the test, and also during the test if so required.

(xi) *Scoring and appraising* - Scoring can be done either manually or mechanically depending on the type of questions and available resources. Appraisal of the test is also called 'item-analysis' which means each item of the test is analyzed to know for its difficulty level or otherwise on the basis of the responses given by the students against the individual items.

(xii) *Using the test results* - The primary usage of the test results is to decide about the promotion of the pupils to the next higher grade. Besides, the results can also be helpful in enabling a teacher to introspect and reorganize the teaching-learning activities. The measurements/ scores in the Geography of a student in achievement test will indicate the meaningful cognitive, affective and psychomotor outcomes and thereby designate the level of mastery of the subject.

Through achievement tests, teachers can be assured that they are measuring students' learning across a wide range of content and reading as well as cognitive processes requiring higher-order thinking. It helps in assisting the teacher to select the most appropriate teaching strategy for creating real geographical knowledge. Teachers are able to determine what topic(theory/practical) is being stressed and also assist in the preparation of tests that reflect what students have assimilated and also limit the amount of time spent on each unit and also consider that no important objective or content area has been omitted.

Model of an Achievement Test Paper of Geography

In this case, any sample of geography question paper can be taken at School levels (Classes IX, X, XI and XII) affiliated to National and State Boards.

Construction of Achievement Test
Content Analysis

Class: X **Subject: Geography**

Units	*Subunits*
Biosphere	• Biosphere • Ecosystem & Ecology • Trophic Structure • Food chain & Food web • Energy Flow

Units	*Subunits*
Soil	• Genesis of soil • Soil profile • Properties of soil • Formation of soil • Type of soils

Instructional Objectives

Unit: Biosphere

Remembering: Students will be able to

- Define ecosystem. (factual)
- Recall the characteristics of ecosystem. (factual)
- Define food chain and food web. (factual)

Understanding: Students will be able to

- Distinguish between food chain and food web. (Conceptual)
- Discuss the phases of energy flow. (Conceptual)
- Interpret different components of biosphere. (Conceptual)

Applying: Students will be able to

- Identify different types of ecosystem in environment. (Procedural)
- Determine the role of different organism found in surroundings in energy flow. (Conceptual)

Analyzing: Students will be able to

- Classify ecosystem. (Conceptual)
- Analyze the trophic structure of ecosystem. (Conceptual)

Evaluating: Students will be able to

- Evaluate the role of different species with respect of ecological balance. (Meta-cognition)
- Justify the interdependence of different species of environment. (Meta-cognition)
- Justify the importance of ecological conservation. (Meta-cognition)

Creating: Students will be able to

- Draw the diagram of energy flow. (Procedural)
- Draw the diagrams of food chain and food web. (Procedural)

Unit: Soil

Remembering: Students will be able to

- Define soil. (factual)
- Recall the equation of soil forming factors. (factual)

Understanding: Students will be able to

- Distinguish between eluviation and illuviation. (Conceptual)
- Distinguish between pedocal and pedalfar. (Conceptual)
- Describe soil forming factors. (Conceptual)

Applying: Students will be able to

- Identify different soils of India. (Conceptual)
- Identify different soils suitable for cultivating different crops. (Conceptual)

Analyzing: Students will be able to

- Classify soil. (Conceptual)
- Divide soil forming factors. (Procedural)
- Analyze zonal, inter-zonal and azonal soils. (Conceptual)

Evaluating: Students will be able to

- Examine the role of soil in human life. (Meta-cognition)
- Evaluate the role of soil in environment. (Meta-cognition)

Creating: Students will be able to

- Draw the diagram of soil profile. (Procedural)
- Draw maps for showing different zonal soils in the earth. (Procedural)

Unit Wise Distribution of Marks

Units	*Item Number*	*Marks*	*% of Marks*
Biosphere	1A.i, ii, 1B.i, ii, 1C iii, iv, 1D i,ii,2 i, ii, 3 ii, 4	26	52
Soil	1A.iii, iv, 1B.iii, iv, 1C i, ii, 1D iii, iv, 2iii, iv, 3 i, iii, iv	24	48
Total		**50**	**100**

Unit Wise Distribution of Marks

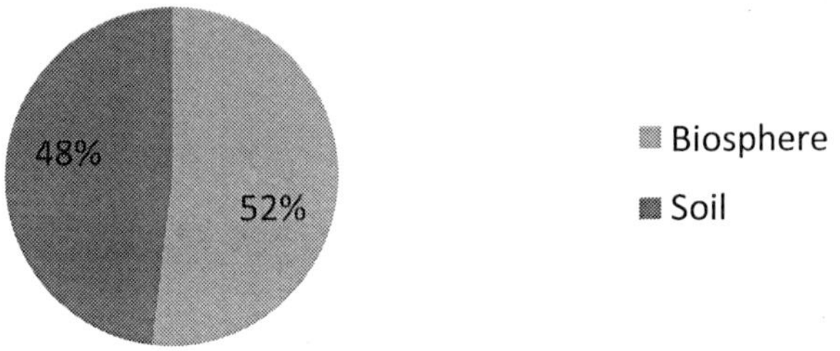

Objectives Wise Distribution of Marks

Objectives	*Item Number*	*Marks*	*% of Marks*
Remembering	1A.i,1B.i,ii,1C iv,1D i	5	10
Understanding	1A.ii,iii, 1B.iv, 1C ii, 1D ii,3i, 4	19	38
Applying	1Aiv, 1Biii, 1Ci, iii,1Diii	5	10
Analyzing	1D iv,2i,3iv	7	14
Evaluating	2iii,3iii	6	12
Creating	2ii,iv, 3ii	8	16
Total		**50**	**100**

Objectives Wise Distribution of Marks

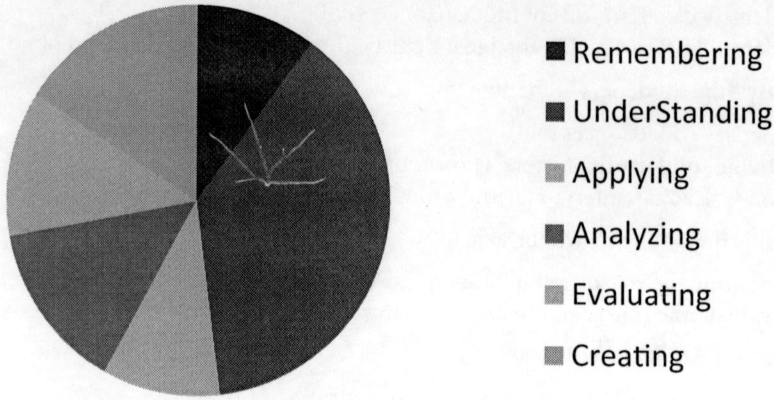

Type of Question Wise Distribution of Marks

Type of Questions	*Item Number*	*Marks*	*% of Marks*
Objective	1A.i,ii,iii,iv, 1B.i,ii,iii,iv,1C i,ii,iii iv,1D i,ii,iii,iv	16	32
Very Short	2i,ii,iii,iv	8	16
Short	3i,ii,iii,iv	16	32
Essay	4	10	20
Total		50	100

Type of Question Wise Distribution of Marks

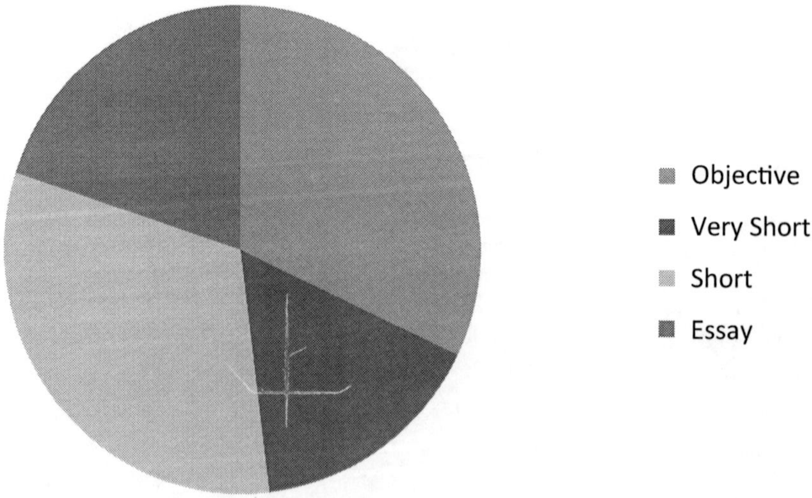

Table of Specification

Objectives	*Remembering*				*Understanding*				*Applying*				*Analyzing*				*Evaluating*				*Creating*					
Units	*O*	*VS*	*S*	*E*	*O*	*VS*	*S*	*E*	*O*	*VS*	*S*	*E*	*O*	*VS*	*S*	*E*	*O*	*VS*	*S*	*E*	*O*	*VS*	*S*	*E*	*TM*	*%*
Biosphere	1A.i (1) 1B.i,ii(2) 1Civ(1) 1Di(1)				*1A.ii(1) 1Dii(1)			4 (10)	1Ciii(1)					2i (2)								2ii(2)	3ii(4)		26	52
Soil					1A.iii(1) 1B.iv(1) 1Cii(1)		3i (4)		1Aiv(1) 1B.iii(1) 1Ci(1) 1Diii(1)				1Div (1)		3iv (4)			2iii (2)	3iii (4)			2iv(2)			24	48
	5				5		4	10	5				1	2	4			2	4			4	4			
Total Marks	5				19				5				7				6				8				50	
% of Marks	10				38				10				14				12				16				100	

* Outside the bracket question or item no. and inside the bracket marks O= Objective Type VS: Very Short Type S: Short Type E: Essay Type

Name of the School

Achievement Test

Class: X **Full Marks: 50**
Subject: Geography **Time: 2 hours.**

Instruction:
i. Students are instructed to write in your own words.
ii. All questions are compulsory.

1. A Fill in the blanks. **1×4=4**

i. _______ coined the term 'Ecosystem' in 1935.
ii. Upper the trophic level _______ the number of consumer.
iii. The influence of parent rock causes _______ soil.
iv. In Birbhum, the high concentration of iron helps to form _______ soil.

1. B Choose the appropriate answer. **1×4=4**

i. Ecosystem is the system of –
   a) Open System
   b) Close System
   c) Quasi-open System
ii. According to 10% law, at each step of ecosystem the actual energy
   a) Increases
   b) Decreases
   c) Remains same
iii. The alluvial soil at the side of Hooghly river bank is a type of-
   a) Zonal Soil
   b) Azonal Soil
   c) Inter-zonal Soil
iv. The regolith is formed due to-
   a) Weathering
   b) Humification
   c) Capillary action

1. C Select true or false statements. **1×4=4**

i. Rendzina soil is found at the Karst region.
ii. The minerals are deposited in the B layer of soil due to illuviation.
iii. The grazing ecosystem is found in oceanic environment.
iv. The energy flow in food chain is multidirectional.

1. D Give answer with one or two sentences. **1×4=4**

i. What is biomass?
ii. Give on difference between producer and consumer.
iii. Where is found sierozem soil in India?
iv. How many minerals are there to form white and gray colour in soils.

2. Give answer within 50 words. **2×4=8**

i. Classify the components of ecosystem.
ii. Draw a flow chart to illustrate trophic levels in ecosystem.
iii. The soil formation takes less time upon permeable rocks- Justify the reason.
iv. Give a diagram of soil catena.

3. Give answer within 100 words. **4×4=16**

i. Differentiate between pedocal and pedalfer soils.
ii. Give a layout of pond ecosystem with a diagram.
iii. Justify the reasons of formation of podzol soils in wet-temperate region.
iv. Analyze the cross-section of a soil profile.

4. Give answer within 200 words. **10×1=10**

Describe the phases of energy flow in ecosystem.

Answer Key

1A i. Tansley
1A ii. Lower
1A iii. Endomorphic
1A iv. red
1B i. a) open system
1B ii.b) Decreases
1B iii.b) Azonal Soil
1B iv.a) Weathering
1C i. True
1C ii. True
1C iii. False
1C iv. False
1D i. Biomass is the weight or total quantity of living organisms commonly referred to a unit area or volume of habitat.
1D ii. The producers generate food for themselves and consumers do not produce anything, instead eating producers, other consumers or both.
1D iii. The sierozem soil in India is found at Thar desert.
1D iv. Mineral sault, limestone and quartz are there to form white and grey colour in soils.

2i.

Components of Ecosystem

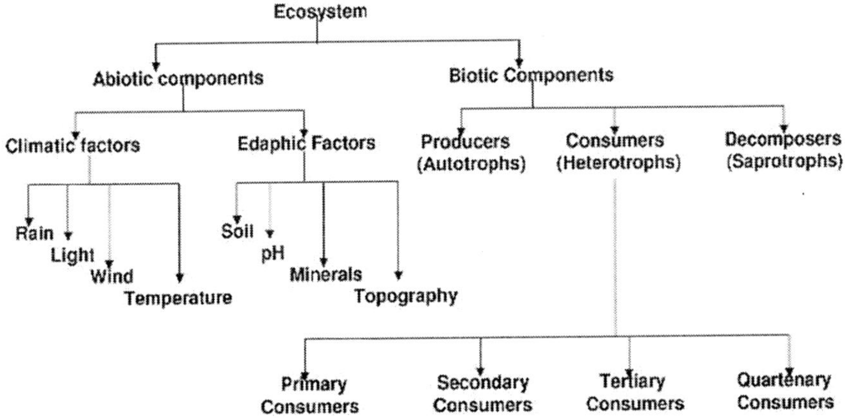

2ii.

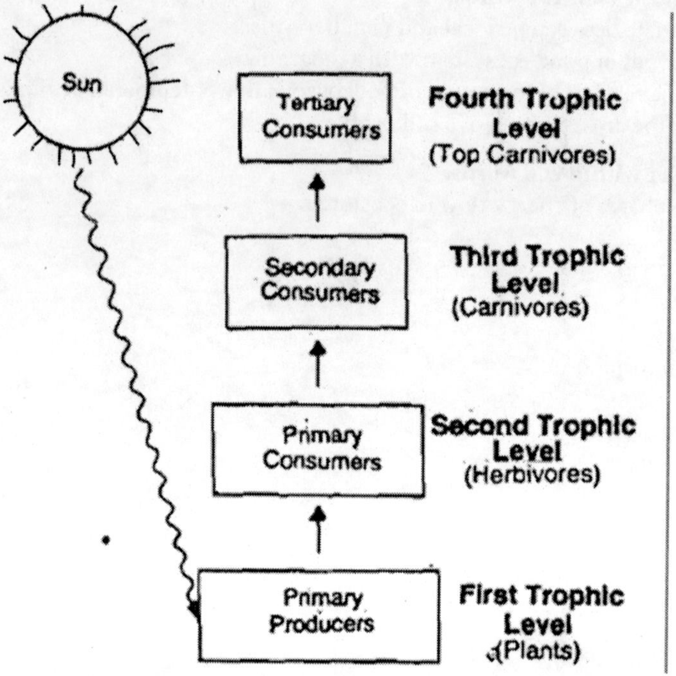
Sun
Tertiary Consumers
Fourth Trophic Level (Top Carnivores)
Secondary Consumers
Third Trophic Level (Carnivores)
Primary Consumers
Second Trophic Level (Herbivores)
Primary Producers
First Trophic Level (Plants)

2iv.

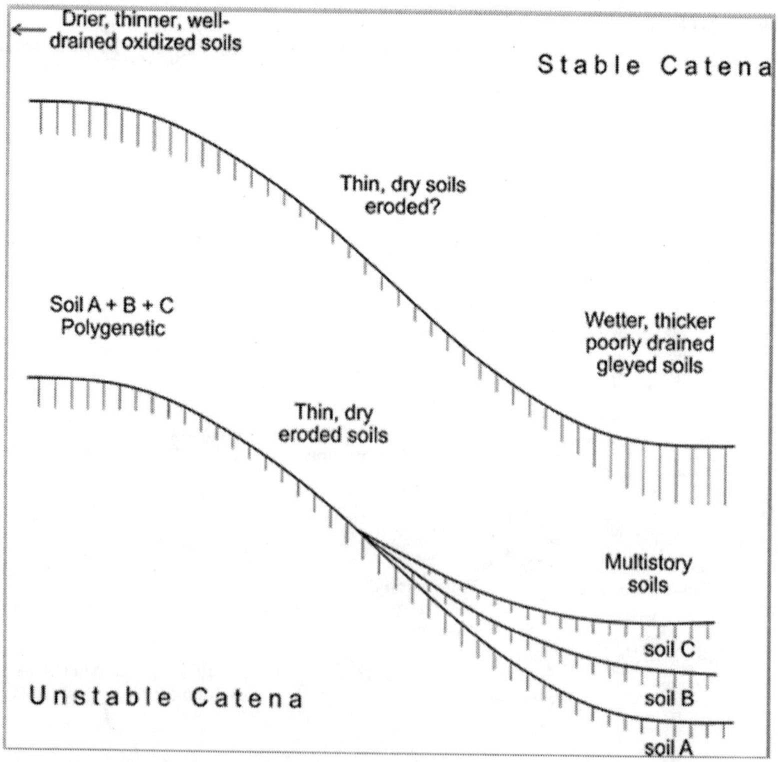
Drier, thinner, well-drained oxidized soils
Stable Catena
Thin, dry soils eroded?
Soil A + B + C Polygenetic
Wetter, thicker poorly drained gleyed soils
Thin, dry eroded soils
Multistory soils
soil C
soil B
soil A
Unstable Catena

3i. **Pedocal**

- Dry, semi arid climate
- Little organic matter
- No mineral leaching
- High limestone content
- parties

Pedalfer

- Enriched with aluminum and iron
- Greater organic matter and leaching
- Found in areas with high temp. and lots of forest cover

3ii.

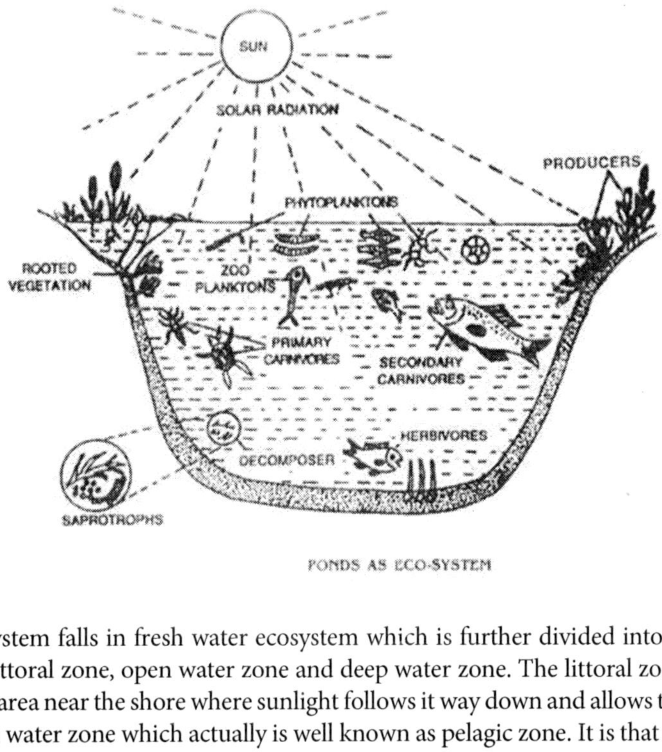

PONDS AS ECO-SYSTEM

Pond Ecosystem falls in fresh water ecosystem which is further divided into three parts which are littoral zone, open water zone and deep water zone. The littoral zone is a zone which is an area near the shore where sunlight follows it way down and allows the plants to grow. Open water zone which actually is well known as pelagic zone. It is that zone where water is neither close to the bottom nor near the shore. The last is the deep water zone which is the lowest level lying above the sea and below the thermocline. These three types can be understood deeply. Some differentiate Pond Ecosystem in other way. It is classified in five types which are salt ponds, garden ponds, freshwater ponds, vernal ponds and underground ponds. Salt ponds are one which are near the sea side and consist brackish water. Garden ponds are those which are artificial in nature. Freshwater pool can form anywhere because of rains. Vernal ponds are those which are created because of depression in the ground and are seasonal. Lastly underground ponds that are created near the rocks.

3.iv A soil horizon is a layer parallel to the soil surface whose physical, chemical and biological characteristics differ from the layers above and beneath. Horizons are defined in many cases by obvious physical features, mainly colour and texture.

O) Organic surface layer: Plant litter layer, the upper part often relatively undecomposed, but the lower part may be strongly humified.

A) Surface soil: Layer of mineral soil with most organic matter accumulation and soil life. Additionally, due to weathering, oxides (mainly iron oxides) and clay minerals are formed and accumulated. It has a pronounced soil structure. But in some soils, clay minerals, iron, aluminum, organic compounds, and other constituents are soluble and move downwards. When this eluviation is pronounced, a lighter coloured E subsurface soil horizon is apparent at the base of the A horizon. The A horizon may also be the result of a combination of soil bioturbation and surface processes that winnow fine particles from biologically mounded topsoil. In this case, the A horizon is regarded as a "biomantle".

B) Subsoil: This layer has normally less organic matter than the A horizon, so its colour is mainly derived from iron oxides. Iron oxides and clay minerals accumulate as a result of weathering. In soil, where substances move down from the topsoil, this is the layer where they accumulate. The process of accumulation of clay minerals, iron, aluminum, and organic compounds, is referred to as illuviation. The B horizon has generally a soil structure.

C) Substratum: Layer of non-indurated poorly weathered or unweathered rocks. This layer may accumulate more soluble compounds like $CaCO_3$. Soils formed *in situ* from non-indurated material exhibit similarities to this C layer.

R) Bedrock: R horizons denote the layer of partially weathered or unweathered bedrock at the base of the soil profile. Unlike the above layers, R horizons largely comprise continuous masses (as opposed to boulders) of hard rock that cannot be excavated by hand. Soils formed *in situ* from bedrock will exhibit strong similarities to this bedrock layer.

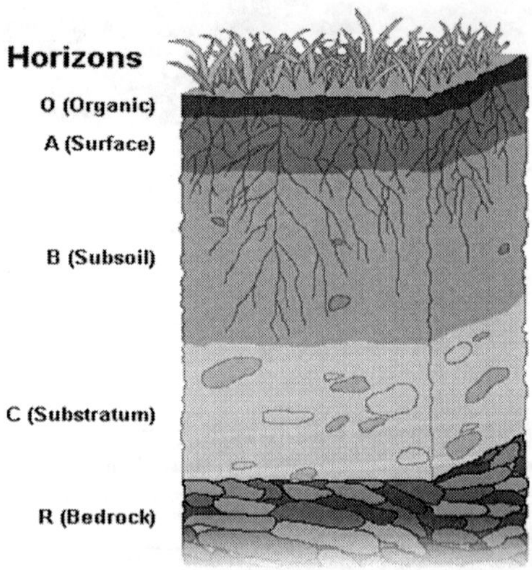

4. The radiant energy is in the form of electromagnetic waves which are released from the sun during the transmutation of hydrogen to helium. The chemical energy stored in the food of living organisms is converted into potential energy by the arrangement of the constituent atoms of food in a particular manner. In any ecosystem there should be unidirectional flow of energy. This energy flow is based on two important Laws of Thermodynamics which are as follows:

   (1) *The first law of Thermodynamics:* It states that the amount of energy in the universe is constant. It may change from one form to another, but it can neither be created nor destroyed. Light energy can be neither created nor destroyed as it passes through the atmosphere. It may, however, be transformed into another type of energy, such as chemical energy or heat energy. These forms of energy cannot be transformed into electromagnetic radiation.

(2) *The second law of Thermodynamics:* It states that non-random energy (mechanical, chemical, radiant energy) cannot be changed without some degradation into heat energy. The change of energy from one form to another takes place in such a way that a part of energy assumes waste form (heat energy). In this way, after transformation the capacity of energy to perform work is decreased. Thus, energy flows from higher to lower level.

Main source of energy is sun. Approximately 57% of sun energy is absorbed in the atmosphere and scattered in the space. Some 35% is spent to heat water and land areas and to evaporate water. Of the approximately 8% of light energy striking plant surface, 10% to 15% is reflected, 5% is transmitted and 80 to 85% is absorbed; and an average of only 2% (0.5 to 3.5%) of the total light energy striking on a leaf is used in photosynthesis and rest is transformed into heat energy.

Energy flow in Ecosystems:

Living organisms can use energy in two forms radiant and fixed energy. Radiant energy is in the form of electromagnetic waves, such as light. Fixed energy is potential chemical energy bound in various organic substances which can be broken down in order to release their energy content.

Organisms that can fix radiant energy utilizing inorganic substances to produce organic molecules are called autotrophs. Organisms that cannot obtain energy from abiotic source but depend on energy-rich organic molecules synthesized by autotrophs are called heterotrophs. Those which obtain energy from living organisms are called consumers and those which obtain energy from dead organisms are called decomposers (Figure 3.7).

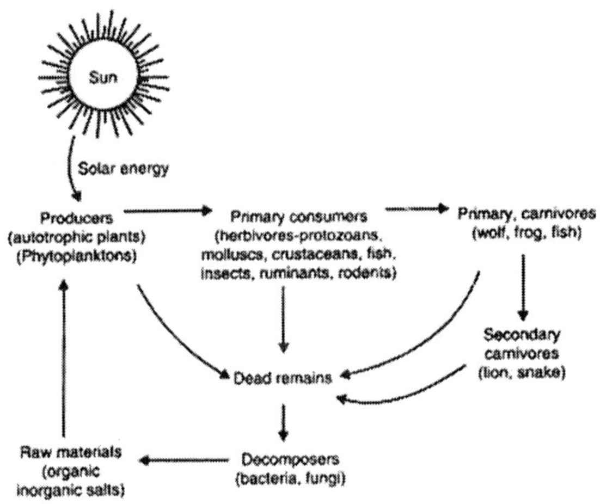

Fig. 3.7. Flow of energy at different levels of ecosystem.

When the light energy falls on the green surfaces of plants, a part of it is transformed into chemical energy which is stored in various organic products in the plants. When the herbivores consume plants as food and convert chemical energy accumulated in plant products into kinetic energy, degradation of energy will occur through its conversion into heat. When herbivores are consumed by carnivores of the first order (secondary consumers) further degradation will occur. Similarly, when primary carnivores are consumed by top carnivores, again energy will be degraded.

19

Simulation Skills: Geography

Skill of Introducing a Lesson

Topic/Unit	*Main Points*	*Components*	*Procedural Details*
Natural Vegetation of India	• Concept and Importance of Natural Vegetation • Types of Natural Vegetation • Measures for Conservation of Forests • Social-Forestry and Agro-Forestry	Securing attention	The teacher will secure the attention of the student by showing a small movie clip demonstrating the forests of Sunderban.
		Assessing motivational level	The teacher will assess the motivational level of the student by asking some questions: ➢ Identify the forest that you have just seen in the movie picture. • Forest of Sunderban ➢ What type of forest is this? • Mangrove forest ➢ Have you seen any other forests either in movies or in pictures or in reality? • Yes ➢ Tell me the name of some other forests that you have seen. • Evergreen forest, Deciduous forest
		Linking with past experience	The teacher will ask questions to relate the previous knowledge of the learners with the unit. For example: ➢ What type of resource is natural vegetation? • Filled resource ➢ Think a little more and then answer. • Flow resource ➢ How does rainfall distribution influence the growth of natural vegetation? • Heavy rainfall, dense vegetation, low rainfall scarcity of vegetation.
		Specifying the points to be taken up for teaching	The teacher will specify the main points to be taken up for teaching: • Definition and importance of natural vegetation. • Types of natural vegetation found in India. • Characteristic features of different types of natural vegetation. • Measures for conservation of natural vegetation. • Concept of social forestry and agro-forestry.
		Using an appropriate device	The teacher can use an appropriate device such as: • A small clipping from the documentary Sunderban: It's not about the tiger only' to attract the attention of the student.

Skill of Questioning

Topic/Unit	*Main Points*	*Components*	*Procedural Details*
Natural Vegetation of India	• Concept and Importance of Natural Vegetation • Types of Natural Vegetation • Measures for Conservation of Forests • Social-Forestry and Agro-Forestry	Precision and clarity of language	The teacher will use precise, clear, relevant, grammatically correct, and accurate language while questioning such as: • What is the importance of natural vegetation? • Why tropical evergreen forests are found on the western slope of the Western Ghats?
		Link with specific objectives	The teacher should ask questions linked with specific objectives. The questions should be like this: ➢ Name some important species of trees found in mountain forests (Remembering). ➢ What is natural vegetation? • The vegetation that grows under natural conditions. ➢ What is the difference between natural vegetation and the vegetation that we grow in our house? • The one that we grow in our house requires care and maintenance but natural vegetation does not require so. ➢ Why do littoral forests develop pneumatophores? (Understanding) ➢ Why conservation of forests is important? (Applying)
		Using the student's response for further questioning	The teacher uses the student's response for further questioning, such as: ➢ Where are tropical evergreen forests found? • Western slopes of the Western Ghats. ➢ Why are they found there? • Because the region receives more than 200 cm of rainfall. ➢ Why does the region receive more than 200cm of rainfall? • Because the moisture-laden wind of the Arabian sea branch of monsoon when ascends along the slopes of the Western Ghats cools and condenses and brings heavy rainfall on the western slopes of the Western Ghats.
		Prompting	The teacher prompts answers to the students in case they can't answer such as: ➢ Why coniferous trees are conical in structure? • The students remain silent ➢ If you place a ball on the top of a mountain will the ball remain there or it will slide down? • The ball will slide down ➢ Where are coniferous trees found? • In mountainous areas ➢ Does snow fall in those regions? • Yes ➢ Now tell me why the conifers are conical in structure. • So that snow can slide down.

Topic/Unit	*Main Points*	*Components*	*Procedural Details*
		Refocusing and redirecting	The teacher uses questions to refocus and redirect specific points such as: ➢ Why evergreen trees are found in the foothill regions whereas alpine vegetation is found in the mountains? • Due to differences in soil types ➢ Is there any relation between temperature and altitude? • With the increase in altitude temperature decreases ➢ Then tell me the reason why natural vegetation will differ. • Due to the difference in temperature

Skill of Use of Teaching Aids

Topic/Unit	*Main Points*	*Components*	*Procedural Details*
Natural Vegetation of India	• Concept and Importance of Natural Vegetation • Types of Natural Vegetation • Measures for Conservation of Forests • Social-Forestry and Agro-Forestry	Relevant to the topic	The teacher adds that aids should be relevant to the topic such as: • A short movie clip demonstrating the mangrove forests of Sunderban. • A pictorial chart on 'Different Types of Natural Vegetation and their Adaptive Mechanism' • A pictorial chart on 'Some important species of different types of natural vegetation'.
		Appropriate to the pupil's level	The chart must be colourful and appropriate to the student's level of maturity.
		Creating interest and thinking	The aids must initiate faster thinking in the students, such as: The students will see the pictures of the adaptive features of different types of natural vegetation and will learn the importance of these adaptive features as they read through the chapter.
		Proper display	The charts will be displayed properly so that the students can see them properly and clearly. The letters in the charts should be bold and they will be written in bright colours against a light backdrop so that it is visible to everyone. The chart should be hung up high so that it is visible from the last bench.

Topic/Unit	*Main Points*	*Components*	*Procedural Details*
		Appropriate use	The teaching aids should be used appropriately. For Example- The movie clip demonstrating the mangrove forests of Sunderban should be shown at the commencement of the lesson to secure the attention of the students. • The chart showing the pictures of different types of natural vegetation and their adaptive mechanism should be displayed throughout so that the student can follow the chart and can easily understand when the teacher explains each type of natural vegetation.

Clips of Blackboard Work done by the Teacher

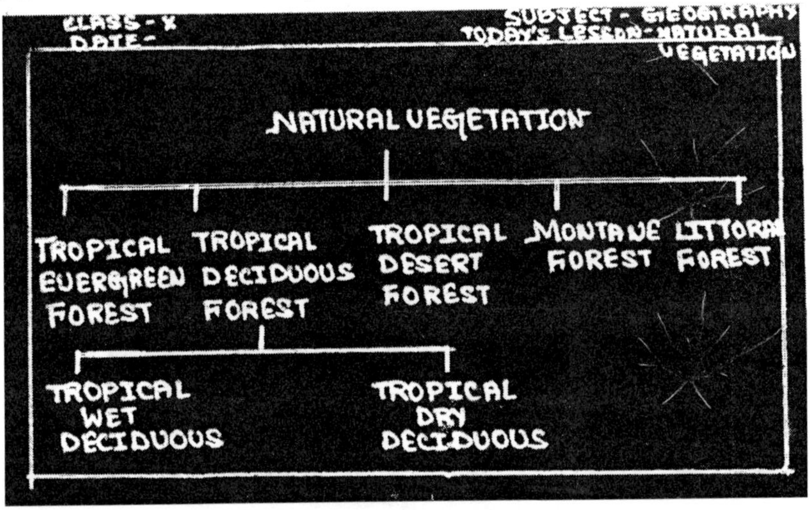

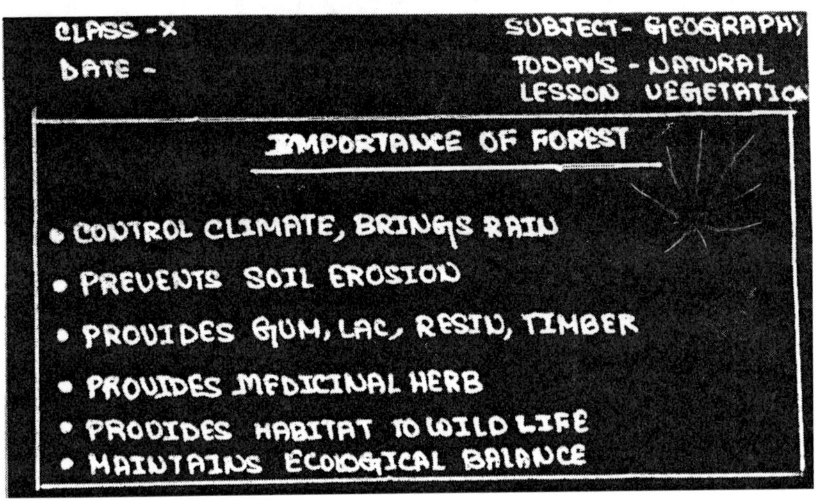

PICTORIAL CHART: TYPES OF NATURAL VEGETATION AND THEIR ADAPTIVE MECHANISMS

TROPICAL EVERGREEN FOREST

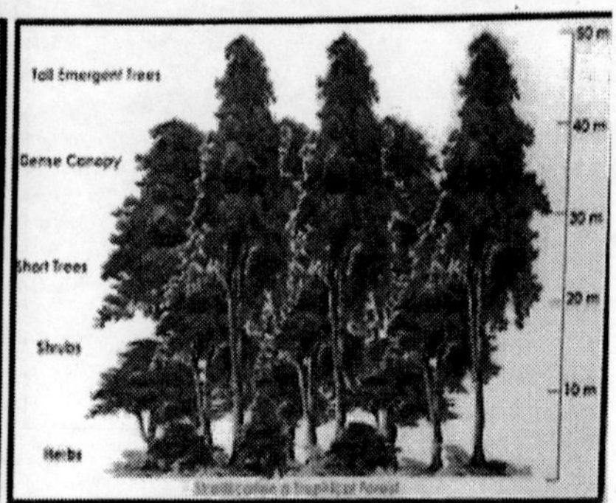

LAYERS OF TROPICAL EVERGREEN FORESTS

HUGE BUTTRESSED TRUNK OF TROPICAL EVERGREEN TREES

CONIFEROUS FOREST

NEEDLE SHAPED LEAVES OF CONIFEROUS TREES

PICTORIAL CHART: TYPES OF NATURAL VEGETATION AND THEIR ADAPTIVE MECHANISMS

TROPICAL DESERT FOREST

ADAPTIVE MECHANISM OF THE XEROPHYTES

TROPICAL DECIDUOUS FOREST

TROPICAL DECIDUOUS FORESTS SHED THEIR LEAVES

MANGROVE FORESTS WITH ANCHORING STEMS

PNEUMATOPHORES OF THE MANGROVES

Skill of Reinforcement

Topic/Unit	*Main Points*	*Components*	*Procedural Details*
Natural Vegetation of India	• Concept and Importance of Natural Vegetation • Types of Natural Vegetation • Measures for Conservation of Forests • Social-Forestry and Agro-Forestry	Positive reinforcement; no negative reinforcement	The teacher should always use positive reinforcement like good, correct, fine, and well done when the student gives the correct answer. This type of behaviour from the teacher increases the self-esteem of the students and motivates them to learn. For example: ➢ Where social forestry is best practiced? • Kerala ➢ Very good The teacher should never use negative reinforcement like bad, very bad, or wrong as it demotivates the students.
		Wrong response weakened or broken	The teacher breaks or weakens the wrong responses of the students in a way that doesn't demotivate the students but redirects them toward the correct responses. For example: ➢ Why shifting cultivation has been banned in the north-western states of India? • Because it leads to land degradation ➢ There is a different reason behind it. Think a bit more and answer. Didn't shifting cultivation destroy the forests? • Shifting cultivation leads to deforestation and for the conservation of the forests it has been banned.
		Providing lead or seeking further information	Instead of scolding the students, the teacher provides clues to the students in case they can't answer a question or give a complete answer. Such as: ➢ Why do the xerophytes have waxy stems? • To limit the loss of water through transpiration. ➢ Think a bit more and give the complete answer. Can you tell me where the xerophytes are found? • In the desert. ➢ Is there a scarcity of water in the desert? • Yes ➢ Now tell me why do the xerophytes develop waxy stems? • In the deserts due to lack of rainfall there is a scarcity of water; hence to limit the loss of water through transpiration the xerophytes develop waxy stems.

Topic/Unit	*Main Points*	*Components*	*Procedural Details*
		Positive non-verbal reinforcement	The teacher should use positive non-verbal reinforcement to motivate the students when the student gives the correct answer to a question. The teacher should appreciate the answer by smiling and nodding.
		Writing pupil's response on the board	The teacher should write the correct answer of the students on the board. This encourages the students and creates a feeling of acceptance among them.

Skill of Illustration

Topic/Unit	*Main Points*	*Components*	*Procedural Details*
Natural Vegetation of India	• Concept and Importance of Natural Vegetation • Types of Natural Vegetation • Measures for Conservation of Forests • Social-Forestry and Agro-Forestry	Relevant examples	The teacher should give relevant examples while explaining a particular topic such as: While explaining the coniferous forests, the teacher will give examples of some trees of coniferous forests (like Pine, Fir, and Spruce) and will show the picture of those trees displayed on the pictorial chart.
		Clear concept	Good examples help in building concepts. The teacher should give relevant examples while explaining a particular topic so that the concept becomes vividly clear to the students. For example, while explaining Tropical Evergreen Forests, the teacher will give an example of the film 'Sabuj Deeper Raja' where the students can see dense evergreen forests.
		Simple and easy	The teacher should give simple and easy examples while explaining a particular topic. The examples should be appropriate to the student's level of maturity.
		Student's interaction	The teacher should encourage student interaction so as to ensure that all the students actively participate in the classroom discussion. For example: After explaining the characteristics of tropical deciduous trees the teacher asks a few questions. ➢ Give an example of a tree that sheds its leaves during winter. • Palash ➢ Name some other trees of tropical deciduous forest. • Teak, Sal
		Correlates with the topic	The teacher should give such examples that correlate with the topic such as: While explaining the conservation of forests the teacher may give example of 'Chipko Movement' conducted by Sunderlal Bahuguna in Uttarakhand where the women embrace the trees to prevent it from being cut down.

Skill of Closure

Topic/Unit	*Main Points*	*Components*	*Procedural Details*
Natural Vegetation of India	• Concept and Importance of Natural • Vegetation • Types of Natural Vegetation • Measures for Conservation of Forests • Social Forestry and Agro-Forestry	**a) Summarization** – When the lesson was introduced, the main points in the lesson were specified. At the end of the lesson, the teacher concludes the lesson with a clear indication that the same points have been covered and a summary of the lesson has to be prepared accordingly.	The teacher will specify the major points of importance, types (by pointing to the learning aid), and measures of conservation of natural vegetation on the Blackboard (writing only *keywords*) in summary. The teacher also distinguishes between Social and Agroforestry prevalent in India for better understandability.
		Linked with the future learning – Indication is given about what may be the implication of the lesson in the next lesson.	The teacher discusses the correlation between natural vegetation and settlement density and also specifies the agricultural and industrial practices associated with it.
		Link with real-life activities – Whether the lesson has any bearing upon the real-life or out-of-school experience and if there is any such link it is to be exactly pointed out at the end of the lesson.	They would be aware 1. To save paper as it's a product of forest 2. To sensitize at school and at home about the importance of trees through programs 3. To be aware of the contemporary issues regarding forests and their fauna, forest products 4. To understand the interdependence of forest and community
		Evaluation – Whether immediate evaluation has been done during closure. This is necessary to assess the success or failure of the lesson.	The teacher asks a few hinge-based questions to check whether the students have consolidated and assimilated the content and are now ready to move on to the next.
		Adequacy of assignment – Whether assignments are given appropriately and whether assignments pertain to the domains of objectives other than cognitive, particularly to the affective domain need to be seen.	The teacher can assign relevant homework on written or project-based activities on an individual basis. Pertaining to the topic taught, group activities can also be conducted and completed within the class hour.

More Praise for *Next*

The biggest climate and development challenges are too big to ignore. Jorisch offers a hopeful glimpse into the inspiring minds behind the innovations that embody our better future.

— **Henk Ovink**, Special Envoy for International Water Affairs, Kingdom of the Netherlands; Sherpa to the United Nations High Level Panel on Water

Avi Jorisch has demonstrated once again that he is one of the most incisive and far-sighted thinkers on our planet, offering us not just a glimpse into what's next over the horizon but a lodestar to where we *should* be going to ensure a better future for us all.

— **Afshin Molavi**, Senior Fellow, Foreign Policy Institute, Johns Hopkins University School of Advanced International Studies

Next offers a thought-provoking narrative: in the coming decades, the world will change to such a degree that today's science fiction is tomorrow's science.

— **Rodolphe Barrangou**, Distinguished Professor, CRISPR Lab Lead, North Carolina State University

Next compellingly builds upon the case for a sustainable existence for our species both on and off planet through pioneering innovation, thrillingly captured in this storied chronicle.

— **Yvonne Cagle, MD, PhD**, Aerospace Mission Specialist

NEXT

A BRIEF HISTORY OF THE FUTURE

NEXT

A BRIEF HISTORY OF THE FUTURE

AVI JORISCH

JAICO PUBLISHING HOUSE
Ahmedabad Bangalore Chennai
Delhi Hyderabad Kolkata Mumbai

Published by Jaico Publishing House
A-2 Jash Chambers, 7-A Sir Phirozshah Mehta Road
Fort, Mumbai - 400 001
jaicopub@jaicobooks.com
www.jaicobooks.com

First published in English by
Gefen Publishing House Ltd., Jerusalem, 2022

Published in arrangement with
RedCell Intelligence Group
2200 Wilson Blvd. Suite 102-310
Arlington, VA 22201

To be sold only in India, Bangladesh, Bhutan,
Pakistan, Nepal, Sri Lanka and the Maldives.

NEXT: A Brief History of the Future
ISBN 978-81-19153-91-6

First Jaico Impression: 2024

Printed by
Nutech Print Services - India, New Delhi

For Eiden, Oren and Yaniv –
the greatest gift the universe has given me.
The future belongs to you.

I don't regret any of my dreams.
My only regret is not having dreamed more.
– Shimon Peres, No Room for Small Dreams

Contents

PREFACE

The Future Is Now

The best way to predict the future is to create it yourself.

– Peter Diamandis, Singularity University

What's Next?

In March 2020, a few weeks before the coronavirus roiled America, my three young sons and I piled into the Washington Convention Center in DC for an AIPAC policy conference. I wanted them to see how technology can help solve the world's most intractable problems, from groundbreaking advancements in water preservation to novel new methods for recycling plastic. You know, the kind of stuff small children love.

As we wandered through the crowd, we bumped into my friend Eli Beer, the founder of United Hatzalah, an Israeli nonprofit that has revolutionized emergency care. There are few people I admire more. Over the past twenty years, Eli has brought together volunteer EMTs – Jews, Christians, and Muslims, among others – to save thousands of lives. One of his best innovations is an Uber-like app that connects anyone who needs emergency care with available EMTs nearby. These volunteer emergency workers often travel by "ambucycle," motorcycles refitted to function as mini-ambulances, which are nimble enough to weave through traffic. In Israel and elsewhere, Eli and company have helped drastically cut down the response time for people in need. (Little did we know that Eli himself would soon be in dire need, as he faced a life-threatening case of COVID-19 himself soon after our meeting.)

My kids gave Eli a hug, and he showed us around his booth. All three boys – Eiden, Oren, and Yaniv – strapped on virtual reality goggles to experience what it's like to work as a United Hatzalah volunteer. As their faces lit up, I felt optimistic not just about the future but about the ability of technology to make the world a better place. "When I get older," Eiden, my oldest, told me, "I don't want to drive a car. I want to ride an ambucycle and save lives."

I couldn't have been prouder. As we walked away, my middle child, Oren, grabbed my hand and pulled me down to his face. "But what's next, Baba?"

He didn't mean our plans for the afternoon.

A few weeks later, my sons and I were quarantined together around a screen in our living room. A SpaceX aircraft was about to launch from pad 39A at the Kennedy Space Center. The boys watched open-mouthed as mission control gave the order: "Three, two, one, blastoff." We tracked the spacecraft continuously for nineteen hours as it hurled itself toward the International Space Station. When the Crew Dragon docked, and the astronauts emerged smiling from their capsule, they made history: it was the first private spacecraft to berth with ISS, carrying cargo under commercial agreements with NASA.

"What's next?!" Oren asked me again. My kids were even more eager for me to tell them what the future is going to look like and what scientists and innovators are doing today to address the earth's scariest problems. I told them stories about *Star Trek*–loving inventors who 3D print spacecrafts in orbit, vegan researchers who replicate the composition and chemical structures of meat in a lab, and mad scientists who save humans from terrible disorders by cutting and pasting genes as if they were letters in Microsoft Word. Then, as much to satiate my own curiosity as theirs, I found myself looking for more. Who are the most impressive innovators out there today, shaping the history of the future?

In this present moment, imperiled by hunger, pollution, and global warming, human beings are at a critical juncture, closer than we've ever been to driving ourselves to extinction. But we're also in an era of abundance – and remarkable wealth – that's fueling the kind of innovation that could allow us to fundamentally change our destiny: curing the sick, feeding the hungry, eradicating poverty, and healing the planet.

Next tells the story of thirteen game-changing innovations that are poised to transform our species from a society of takers to a society of givers. Call it Humanity 2.0. Each venture featured in the following pages – together with the ideas and doers behind them – is already having an outsized impact on human history. In the coming decades, what's next will be dictated by our growing ability to leverage exponential thinking to bring about truly extraordinary innovations that are far beyond our current imagination – technologies that will, if used for good, allow us to make the world a far better place.

An Eruption of Disruption

More than half a century ago, long before the invention of the personal computer, engineer-turned-entrepreneur Gordon Moore received a letter from the editor of a trade publication called *Electronics*. Moore, the cofounder of a then unknown semiconductor firm called Intel, was delighted to learn that the magazine wanted him to write a piece for their thirty-fifth anniversary. The topic: how changes in the circuits that power computers would impact society. Moore was so excited, he scribbled a giant note to himself at the top of the letter: "Go-Go."

"I find the opportunity to predict the future in this area irresistible,"

Moore responded, "and will, accordingly, be happy to prepare a contribution."[1]

What Moore came up with would radically change the tech we use and the way we live. The power of microchips running our computers, he concluded, will double every year. And as he pondered what that might mean for the world, he also managed to predict the invention of home computers, cellphones, self-driving cars, and smartwatches – all of which would become cheaper over time. As *New York Times* columnist Thomas Friedman once joked, "The only thing he missed... was microwave popcorn."[2]

By 1975, Moore's hypothesis had already been proven true, but he believed the pace of growth was accelerating, so he updated his prediction. "Moore's Law," which posits that the power of microchips will double roughly every two years while their cost remains about the same, has held up ever since.

We feel the impact of this phenomenon every day. It's why our technology is almost unrecognizable compared to what we were using just two decades ago. Smartphones, social media, and GPS have all become commonplace in the past twenty years. And it's not just about the gadgets. We've seen a true revolution in science and technology, from crops genetically engineered to yield more food to advances in microfinance that have helped lift millions out of poverty.

Ray Kurzweil knows a lot about these changes. He's not only Google's director of engineering, he's the preeminent inventor of the flatbed scanner and reading machines for the blind. Using Moore's Law, he foretold some of the most amazing technological changes in the modern world. In a 2010 report on the 147 predictions Kurzweil made over thirty years, 115 turned out to be entirely correct, twelve essentially correct, seventeen partially correct, and just three wrong. Among other things, Kurzweil predicted the fall of the Soviet Union, computers with no moving parts, facial recognition software, voice-activated typewriters, robotic prostheses that allow paraplegics to walk, and a fully mature cloud-computing network.

In his 1999 book *The Age of Spiritual Machines*, Kurzweil used Moore's framework to analyze human history, arguing that exponential growth isn't a recent phenomenon, but something we've seen for thousands of years. He called it the Law of Accelerating Returns. By the nineteenth century, he wrote, technological change was happening at a rate faster than the previous nine hundred years combined – and included innovations such as the steam engine, the electric telegraph, and gas lighting. The first few decades of the twentieth century saw more technological advancement than the entire previous century – from cars and airplanes to radio transmission.

Kurzweil makes an utterly convincing case that the rate of global technological progress will continue to be exponential. "We won't experience 100 years of progress in the twenty-first century," he says, "it will be more like 20,000 years of progress."[3] He predicts the next few decades will see radical technological changes cut from the pages of science fiction novels.

According to Kurzweil, humans will be able to cure most diseases by the end of this decade. We'll be able to reverse-engineer the brain and use nanobots to treat debilitating illnesses, including neurological disorders like Parkinson's and Alzheimer's. And as a result, our lifespans will increase dramatically.

His predictions about energy have equally far-reaching implications. "In 2030 we will have [total] renewable energy," Kurzweil says, "and it will be inexpensive."[4] Solar, he maintains, will satisfy the total demand of humankind, likely meaning cleaner air and fewer wars.

His most shocking predictions involve virtual reality and artificial intelligence. Kurzweil believes that by 2029, computers will exhibit intelligent behavior equivalent to or indistinguishable from that of humans – including the capacity for love, romance, and humor. Artificial intelligence will eventually respond to the personalities and preferences of human users, and we'll build real human-computer relationships. But this is also the moment that AI surpasses us, as super intelligent machines gain the ability to conceive of ideas no human has ever thought of.

By the early 2030s, Kurzweil thinks we'll be able to copy a human brain and reproduce it in electronic format. Humans at this point will no longer need flesh, blood, or bones, and we won't die in the traditional sense. We'll just 3D print fresh organs. "When we change from an older computer to a newer one, we don't throw all our files away," he says. "Rather, we copy them and reinstall them on the new hardware."[5] The same will go for our bodies.

Even greater changes will come by 2045, though we can hardly comprehend them, because our brains are wired to believe that the future will evolve linearly and predictably, rather than exponentially. By that year, Kurzweil famously posits, we'll have achieved what's known as singularity, a period when humans will multiply their effective intelligence a billion-fold by merging our brains with artificial intelligence. This is the point in time when human civilization experiences unforeseen changes due to uncontrollable and irreversible technological growth – which will fundamentally impact how humans learn, work, recreate, and wage war. "This leads to computers having human intelligence, humans putting them inside our brains, connecting them to the cloud, expanding who we are," he says.[6]

That Kurzweil's reassuring predictions sound more like *Star Trek: The Next Generation* than *The Terminator* is not lost on futurists like Nick Bostrom, who believe that when singularity arrives, there's a strong possibility that machines will either enslave or destroy human beings. Bostrom envisions an AI that will "annihilate Earth-originating intelligent life or permanently and drastically curtail its potential."[7] The Rand Corporation, a US defense think tank, warns that the use of AI in military applications could give rise to a nuclear war by 2040.[8]

Even Elon Musk, the founder of SpaceX and Tesla, two of the

most innovative companies of the twenty-first century, has called artificial intelligence "more dangerous than nukes." He, too, believes the rate of improvement in AI is exponential, and that it's "capable of vastly more than people really understand."[9] As he once put it on Twitter, "competition for AI superiority at a national level [is the] most likely cause of World War III."[10]

Whether or not we're headed toward utopia or dystopia – or something amorphous in between – will rest on the decisions humanity makes over the coming few years.

The Paradox of Progress

As a species, we have thousands of years of experience grappling with new technology and making a mess of it. In the Bible, Genesis tells the story of early man using bricks to climb higher and higher into the sky, hoping to achieve God-like status. Seeing that society has lost its way, God decides to confound human speech so we can't understand each another, then scatters us across the globe. Or so the story goes. Either way, the moral of the Tower of Babel remains: human beings shouldn't play God. And if even a fraction of Kurzweil's predictions come true, things are about to get extremely biblical.

The past fifty years alone have put the promise and perils of technology in stark relief. Our lifespans have more than doubled, per capita income has tripled nearly everywhere, and the cost of food and energy has dropped thirteen- and twenty-fold, respectively. But our world has also changed at such a rapid pace that we've struggled to adapt, and in the trade-off for progress, we've been all too willing to accept or ignore the more sinister ramifications. Using genetic engineering to cure diseases also gives drug cartels the ability to synthetically alter organisms and create illegal narcotics. The same platforms that democratize free education have also taught criminals to rob banks with a single keystroke. The computer technology that has allowed us to get from one place to another autonomously will also inevitably allow a terrorist to remotely coordinate a suicide attack.

In 2020, social media allowed us to stay connected during a pandemic and, among other things, was instrumental in documenting police abuses, but it also continued to fester with propaganda and misinformation, which if left unchecked could portend a future of social engineering on a grand and terrifying scale. We also continued to allow the unfettered collection of our personal data, which is bought and sold to ends we don't yet fully understand and is all but guaranteed to have far-reaching implications for society. And the technology many countries used to stop the epidemic in its tracks – connected wearables, location tracking, facial recognition – was also abused to invade our privacy, break up protests, and racially profile.

"The ability of one to affect many is scaling exponentially," says Marc Goodman, author of *Future Crimes: Everything Is Connected, Everyone Is Vulnerable and What We Can Do about It*. "It's scaling for good and it's scaling for evil."[11]

But compared to the existential problems the earth is facing, the downsides of technology are the least of our worries. In 2020, the Paris-based research organization Future Earth released a report on climate change that combined the insights of more than two hundred leading scientists from more than fifty countries. They concluded that the challenges confronting humanity – extreme weather events, the decline of life-sustaining ecosystems, food insecurity, and dwindling stores of fresh water – compound one another. On their own, they're devastating enough, but taken together, they may wind up destroying our cities, our countries, our planet, and ultimately us.

Extreme heat waves increase global warming by releasing planet-warming gases from natural resources while intensifying water and food shortages. And the loss of biodiversity weakens our natural and agricultural systems, putting food supplies at even greater risk. "Our actions in the next decade," says Amy Luers, the group's executive director, "will determine our collective future on Earth."[12]

There's agreement across the board that the time to act is now. "We have only a very small window," says World Economic Forum president Børge Brende. "And if we don't use that window in the next 10 years, we will be moving around the deckchairs on the Titanic."[13]

We Choose to Go to the Moon

To meet the moment, we're going to have to take more moon shots than ever before. When President John F. Kennedy announced the Apollo program in 1962, landing a man on the moon seemed all but impossible. And if anyone was going to do it, all bets would have been on the Soviet Union, which had already launched the world's first satellite, Sputnik, and the world's first man, cosmonaut Yuri Gagarin, into orbit. America was losing the space race, and Kennedy wanted our astronauts to do something earth-shattering, to show people all over the world how remarkable the United States was – and by implication our system of capitalism and liberal democracy. "We choose to go to the moon in this decade and do the other things," Kennedy famously declared, "not because they are easy, but because they are hard."[14]

Not even seven years later, on July 16, 1969, Neil Armstrong, Buzz Aldrin, and Michael Collins blasted off into space on the Apollo 11 spacecraft. With more than half a billion people watching on television, when Armstrong climbed down the ladder onto the Sea of Tranquility, he couldn't have said it better: it was one small step for man, one giant leap for mankind.

The Apollo program remains one of the greatest mobilizations of resources and manpower in human history. The cost was about $22 billion, or $260 billion in today's dollars.[15] But it was worth the benefits – and not necessarily for the reasons Kennedy intended. NASA had to develop the world's biggest rocket, along with the smallest, fastest, and most nimble computer. The device was among the first to use integrated circuit computer chips and ultimately helped accelerate the silicon chip revolution. Kennedy's moon shot also jump-started the development of a high-speed data network, spacesuits, space food, heat shields, and numerous fireproof materials, among other products.

Well after the mission was complete, the impact of the investment continued to reverberate in new technologies, including the CAT scanner, originally developed to enhance images of the moon; the ear thermometer, first used to detect the birth of stars; and even invisible braces for teeth, which got their start as a material that could detect cracks in aircraft.

The world has changed dramatically since the Kennedy era. And while the United States is still considered to be a great problem solver, the federal government is no longer at the forefront of innovation. Few expect government spending on moon shots to increase in the near future, in part due to partisan gridlock in Washington. Fortunately, venture capitalists and entrepreneurs have taken an increased interest in moon shots, sometimes quite literally: with their help, the United States has announced it intends to send humans back to the moon by 2024, if not sooner. Water has now been found on this celestial body, which will deeply impact the future of lunar bases – both as a source of drinking water and fuel – and further the exploration of deep space.

Governments are no longer the only entities that can play a meaningful role in tackling seemingly intractable global challenges. Corporations, research institutions, and DIY innovators with few resources and little manpower, alongside a generation of techno-philanthropists like Bill Gates, who are spending billions of dollars of their own personal wealth, are now capable of achieving moon shots on their own.

If we're going to meet the challenges of this century – from climate change to food scarcity – governments, companies, and entrepreneurs from around the world will need to work together and use the type of bold, creative thinking that got us to the moon. As Astro Teller, the director of Google x, a research and development facility that creates radical new technologies, explains: "Moonshots live in the gray area between audacious projects and pure science fiction."[16] It is, he asserts, "almost more an exercise in creativity than it is in technology."[17]

Many are already embracing this mindset. Among them are Peter Diamandis, futurist and chairman of the X Prize Foundation, a nonprofit that funds public competitions to encourage technological development that benefits humanity. In 2008, he and Kurzweil founded Singularity University, a school that focuses almost exclusively on "moon shot thinking," where the world's greatest scientists, inventors, engineers, and technologists come together with entrepreneurs, investors, and philanthropists to help solve the world's most pressing problems.

The university doesn't strive to achieve singularity, nor does it give out degrees. Its goal is to teach students about advances in biotech, nanotech, artificial intelligence, robotics, neuroscience, and energy systems. More than a million people from more than 150 countries have attended the school's programs since it was founded. (I had the privilege in 2018.) And it not only gives people the knowledge they need to think boldly about the future, but in many cases access to the money they need to make their ideas a reality. Everyone who takes part is asked to focus on a central question: How can I positively impact the lives of at least one billion people and solve a grand global challenge?

Too Big to Ignore

Interviewing the innovators in this book and diving deep into their stories fueled me with a tremendous sense of faith and optimism that we'll be able to conquer the daunting obstacles that lie ahead. Today, advances in air travel and the internet have shrunk our world to the size of a fist: we now wield more processing power in a single smartphone than the entire Apollo 11 guidance system. There's always the risk of stumbling toward dystopia, but as Kurzweil says, "This is by far the best time in human history."[18]

The transformation the world is currently undergoing is profound and will far supersede the impact of the Industrial Revolution. Above all, three mega-trends are the driving forces behind the issues tackled in *Next*: the growth of a global middle class, rapid urbanization, and increased physical and technological connectivity. By the end of the decade, the world's population will reach 8.5 billion, the majority of which will be part of the economic middle class. And though half the world's population is now considered urban, by 2030 a full two-thirds of it will be. Many will live in a mindboggling forty mega-cities with populations of more than ten million people each.

"These powerful economic drivers will continue to dramatically transform our world over the next several decades," says Afshin Molavi, senior fellow at the Johns Hopkins SAIS Foreign Policy Institute, "lifting millions more out of poverty, reshaping global trade patterns, altering geopolitical alliances, creating new centers of innovation, and posing new and previously unimaginable challenges to societies and states."[19]

The innovations featured in the chapters that follow address thirteen critical areas outlined by the United Nations as key Sustainable Development Goals: space, learning, shelter, the environment, hygiene, medicine, disaster resilience, energy, prosperity, food, water, governance, and security. Many of the people behind the innovations were inspired by the likes of Kurzweil and Diamandis. Some of these visionaries have attempted to solve problems so big, most of us wouldn't know where to begin. *Next* offers readers a look at the key decisions – and failures – behind the process of turning an idea into a real-world breakthrough. And though each of their success stories is unique, collectively they reveal a blueprint for what it takes to tackle the impossible. All have required moon shot thinking.

Among them are Jun Sato, one of Japan's most renowned architects, who explained to me how his country builds structures to withstand natural disasters – and who has become a global pioneer in both seismic engineering and society-level planning in the bargain. During our interview, a magnitude 5.4 earthquake hit. Sato shrugged it off. "That wasn't so bad," he laughed. "The really big one is yet to come."[20]

You'll hear from Toomas Hendrik Ilves, the former president of Estonia, who helped transform his country into the world's first fully digital society, providing hundreds of social services from the cloud, including voting. From the Netherlands, where a significant portion of the population lives below sea level, you'll learn about the colossal efforts of water specialists like Henk Ovink, who are showing the rest of the world how to keep rising oceans at bay. And you'll read about revolutions already taking hold across the spectrum, from the electric and autonomous vehicles reshaping transportation to the difference-makers like Salman Khan reinventing education and Muhammad Yunus combating poverty through microfinance.

You'll sink your teeth into lab-grown meat that could save us all and get a closer look at earth-shattering technologies like CRISPR gene editing, which are set to transform medicine – and nearly everything else. You'll also discover how two solar panels, a refrigerator, and a simple water pump connected to a computer thousands of miles away can entirely transform a starving African village, and how Sivan Yaari is proving it, one project at a time, impacting the lives of millions. And you'll find out how a tiny wrench produced by an experimental, zero-gravity 3D printer could soon unlock a quantum shift for the human race, on our own planet and beyond.

These technologies and the new realities they present to humanity are *already* here now – they are firmly embedded in some places, and spreading. All of these ideas offer a glimpse at what's next for humanity. But above all, if I've done it right, they will also give readers hope about our lives. About our future as a species.

As the science fiction writer William Gibson once put it: "The future is already here. It is just not evenly distributed."[21]

CHAPTER 1

Space: Print Me Up, Scotty!

The Earth is the cradle of humanity, but mankind cannot stay in the cradle forever.

– Konstantin Tsiolkovsky, from a 1911 letter

International Space Station over Miami, Florida (NASA)

Big Dang Deal

In November 2014, Barry Wilmore, a NASA astronaut, looked out the small window of the International Space Station and worried about what he was going to tell his boss. Weeks earlier, he and two Russian cosmonauts, Aleksandr Samokutyayev and Yelena Serova, had arrived at this outpost about 220 miles above Earth – the only off-planet home for humankind. Almost immediately, they began conducting experiments ranging from disaster relief to how the human eye changes during space travel. As they worked, however, one of Wilmore's tools – a ratcheting wrench – drifted away in the spaceship.[1] "It's not uncommon for things to disappear in space," he says. "You just don't have gravity keeping stuff in place."[2]

If an emergency occurs or something breaks down, the consequences of losing tools can be calamitous. Which is why the space station spends millions on spare parts and backups, most of which will never be used. Unfortunately, the astronauts on his mission could not find another wrench.

Frustrated, Wilmore radioed mission control back on Earth. The NASA engineers on the other end of the line were concerned, but they quickly came up with a solution. They contacted a company called Made in Space, a California-based startup that was trying to prove that 3D printing could work without gravity.

A few weeks prior, NASA had sent one of their devices to the astronauts on a cargo spacecraft. The timing was fortuitous.

Invented by Charles Hull in 1984, 3D printers are similar to their ink-and-paper predecessors. The difference is they can make three-dimensional objects. "The ink is now plastic," says Jason Dunn, Made in Space's cofounder. Or an acrylic-based material known as photopolymer. When UV laser beams hit this material, the portion exposed to light turns into a solid piece, which the printer molds into a 3D shape. As Dunn explains: "We're melting it just like a hot glue gun out of a nozzle."[3]

Thanks to 3D printing, companies can now test designs without having to make a huge upfront investment. Today, industries across a broad range use this technology, from construction to aerospace. In space, 3D printing could revolutionize travel, telecommunications, and other industries. But the devices require gravity – a problem the company knew they had to solve. If every single belt, cog, and pulley isn't secured, once you remove the downward force ever-present on Earth, the printing mechanism fails or leads to strange results.

By the time Willmore lost his wrench, Dunn and company had been developing their printer for years. If they succeeded, the tool would be the first designed on Earth and made in space.

Back at Silicon Valley's Ames Research Center, the company's engineers produced several wrenches to ensure they met the proper specifications. After five days of work, mission control sent the digital file to the space station. After two hours of printing, mission control asked Wilmore to go over to the box-like device about the size of a small microwave. He opened the door and pulled out a wrench. "[It's] a game-changer," he said. "Or as we say in Tennessee, it's a big dang deal."[4]

One small tool for human beings, one huge leap for humankind.

The Vomit Comet

Dunn's interest in space began when he was a child growing up in Florida. He was fascinated with exploring the world, from the oceans on Earth to the planets beyond it. The first time he saw the Space Shuttle launch from the nearby Kennedy Space Center at Cape Canaveral, he watched what looked like a little star moving up in the sky and was amazed that there were people on it disappearing into space.[5]

As he grew up, Dunn's interest in space increased. At the University of Orlando, he studied aerospace engineering and worked on NASA's shuttle program through an internship at Boeing. He got involved in a college club called Students for the Exploration and Development of Space, where he met Peter Diamandis, the Singularity University cofounder, who became his mentor.

In 2010, NASA announced it was going to stop flying into space. "I felt a little heartbroken and very confused," says Dunn, "because the whole reason I was doing this was to work on the space shuttle."[6] But Diamandis encouraged Dunn to join the Singularity University graduate studies program in Mountain View, California. There, Dunn met two entrepreneurs – Aaron Kemmer and Mike Chen. All three had a lot in common. They were obsessed with *Star Trek* and space, and they had pursued college degrees in either engineering, computer science, or philosophy.

Kemmer played a major role bringing the cofounders together. He grew up in the Florida panhandle and watched space shuttle launches from his home. He was mesmerized by the shuttles, by the fire, the smoke, and the chance to explore life beyond Earth. And from an early age, he knew he wanted to be an astronaut. He remembers jumping up and down on his bed when he was five years old, holding his Hot Wheels Space Shuttle. He tripped and fell and cut open his forehead, but wouldn't let go of the toy. "That's when I got my Harry Potter scar," Kemmer says. "Going into space was seared into my forehead."[7]

As he got older, his obsession with space didn't wane. And after graduating from the University of Florida and starting a news aggregation website, he kept thinking about what he could do that was bigger than himself – and for him, that meant a space venture. When he was twenty-five, he enrolled in Singularity University, hoping to learn all he could from Diamandis.

When he met Dunn and Chen in the program, he quickly realized they all had a common interest in 3D printing and its potential in outer space. NASA, they learned, had experimented with 3D printing in the 1990s, but the agency decided the technology wasn't ready. That changed in 2005, when Congress designated the US segment of the International Space Station as a national lab to encourage experiments that would improve life on Earth.[8]

After finishing the Singularity program, the three men decided to start a company along with Michael Snyder, a close friend of Kemmer. They alternated between calling themselves the Teenage Mutant Ninja Turtles (each one was assigned a character based on their personalities) and the four Jedi.[9] As they decorated their office, they filled it with science fiction posters from *Star Wars*, *The Tholian Web*, and *Mirror Mirror*. They also have a life-sized cardboard cutout of Mr. Spock in the reception area. "We are creating the technology of *Star Trek*," says Dunn, "building the replicator and transporter all in one with 3D printers."[10]

Shortly after forming their company, Kemmer and Dunn traveled to Alabama to NASA's Marshall Space Center where they pitched the idea of putting a 3D printer in space. It went well, and NASA eventually connected them to academics and government agencies, and let them conduct experiments on its reduced gravity aircraft. Known as the "Vomit Comet," it's a converted passenger jet that flies in the same way a roller coaster moves. With every nosedive, passengers get about twenty seconds of weightlessness. This allowed them to start testing a variety of 3D printers to see if they would work outside Earth's gravity.

Made in Space quickly got to work. From July through September 2011, the company performed more than four hundred tests on ten flights. "The feeling of zero gravity is impossible to explain," says Dunn. "It's entirely new. Outside of when you're an infant, you don't really have any new sensations or feelings."[11]

On the first nine flights, the company's printers failed. Because of the anti-nausea pills and the new floating sensation, the team had to automate every part of its experiments. Making real-time decisions was impossible. At first, everything the team printed turned to mush. Heat transfers differently on Earth than it does in space – it rises in gravitational environments, which is obviously impossible where "up" doesn't exist – and that impacts the melting and cooling of the plastic.

Made in Space had one chance left, and it had to design a new printer to make it work, an arduous task. The team spent days redesigning the device. The four eventually decided to try adjusting the printer's surface tension to make the layers of plastic stick together. They boarded the flight far from certain that their work would pay off. NASA, in the face of hitherto unproductive test flights, was not going to give them any more flight time. "We were down to our last chance to ever figure this out," Kemmer recalls.[12]

The pilot counted down from ten, then pulled back on the joystick controlling the aircraft. It climbed a kilometer into the sky. Then the pilot cut the engines. The plane began to drop. The team fired up the printer. Goo oozed from the device. The team waited a moment – and then it happened: the goo hardened. Everyone pumped their fists. "It was this amazing moment," says Kemmer.[13]

The tests proved that Made in Space had created a product that could actually work in zero-gravity environments like the moon or the surface of Mars. "In my heart, I always knew these were the guys to figure this out," says Yvonne Cagle, a NASA astronaut and the Singularity University instructor who served as their liaison to the space agency. "They had the passion, the drive, and determination to solve this challenge no matter what."[14] NASA soon offered them a contract to build a 3D printer that the agency could send to the space station. And after two years of trial and error, Made in Space's dream became a reality with Willmore's wrench.

Today, the company's devices can help reduce the risk of missions by fixing tools, or creating new ones in real time. And in the future, Dunn says, humans will be able to 3D print larger objects at the space station. That could mean we'll be able to travel deeper into the universe and live full-time in space or on other planets. As Dunn puts it: "We need to make it so that going to space is as safe as getting on an airplane."[15]

Life beyond Earth

In 1998, the United States, Russia, Japan, Europe, and Canada created the International Space Station, a research laboratory where astronauts study how to launch long-term missions to the moon and Mars, among other things. The driving force behind this project is to preserve life on Earth and build colonies in space.

It's an important mission – and one that could be existential. Today, most experts believe there are not enough resources on our planet for every person to have a standard of living equal to those in the Western world. If we fail to explore planets beyond Earth, we may become extinct. "There's an intrinsic limit to how much energy production we can produce and consume, how much CO_2 we can keep putting into the atmosphere, how much food we can produce as long as we are a one-planet species," says Dunn. "It's not about leaving Earth, but expanding and saving it."[16]

Already, entrepreneurs and governments see the moon, asteroids, and other planets as vast, untapped resources that humans can use to sustain ourselves – from water and gold to cobalt and iron. "The expansion of humanity into space will unlock unlimited potential for economic and population growth from the billions to trillions," says Bernard Kutter, the chief science officer of United Launch Alliance, a US space service provider that is a joint venture between Boeing and Lockheed Martin.[17]

Ramin Khadem, the former chairman of the board of trustees of the International Space University, agrees. "The moon is our eighth continent," he says. "Among its many important elements, it has water ice in its permanently shaded regions that humans will eventually use not only for their habitat and sustaining life beyond planet Earth, but also for creation of propulsion fuel to explore the universe beyond."[18]

Some of the top minds in the field are working to permanently settle humans beyond Earth. Elon Musk's company, SpaceX, wants to turn Mars into a vibrant green home. Jeff Bezos's Blue Origins believes humans can live in free-floating, cylindrical space colonies.

When we will leave the cradle of planet Earth remains to be seen. Some, such as Brian Weeden, the director of program planning for Secure World Foundation (a US-based organization dedicated to the sustainable use of space), notes that there are significant challenges ahead, from building the structures needed to sustain us to understanding how space may impact the human body.[19] But most experts agree that deep space exploration is not a question of if, but when. "What was impossible only a decade ago," says Yonatan Winetraub, cofounder of SpaceIL, the venture that launched the Bereshit robotic lunar lander and lunar probe, "is now a reality. If you dream big, technology will eventually catch up."[20]

Right now, the main obstacle is cost. Space travel is incredibly expensive. NASA spent about $100 billion to build the International Space Station, and it spends $3 to $4 billion a year to maintain it.[21] Meanwhile, a rocket to the moon costs as much as $1.6 billion per launch.[22] All of the private companies on the market are focused on cheaper ways to leave Earth – from larger or reusable rockets to solar propulsion.

Made in Space's cofounders began their journey by asking a different question: What if we didn't need to launch anything at all?[23] Both Kemmer and Dunn envision a future in which 3D printing will manufacture an entire spacecraft at the station, eliminating the constraints of traveling from Earth. "By taking one tiny step, they have fundamentally caused a quantum shift in thinking," says Chris Stott, a space entrepreneur. "This completely changes the equation for the human race."[24]

If we want to live on the moon or Mars, we'll have to learn how to manufacture basic items beyond Earth. Aside from creating tools and vehicles, Made in Space has started working on future space colonies by 3D printing bricks made of a special polymer that simulates regolith – the fine powdery material found on the surface of the moon and Mars.[25] Eventually, the company will begin to experiment with the actual material. By sending 3D printers to the moon, Mars, or anywhere else in the universe, NASA could build roadways, landing pads, and other structures humans need to sustain themselves – all from Rego brick. "Most of the technology is actually here, right now," says Michael Potter, director of the award-winning space documentary film *Orphans of Apollo*.[26]

Space and satellite expert Carol Goldstein agrees. "The combination of technology advances across a number of disciplines, including rockets, robotics, communications, and 3D printing technologies, to name just a few, makes it possible to see building a permanent presence off-planet in the foreseeable future."[27]

Over the next ten years, Kemmer predicts that space will also be home to factories that orbit the earth, much like oil platforms floating in the ocean. And that could mean big business. Morgan Stanley, Goldman Sachs, and Bank of America have estimated that by 2040, space-related products and services will reach $1.1 trillion.[28] "This venture represents the . . . next huge industrial revolution," says Peter Garretson, a former US Air Force officer and now a senior fellow at the American Foreign Policy Council.[29]

Universal Quality of Life

This revolution can also help improve life on Earth. Today, there are more than 50.1 billion devices online worldwide, used by almost half the world's population. Some analysts predict the amount of data we use will grow by 61 percent to 175 zettabytes by 2025.[30] If we stored this amount of data onto Blu-ray Discs, we would have a stack that goes to the moon twenty-three times.[31] This increase in data also means we need new equipment to handle the additional bandwidth.

One of the strongest candidates: ZBLAN optical fiber, which is used to build high-speed internet, transoceanic telecommunications, and medical devices.[32] When these fibers are produced in a zero-gravity environment, they are ten to one hundred times more efficient.[33] Starting in 2018, Kemmer, Dunn, and company began manufacturing ZBLAN in space. Eventually this business could generate billions in revenue. "The new space economy is the catalyst that will fundamentally transform life on Earth," says Stephen Eisele of Virgin Orbit, a Richard Branson–owned company that provides launch services for small satellites. "We are on the brink of opening space for everyone."[34]

And yet companies should proceed cautiously in all their space-related endeavors. Without thoughtful planning, unchecked industry and mining in space could be harmful and tarnish the natural beauty of planets, moons, and the universe. Humanity's record of protecting our planet is abysmal – from pollution to deforestation. But space offers us a second chance to change our relationship with nature.

"Our solar system, the moon and planets are pristine," says Dr. William R. Kramer, a lecturer at the International Space University and retired official at the US Fish and Wildlife Service. "Before we begin to settle, exploit, and dramatically alter these places of stark beauty and incredible scientific value, we would benefit from considering what is worth protecting. Should we repeat the same mistakes we've made on Earth?"[35]

Kemmer, Made in Space's cofounder, agrees. He wants his company and others to help humanity safely explore the universe without repeating the mistakes our species has made on planet Earth. "It's a unique challenge, a unique opportunity," he says – and one that other experts believe will save humanity.[36]

As Christian Sallaberger, the CEO of Canadensys Aerospace Corporation and chairman of the board of the International Space University, told me: "It is the first time... our species has come together to design things that benefit our entire society and improve the quality of life for all on Earth."[37]

CHAPTER 2

Learning: The Internet Academy

No schooling was allowed to interfere with my education.

– Grant Allen, *Rosalba: The Story of Her Development*

Salman Khan at 2011 Ted Conference (Steve Jurvetson, CC BY 2.0)

Tiger Cousin

In August 2004, Salman Khan was living in Boston and working for a hedge fund when he and his wife, Umaima Marvi, hosted a family visit that would change his life – and the lives of millions around the world.[1] That weekend, Khan's twelve-year-old cousin Nadia Rahman visited from New Orleans and shared that she was having trouble with pre-algebra. Her school had placed her in the less advanced class, and Khan didn't think her parents realized the repercussions – that it might set her on a path of underperformance and low expectations. "I want to work with you," Khan told his cousin, "if you are willing."[2]

Khan understood the importance of getting an education; until much later in life, he didn't have the opportunity to attend good schools. Khan was raised in Metairie, Louisiana, a town best known for electing white supremacist David Duke to the state legislature.[3] His parents – both from Bangladesh – divorced in the 1970s, and his father left when Khan was two. They only met once more, a decade later, before his father died of a heart attack.[4] "He never sent child support," recalls Khan, "because he didn't have money despite being a doctor."[5] To feed her family, Khan's mother worked at convenience stores.[6] They were poor, he says, but they always had food on the table.

Khan attended Grace King High School, an unexceptional but diverse public school that influenced his approach to education. Most of the attendees were minorities, half were poor, and "a few classmates were fresh out of jail," he recalls. Others were college bound and took advanced placement classes.[7] Part nerd, part punk-rock rebel, Khan sported wild hair and pierced ears but was known for his keen mathematical mind. He was so sharp that he even got a perfect score on the math portion of the SAT and became his high school's valedictorian.[8]

For college, Khan was accepted to MIT, where he earned degrees in math, electrical engineering, and computer science. (He also became the lead singer of Malignancy, a heavy metal band.)[9] Because his family was poor, he qualified for financial aid but still had to take out $40,000 in loans to cover the rest.[10] He later got his MBA at Harvard, where he claims his main goal was to get married. "I'm dead serious," Khan says. "Silicon Valley in the late '90s was the absolute worst place to find a wife or a girlfriend."[11]

He succeeded – and met Marvi, who was attending Harvard Medical School and later became a physician. When the two hosted Nadia that weekend in 2004, Khan wanted to see her succeed just as he had. When she returned to New Orleans, he started tutoring her by phone, carving out time between his work and her soccer practice. "I became what I like to call a 'tiger cousin,'" Khan says.[12] Two months later, she had made so much progress that Khan called the school and asked them to test Nadia again. After the exam, her teacher moved her into more advanced math classes.[13]

Soon, Khan's extended family learned about Nadia's success – and they wanted help, too. "I found myself working on the phone with about fifteen cousins," he says.[14] Most of their basic math skills were poor – even when it came to simple division. Khan was shocked and wanted to help them improve. But it was hard working with so many different family members at once. To make it easier, he created a website to allow his students to practice problems and track their progress.

About two years later, Khan explained to a friend what he was doing for his family.

"How are you scaling your lessons?" the friend asked.

"I'm not."

"Why don't you make some videos of the tutorials and post them on YouTube," the friend suggested.

"That's a horrible idea," said Khan. "YouTube is for cats playing piano. It's not for serious mathematics."[15]

The more Khan thought about it, however, the more he was

inclined to take his friend's advice. In November 2006, he started recording YouTube videos out of his closet, "a kind of monk's cell," he recalls, "without distractions or the temptations of too much comfort."[16] His only equipment: a desktop, a microphone, Microsoft Paint, some free photo software, and a very basic recording camera.[17] The total cost: $1,100.[18]

Khan quickly discovered that his cousins liked him far more on YouTube than in person. They found his explanations more valuable when they could have him on demand. "The worst time to learn something," he says, "is when someone is standing over your shoulder going, 'Do you get it?'"[19]

Others felt the same way. Not long after he began posting videos, Khan learned that many people who weren't related to him were watching his videos, too. By 2008, thousands of people were watching his videos every month.

He had stumbled into a new career – entirely by accident.[20]

It Was Very Strange

Every year, America spends $1.3 trillion on education – more than any other nation.[21] And yet only about a quarter of fifteen-year-olds in the United States can use math in their daily lives. About a fifth of them don't have basic competency in science, according to the US President's Council of Economic Advisors.[22]

Khan wanted to change these unfortunate statistics, but doing so would require him to quit his job at the hedge fund and devote himself full-time to his online learning platform. He was torn. "I was very happy and satisfied," Khan recalls.[23] He loved his job and his boss and was doing well financially. But in 2009, he quit and created a nonprofit, using his savings to expand his video operations. The experience was terrifying but oddly liberating. "It was very strange for me," he said, "to do something of social value!"[24]

By then, his platform, Khan Academy, was attracting an even larger audience, in part because he allowed students to work at their own pace. Users could only move to the next level after mastering the previous one.[25] Students wrote to Khan, thanking him for his help; because of it, they were able to pass difficult subjects such as algebra and trigonometry. Teachers sent him emails saying he had saved their students from dropping out of school.

Khan's concept was simple but has proven to be effective. Studies show that everyone learns at a different pace. Some grasp a subject faster and are able to rapidly move to the next subject, while others lag behind and struggle. When children are given the appropriate time to master concepts, they will eventually move on. But most classrooms test kids at the same time, at the same pace. "It is not just about zeroes and ones," says former astronaut Nicole Stott, an advocate for STEM (science, technology, engineering, and mathematics). "The Khan Academy opens the world of knowledge to learners of all ages, giving them critical tools and the confidence to learn any subject, anytime they want."[26]

Khan's goal isn't to replace teachers, but to help them and their students inside or outside school. "[Khan Academy] can be a useful tool in the hands of a thoughtful teacher, as long as the ultimate goal is not to deliver information to a passive group of students, but instead to spark their critical thinking and build their problem-solving skills" say Aleta Margolis, founder and president of Center for Inspired Teaching, a nonprofit based in Washington, DC.[27]

He's also hoping to change the way people learn, whether they live in Minnesota or Brazil. Khan doesn't believe students fail math because it's too difficult or because they aren't bright. He likes to provide the following example: Two students in seventh grade learn exponents, then take a math exam. One gets a 70 percent and the other 80 percent. Both children get passing grades, and the teacher moves on to teach the next-level subject, negative exponents. But doing so becomes much harder, because they never learned the previous lesson. The same goes for algebra and calculus. Each lesson builds off the previous one, so if students fall behind, and no one is helping them, they never catch up.

"There are not enough skilled teachers in the world to meet the demand for quality STEM teaching and learning," says Robert Murphy, a former senior Rand Corporation education technology expert. "Especially in emerging economies, rural areas in the developed world, and among refugee populations. Khan Academy provides a free, effective instructional program with timely targeted feedback for learning that can be used to help fill these major gaps, particularly when there are no viable alternatives."[28]

By the end of 2009, Khan had a growing audience and had gotten some press, but he still hadn't secured major funding. His savings were nearly gone, and he had no idea how to continue to expand his audience and make money.[29] He would often meet with venture capitalists who loved the idea but challenged his basic premise that the videos should be free. "This you need to pay for," they would say. "You need to put in ads."[30]

But filling his platform with ads didn't seem right to him. If kids wanted to learn algebra, they should be able to do so without any distractions. Khan felt somewhat delusional for being so stubborn, considering he was still recording videos out of a walk-in closet. But he refused to budge. "It was stressful," Khan recalls.[31]

He was getting five- and ten-dollar contributions, but that wasn't enough to sustain him. He was about to give up. "I'd realized I had less tolerance for digging into my savings than I thought," he recalls. "[There is] nothing like burning $5,000 a month out of savings while having a toddler in the house to put a strain on marriage."[32]

And then something miraculous happened.

It's Raining Math

In early 2010, just as he was thinking about closing down the academy, Khan got a PayPal donation through his website for $10,000. He emailed the donor, Ann Doerr, to thank her, and when she learned he knew her husband, John, a well-known Silicon Valley investor, she invited him for lunch. She was surprised to learn he was struggling. Because he had been on a number of major TV news shows, "[I] assumed he was set financially," Doerr recalls. "Then I found out I was his biggest contributor."[33]

As they ate lunch, Khan shared that he was about ready to quit and get a "real job." Twenty minutes after they said goodbye, Khan pulled into his driveway. He looked down at his phone and saw a text from Doerr saying she was going to send him $100,000 right away. "I almost crashed into the garage door," he wrote.[34]

A few months later, in July 2010, Khan received a text message from Doerr, who was attending the Aspen Institute Ideas Festival. "Bill Gates is talking about your stuff on stage," she said. Khan was surprised to learn that Gates used the academy to teach his children. He immediately downloaded the video and promptly "shit a brick," Khan recalls.[35] There was Gates, the world's wealthiest and arguably most charitable man, telling the audience that he was using Khan Academy, a relatively unknown platform, to teach his children math and science. "Today if you are motivated to learn," he said, "this is an amazing time for everyone... this one guy is doing some unbelievable fifteen-minute tutorials.... I think you just got a glimpse of the education of the future."[36]

Suddenly, the money started pouring in. Gates followed up with an in-person meeting in Seattle and gave Khan several donations totaling $5.5 million. Others were just as generous. Google gave Khan $2 million.[37] The O'Sullivan Foundation gave him $5 million. Reed Hastings, founder of Netflix, donated $3 million. Intuit cofounder Scott Cook and his wife donated $1 million; and Mexican business magnate Carlos Slim (then the richest man in the world) gave him $317 million.[38]

As Khan put it: "It felt epic."[39]

Teachers Were Blown Away

With the money from all those donations, Khan quickly started hiring engineers, programmers, and developers to build out the back end of the website and provide teachers and parents with data on what students were accomplishing.[40] "Teachers were blown away," says Ben Kamens, one of the programmers, "every time. They were completely shocked, as if this had never existed before."[41]

And while Khan continued producing all the math and science videos, he hired a team to create lesson plans for dozens of other subjects – from world history to SAT prep. Today, the academy has delivered well over several billion combined lessons and exercises.[42] The official platform is available in eighteen languages, and outside volunteers have translated part of the content into an additional twenty-six languages.[43] Khan Academy has even partnered with famous institutions such as the Metropolitan Museum of Art, the British Museum, and the American Museum of Natural History, providing them the platform to deliver their own educational content. "Over a billion people around the world now have access to a high-quality free math education," says Supriya Tripathi, a New Delhi–based Hindi content creator for the Khan Academy. "The fact that this content is available in so many languages fundamentally changes the destiny of an untold number of global citizens who want and deserve better lives."[44]

In July 2018, Khan Academy launched Khan Academy Kids, a free mobile app primarily for children ages two to six. With society's youngest students exposed to a significant amount of screen time, researchers from the Early Academic Development Lab at the University of Massachusetts-Amherst found that Khan Academy Kids boosted early literacy skills across the socioeconomic divide. "This app nailed it," says Professor David Arnold, who led the study. "It is engaging, targets the right subject matter, and scaffolded the material so that it wasn't too easy or over the kids' heads."[45]

Yet there are educators who are critical of Khan's approach, who think he's simply feeding children the educational equivalent of Chicken McNuggets.[46] Gary Stager, a veteran teacher who has worked with students on six continents, is one of them. He believes that criticizing students' performance without questioning what they are being asked to perform does little to prepare them for an uncertain world. "Sal Khan may be helping students do marginally better at an unchallenged traditional school curriculum. That's fine, perhaps even laudatory," he says. "[But] since Kahn is profoundly incurious about how learning occurs or about the history of education, he reverts to a toxic 'only I can fix the education system' worldview shaped by those who got lucky on Wall Street or in Silicon Valley."[47]

Karim Ani, founder of Citizen Math, agrees. "Khan Academy bills itself as providing a 'world-class education.' It is a helpful resource, but hardly transformative. It offers exactly the same experience teachers have provided for generations; the only difference is that it's online."[48]

Sylvia Martinez, the coauthor of *Invent to Learn: Making, Tinkering, and Engineering in the Classroom,* feels similarly, adding that any discussion on Khan's platform must start with the question "How do you believe people learn?" "The things they're doing are really just rote," she says. "Students 'fumbling around' is actually where the learning happens, and there's no shortcut for this process."[49]

And yet Khan claims that his program allows teachers and students to do more fumbling around, precisely because it frees up time for creative activities during the school day, including art, games, and collective brainstorming. "It'd piss me off, too," says Khan, "if I had been teaching for 30 years and suddenly this ex-hedge-fund guy is hailed as the world's teacher."[50]

Gates is even more blunt. "It's bullshit," he says of the criticism against Khan. "If you can't do multiplication, then tell me, what is your contribution to society going to be?"[51]

Both Gates and Khan acknowledge that for subjects such as writing and history, there is no easy way to automate learning. So what happens if a student becomes so advanced in math but is still functioning like a normal kid when it comes to writing, history, and social studies? There is no good answer – at least not yet. But Khan hopes to figure it out – and believes that in the future, people will look back at education today and think of it as the Stone Age.

Nothing Seemed to Work

In the fall of 2017, after a battery of tests, I learned that my six-year-old son, Eiden, had a serious learning disability. Professionals recommended I sign him up for an extensive tutoring program, which I did. But part of me struggled to accept the diagnosis (and as it turned out, he simply had attention deficit disorder, a far more benign disability than they originally thought). And so a few weeks after his initial diagnosis, while I was at a conference in San Francisco, I had a conversation with a colleague about Eiden's struggles that changed the way I thought about education.

As we boarded a ferry and stared at the sun setting over Alcatraz Island, he told me his daughter had just finished calculus.

"Oh, yeah?" I responded. "How old is she?"

"She's twelve."

I couldn't believe it. Most kids in America don't learn calculus until they're seventeen. And that's when he explained she had been doing online learning with Khan Academy – and that it was free.

Up to that point, I had never heard of Khan or his platform, but when I got back to Washington, DC, I downloaded the app and tested my son. He really was about two and a half years behind his peers in math.

We decided to try the program to improve his skills. He started working diligently, at first under my supervision, and after a few weeks, completely on his own. Within three months, he had caught up with his friends. After a year, he was a full grade ahead of his class. What really turbocharged Eiden's learning was the ability to watch the videos as many times as he needed to understand the concepts, then take tests that didn't allow him to move on until he got everything right. Not only did his self-confidence improve, he had learned one of the most valuable life lessons: there is nothing he couldn't teach *himself* as long as he applied determination and grit. He quickly became interested in the other subjects Khan Academy offers, including history, economics, and art.

I was so excited about his success, I started having my other kids take the courses. Khan Academy proved especially important as we all went into lockdown during the spring of 2020, as the coronavirus pandemic spread. The virtual learning program my children's school offered was good, but far from perfect. I supplemented it with Khan courses to make sure they continued to learn. Doing so was incredibly hard, but yielded tremendous results. It was clear that this was a wonderful and important supplement to – though not a replacement for – traditional classroom education. But in the long term, Khan's courses had transformed Eiden from a poor student into someone well above average.

Other Khan students may experience similar transformations – and at a critical time. In 2020, as the world battled COVID, many classrooms around the world attempted "Zoom school" with various degrees of success. In the United States, research shows that this switch wiped out academic gains for many students and widened racial and economic gaps.[52] Those using the Khan Academy appeared to do better, especially among minority groups who historically have had fewer opportunities. Survey data shows that African Americans, women, and economically disadvantaged students were among the heaviest users of the Khan Academy's official LSAT program, and there is at least anecdotal evidence of their success as a result.[53] "This achievement demonstrates Khan Academy's "groundbreaking commitment to helping [minority] students prepare for the LSAT," says Kent Lollis, vice president and chief diversity officer of the

Law School Admissions Council, a Pennsylvania-based nonprofit that promotes quality, access, and equity in law and education.[54]

Nadia, Khan's niece, also succeeded as a result of his classes. More than a decade after their tutoring began, she has grown up to be an impressive young woman. In 2014, she graduated from Sarah Lawrence College and later received two master's degrees from the Icahn School of Medicine at Mount Sinai and the New School. Today she is a research coordinator at the New York University Langone Medical Center.[55]

As Khan wryly puts it: "So far, so good."[56]

CHAPTER 3

Shelter: Let There Be Light

Namaste: The inner light in me
recognizes the inner light in you.

– Sanskrit greeting

Sivan Yaari bringing solar energy panels to African village in Karamoja, Uganda (Lior Sperandeo, CC BY-SA 3.0)

Heart of Darkness

As Sivan Yaari's white SUV rumbled over the dirt road into the tiny village of Karamoja, Uganda, the horror started to spread through her limbs. It was March 2017, and for weeks, Yaari had read the reports of death and starvation, as a quarter of Uganda's forty million people faced the specter of acute famine. As the head of Innovation: Africa, an aid organization that brings water and electricity to remote areas of the continent, she was no stranger to suffering. But this sight was unlike anything Yaari had encountered before.

When she and her team arrived, they were confronted by emaciated children caked in mud. Rail-thin women who had bound their stomachs with rope to fight off hunger pains were wailing. People of all ages were on their hands and knees drinking cow's blood out of a soiled bowl for sustenance. No rain had fallen in three months, and the region was in the grips of catastrophic drought. Many of the villagers hadn't eaten in days.

Yaari and her staff gave the people whatever food and water they had. Then they got down to work.

Within days, they returned with truckloads of food and equipment. To the cheers of hundreds of villagers who came to help, they unloaded bags of beans and corn and enough solar panels to provide desperately needed energy to pump enough water for six surrounding villages. Heavy drilling machines also rolled in on big blue rigs with pipes long enough to reach an aquifer 130 feet down. And when the fresh water began to pump through the newly installed taps, throngs of men, women, and children erupted in joy.

"God bless you," they shouted as they lined up to hug Yaari and her team. Drinking water from a tap was unlike anything they had experienced before – it was the first time the village had access to free-flowing water. The singing and dancing lasted for days.

"It really warms your heart," says Yaari, "to see how such a simple solution can change the lives and futures of so many people."[1]

Few of us will ever be in a position to make such a direct and singular impact on the lives of so many. Sivan Yaari has done it thousands of times.

Where Is Madagascar?

As unlikely as it sounds, it was in the service of fashion, not people, that Yaari's relationship with Africa first began more than twenty years ago. Yaari was just nineteen and fresh off a tour with the Israel Defense Forces in 1998 when she got the job she now credits with changing her life. Though born on the outskirts of Tel Aviv, Yaari had spent her teenage years in France, and her fluency in the language made her a perfect candidate for a position in quality control at a Jordache jeans factory in Madagascar. When she interviewed in Tel Aviv with the company's cofounder, Rafi Nakash, she had to ask him where the country was. He had barely said "Africa" before she offered an enthusiastic yes.

Yaari grew up in a family of modest means, but it wasn't until she got to Africa, she says, that she discovered what it really means to be poor. When she arrived in Madagascar, she was deeply unsettled by the number of people – and children in particular – who were without shoes and clothing, looked sick and malnourished, and who didn't have access to clean water, basic medical facilities, or schooling. She visited a medical clinic in the countryside where she saw scores of people waiting outside for help and getting none. When she asked if there was anything she could do, the nurse said thank you, but no. There was nothing to be done. "I don't have vaccines or medication," the nurse told her. "They all went bad, because we don't have electricity or refrigeration."[2]

Yaari quickly came to understand that the lack of electricity created even deeper problems. Her colleague took her to a nearby secondary school, but by the time she got there, it was after dark. The classrooms were full, and the students were using gas lamps. The children from the best-off families sat in the front rows near the light, because they were the ones paying for the kerosene. Those less fortunate sat in the back in near darkness.

The dearth of light was also a major issue for nurses in the village, who were doing things like delivering babies with one hand while holding kerosene lamps in the other. It came as no surprise to anyone, Yaari says, that so many mothers and babies didn't survive.[3]

When Yaari began traveling to Jordache's six other factories around Africa, she found the same despair wherever she went. "It took me time to understand why this was happening," she says,[4] but one thing seemed clear. "If we could just bring energy, it would be so much better."[5]

Yaari was still in her early twenties, and after working in Africa for a few more years, she moved to New York to get a degree from Pace University and a master's in international energy management from Columbia's School of International and Public Affairs. But Africa was never far from her mind.

As a grad-student volunteer for the United Nations' Energy Bureau, she traveled to Senegal to inspect how machines were being used in villages to grind wheat and corn – and concluded it did not work very well at all. Because villagers had no money for electricity or the diesel fuel it takes to run machines, she says, "I understood I needed to find a completely different solution."[6]

Yaari consulted with her professor, Philip LaRocco, who specializes in renewable energy.[7] "It is quite simple," he told her. "All you need is two solar panels."[8]

Just Add Water

Approximately half the citizens of Africa are without electricity, but what the continent lacks in power, it makes up for in sunshine. A single pair of solar panels can generate enough energy to power an entire medical center, twelve light bulbs, and even a refrigerator. They're also low maintenance, easily installed, and can last for thirty years. Nearly every roof in Israel has a solar panel. Yaari wondered if the same could be done in Africa.

By 2008, Yaari had set up her first project in the village of Kidigozero, Tanzania. For $4,270, she was able to install two solar panels and a refrigerator in the local clinic. On the very first day the lights came on, residents were being vaccinated. For many, it was also the first time they had seen light at night. Within a few weeks, a doctor agreed to move to the village.[9]

"It is so simple," says Yaari. "And yet it makes such a big difference. I felt like we had to do more of it, village by village. Until we cover the entire continent."[10]

Back in New York, she officially registered Innovation: Africa (originally called Jewish Heart for Africa) as a nonprofit to bring Israeli solar and water technology to the continent, and immediately started raising funds.

"It was as if she was struck by lightning," recalls LaRocco.[11]

By 2009, Yaari had installed her second set of solar panels at a primary school in Putti, Uganda, which had just ten classrooms for eleven hundred students. The lights enabled the children to complete their homework at night and study for national exams. But there was a problem: the school couldn't afford to pay for new light bulbs when they burned out. The only sustainable solution was to find a way for the villagers to raise the money themselves.

Though many in Putti didn't have shoes or clothes, almost everyone had mobile phones. "There is one guy who collects all of the cell phones and walks eight miles to another village where there is also no electricity, but there is a car," says Yaari. "People were paying ten cents to use the car's battery to charge their cell phones."[12]

To help schools, medical centers, and orphanages raise money, she decided to let people use the solar system to charge their phones for ten cents a shot. Pretty soon, Yaari recalls, "They had more money than they needed."[13]

All of Innovation: Africa's installations are now maintained through this pay-per-use model, which generates the funding required for villages to replace light bulbs and batteries – and is infinitely replicable and scalable.

But the biggest lesson Yaari learned from her early work in Putti was much more elemental. When she returned a few months after the installation to check on the school project, the headmaster told her they weren't using it. "The students are too weak to walk to school," he explained. "There is no water." Putti was experiencing severe famine.[14]

Yaari learned that the average African woman spends over three hours a day collecting and carrying water. This is 1,095 hours per year, or the equivalent of 136 full American workdays. And the water they do manage to bring home is often unsanitary, causing more than 80 percent of the developing world's deaths and diseases. Women and children in Putti were fetching water instead of working or going to school, standing on long lines for the privilege of digging into holes with their hands to fill jerricans with filthy, disease-ridden water.

Without electricity, access to basic education and health care becomes nearly impossible, but far more fundamentally, it also means you can't pump water from the ground. Electric light and refrigeration alone was never going to make a dent.

"I realized we were fixing the wrong problem," she says. "People were weak and unhealthy because the water was making them sick."[15]

Yaari had always assumed that Africa didn't have much water, but she soon found out it has plenty, in deep underground aquifers. It's directly beneath the feet of those who need it most, anywhere from sixty-five to 650 feet down, all across the continent.

"What people need is energy to pump it up," she says.[16]

In 2009, Yaari returned to Putti to carry out what would become the first of hundreds of Innovation: Africa aquifer projects, all of which follow the same basic playbook. Yaari hires a geologist to survey where and how deep to drill, then brings in a drilling company and local contractors to do the rest, which includes installing a pump and water tank and working with the community to decide where to locate the taps.

After the Putti project was complete, the villagers were able to pump twenty thousand liters of clean, safe water per day to use for cooking, drinking, and livestock. And in the aquifer projects that followed, Yaari decided to install an extra tank that would provide water to drip irrigation pipes, a technique developed in Israel in the mid-1960s that's far more effective than traditional flood irrigation and sprinkler systems, regardless of the plants or location. Drip irrigation uses just a third of the water and doubles the crop yield.

With electricity, clean water, and drip irrigation in place, the villages touched by Innovation: Africa began to thrive. Children were able to bathe and receive an education; adults were able to use the water to start businesses, earning income that paves the way toward healthier, happier, and financially independent lives. Many started by selling the extra fruits and vegetables they produced; others made bricks or launched bakeries.

"What amazed me was the entrepreneurial spirit in the village," recalls Yaari of what she discovered in Putti. "We saw water as the color blue. They saw it as green."[17]

Cloud Backup

To date, Innovation: Africa has completed over three hundred solar and water projects, transforming the lives of more than two million people in ten African countries, including Cameroon, Ethiopia, Malawi, Tanzania, Senegal, South Africa, the Democratic Republic of the Congo, and Uganda. And with each project, the organization has proven that it takes the "innovation" part seriously.

IA's chief engineer, Meir Yaacoby, created a custom-designed monitoring system that collects data from each solar array they install, which is uploaded to a server that can be accessed anywhere in the world. The technology allows Yaari's Israel-based team (and her donors) to stay up to date on the water output and energy consumption at every site. It also uses artificial intelligence and advanced algorithms to predict problems before they start. When there's an issue at any location, Yaari's team in Tel Aviv and local project managers receive an alert on their phones.

"It's off-grid, remote monitoring, so, at any point, we are able to know how much water we're pumping into every village," says Yaari. "If something breaks, meaning a pump hasn't pumped water in 24 hours, we are notified by the system."[18] The platform won the United Nations Innovation Award in 2013.

"IA's impact far surpasses much larger, less focused NGOs," says Gil Haskel, head of Israel's Agency for International Development Cooperation. "The secret to their success is the focus on specific technologies that transform lives."[19]

IA has also launched a new technology they call the Energy Box, a single container that includes everything needed to bring a school or medical center on line, including LED bulbs, batteries, and a human-machine interface (HMI), which serves as a dashboard that connects a person to the device, along with the remote management system.

IA has done all of this with a modest annual budget, all of which is raised through private donations. "One hundred percent of it goes to Africa – and to the Israeli companies whose technology and equipment are purchased," says Yaari. "Not a penny of it is spent on the running of the NGO. All our salaries and overhead are paid through foundations."[20]

"Yaaris model blows me away," says billionaire businessman and philanthropist Natie Kirsh, who has invested both his time and money in helping the organization scale. "It will get it done."[21]

IA's projects not only save lives, they also create jobs for locals who are trained by Israeli engineers to do the work in their own communities. And above all, the efforts teach local community leaders, project managers, and engineers how to be self-reliant and maintain the technologies on their own.

"Africa has among the highest levels of global inequality around the world," says Stephen Koseff, the former CEO of Investec, Africa's largest investment bank and a member of IA's board. "IA is making a meaningful dent at reducing this disparity and bettering the lives of millions."[22] David Arison, fellow board member and VP of global business relations at Miya, agrees. "In my experience, I have not encountered any other person alive who has made more of an impact on the quality of life in African villages."[23]

Yet Yaari still feels she has not done enough. "It is only a drop in the ocean," she says.[24] There are currently over 600 million people in Africa without energy, and 350 million people searching for water every single day. And the problem is only going to get bigger.

It took 200,000 years for the earth's human population to hit one billion at the turn of the nineteenth century – and just two hundred more for us to multiply to the 7.9 billion we number as of this writing. In just thirty years, that total will increase by two billion more, and half of them will be born in Africa – competing for the same food and water.[25] Yaari plans to reach at least a thousand villages by 2025. We're just going to need a lot more Yaaris.

"The solution is simple because the technology exists, and the impact is priceless," she says. "It just has to be done. The simplicity of the solution is what keeps me going."[26]

CHAPTER 4

Environment: Water World

God created the world, but the Dutch created the Netherlands.

– Dutch saying

Henk Ovink (left) with Peruvian politician and engineer Reynaldo Hilbck Guzmán (right) in a helicopter flying over the Piura River after the flooding (Embassy of the Netherlands, Lima, Peru)

New Amsterdam's Looming Water Crisis

Six weeks after Hurricane Sandy ravaged the New York region in 2012, Shaun Donovan visited the Netherlands to learn how the tiny country so successfully keeps the ocean at arm's length. Then secretary of housing and urban development, Donovan had been tapped by President Obama to lead a task force to rebuild in Sandy's wake, but also to explore radical new ways to address America's insufficient infrastructure in the context of climate change. His guide was Henk Ovink, the Netherlands' director general of spatial planning and water management, who drove Donovan around in a small minivan, opening his eyes to a Dutch water-management system that's the envy of the world. For the first time, Donavan understood what was possible when a society's engineering and culture actually merge for the greater good.

When Donavan's plane landed back in Washington, he saw an email from Ovink that was short and to the point: "I hope this isn't too forward, but could I come work with you?" Donovan replied before the plane reached the terminal. "You're just forward enough," he wrote. "When can you start?"[1]

By the end of this century, scientists predict that the ocean will rise by as much as four feet across the globe. In New York City, according to many climatologists, that could mean storm surges of twenty-four feet. Miami Beach will almost certainly be entirely under water. But far sooner than that – within the next five years – almost two-thirds of the world's population will already be living in coastal areas vulnerable to increased ecological problems, including harsher storms, flooded deltas in winter, parched deltas in summer, and salt water infiltration of underground aquifers.[2] Large parts of Vietnam will be underwater at high tide, with over twenty million people affected. Bangkok will face massive water threats, as will most of Shanghai, one of Asia's most important cities. India's financial capital of Mumbai is at imminent risk of being completely wiped out by water, along with Alexandria in Egypt and Basra in Iraq.[3] The list goes on.

"Time is running out," says Stefan Rahmstorf, professor of physics of the oceans at Potsdam University. "It is time to end fifty years of dithering and finally act decisively to avert a looming planetary catastrophe. Otherwise, rising sea levels will flood large coastal cities and destroy entire island states."[4]

Henk Ovink agrees. "We will feel the impact of climate change all over the world," warns the planet's leading flood expert, "most profoundly through water."[5]

In August 2013, Shaun Donovan and the Hurricane Sandy Rebuilding Task Force issued its report, which leaned heavily on the Dutch "resiliency" model and Ovink's policy recommendations. These included building dunes and undersea barriers, overhauling wastewater treatment plans and the weakest links in the electrical grid, constructing storm-surge bulwarks, and building purpose-designed affordable housing.

"Water, floods, and climate change have not been a policy issue in the US," observed Ovink. "Your vulnerability to massive rains, stronger hurricanes, and longer droughts make water *the* policy issue of today."[6]

Malcolm Bowman, head of the Storm Surge Research Group at the State University of New York at Stony Brook, frames the matter even more urgently. "My middle name is Noah," he says. "The Flood's coming, you'd better build an ark, get everybody aboard."[7]

Swamp People

The Dutch have been battling the sea for centuries. Two thousand years ago, the area's first inhabitants were drawn to its peat swamps. The moist forests have waterlogged soil that prevents dead leaves and wood from fully decomposing, resulting in abundantly rich clay. Roughly the size of Connecticut and Massachusetts, the Netherlands literally means "lower countries," with half of its flat topography at less than three feet above sea level – and 27 percent fully below it. More than 60 percent of the tiny country's population of seventeen million lives in this area. Dutch water authorities are mandated to ensure that the risk of death as a result of flooding never rises above one in 100,000. (By comparison, your chance of dying on an airplane is one in eleven million.) If a mega-storm ever breaks through, it could flood thousands of square miles and paralyze one of the most important economies in the world.

"There is nothing inherently distinct about the Netherlands," says Jos Dijkman, a retired member of the Deltares Water Institute. "What is special is that due to the setting of the country – one-third below sea level – there is a high societal priority for flood protection, and unlike most of the world, we abide by the very real law of 'thou shalt take into account sea-level rise.'"[8]

For the last decade, the Dutch government has worked with the Intergovernmental Panel on Climate Change, which won the Nobel Prize in 2007, to figure out what it will take to climate-proof their country for the next two hundred years.

The group predicts that the North Sea will rise by almost a foot and a half by 2050, between two and four and a half feet by 2100, and up to thirteen feet by 2200. The government takes no part in the debate on whether climate change is real or not. It simply accepts that water is rising, that extreme weather conditions cause massive flooding, and the possible effects of both. "It is an adaptive strategy because we are not certain what will happen in the next decades," says Jan Hendrik Dronkers, director general at the Dutch Ministry of Infrastructure and the Environment. "The secret of our new program is that if the sea level changes, we expect we will be able to deal with that."[9]

In 2007, the Dutch parliament put together the Delta Committee to implement that goal. Their plan calls for aggressive measures, including extending the coastline, building massive surge barriers, and fortifying levees. The commission estimates it will cost $1.5

billion annually for the next hundred years to defend the country and its people against high water.[10]

And perhaps more than any other country on the planet, Holland is fully prepared to make the investment. For the better part of the last century, it has been building and refining complex systems to protect its population from the regular havoc caused by the ocean and flooding, which once destroyed so many of its homes and crops. When it comes to stemming the tides of nature, the country already leads the world. And for the most part, it owes its position to the vision of one man: Cornelis Lely.

The Mother of All Dikes

Schoolchildren around the world are taught about the fictional Little Dutch Boy who saved his country by putting his finger in a leaky dike. The boy in the Mary Mapes Dodge novel *Hans Brinker, or the Silver Skates* stays up all night in the cold so the adults can make the necessary repairs in the morning. The Dutch built dikes for centuries, until it became clear to just about everyone that they couldn't go on simply plugging leaks. As Holland became an international powerhouse for goods, commodities, and shipping, they knew they needed bigger ideas and more effective, long-term solutions to keep the water at bay.

In 1913, Holland's minister of transport and water management, civil engineer Cornelis Lely, put before the Dutch cabinet a revolutionary plan to permanently close off the Zuiderzee Bay and protect it from floods with a massive hybrid structure known as a dike dam. The cost of building it would be roughly equal to the country's entire annual budget. But to help the project partially pay for itself, Lely's design called for the land inside the area of the dam, known as a polder, to be pumped dry to mine its clay and then be used for extensive agriculture.

For years, lawmakers weren't interested. But two major events eventually changed their minds. World War I made food scarce for millions of Dutch and Europeans, and vast amounts of farmland were needed to grow more. (Building the dike dam would eventually turn Holland into the second-largest exporter of food in the world.) But more importantly, a huge flood hit the Netherlands in 1916, breaking many of its dikes and destroying massive numbers of homes. Several Dutch counties went bankrupt trying to repair the damage, and government officials were reminded once again how vulnerable the country was to flooding.

Two years later, the Dutch parliament passed the Zuiderzee Act, which sought to protect Holland from the potentially calamitous effects of the North Sea, increase food supply through the development of new agricultural land, and convert the Zuiderzee into a freshwater lake to help manage water. Effectively closing off the entire bay, the Zuiderzee Works became one of the largest land reclamation and water drainage projects in history, with its main dam and enclosure dike, the Afsluitdijk, being the largest ever built until 2006. The American Society of Civil Engineers proclaimed the project to be among the seven wonders of the modern world.[11] "It is an extraordinary feat of Dutch engineering," says Bas Jonkman, professor of hydraulic engineering at Delft University of Technology. "Along with other impressive structures on Earth, like the Pyramids of Giza, [it] can be seen from space."[12]

Constructed from 1927 to 1932, the Afsluitdijk is so huge and ambitious in scope, it's almost impossible to imagine how humans were able to build it. Starting in four locations, tens of thousands of workers began by creating two giant islands in the middle of the sea, dropping twenty-three million cubic meters of sand and thirteen million cubic meters of glacial sediment via ship every day. Heavy stones were then dropped on the inland side and boulder clay placed on the ocean side. Brushwood mattresses, in turn secured by boulders and old concrete, kept the material in place. Both a dike and a dam, with water on both sides, the Afsluitdijk was raised twenty-five feet above sea level with sand and clay, over which grass was planted. A road three hundred feet wide and twenty miles long was then built on top, connecting Northwest and Northeast Holland. The dam also turned the southern sea into a lake, named the Ijsselmeer for the river that feeds into it. Sluices were built to let the water flow out from the lake into the ocean, converting the seawater into freshwater.

The next step in Lely's plan was to create polders, a word that comes from the Old Dutch for "dry land," which has been adopted by thirty-six languages around the world. Poldering is an engineering marvel whereby water is drained from a large body of water, uncovering fertile land below it. The first polders were created in Holland in the eleventh century – and by 1961, with more than four thousand polders, roughly half of the country's total land area was made up of ground reclaimed from the sea. They even have a saying for it: "God created the world, but the Dutch created the Netherlands."

According to Cornelis Lely's instructions, after portions of the Ijsselmeer were dammed off, engineers built ten miles of dikes around what was once the southern sea and pumped the water out using windmills. The best land was used to grow various grains such as rye, wheat, barley, and oats, while the less fertile land was forested.

The Afsluitdijk was finally put to the test in 1953, when one of the most severe storms in history hit the country. In the south, where the dikes were less sophisticated, 340,000 acres were flooded, forty-seven thousand buildings were damaged or destroyed, thirty thousand livestock were lost, and almost two thousand people were killed. In the area protected by the Afsluitdijk in the north, there was no damage. On a single night, the massive investment had paid off in lives and money saved. The area was no longer vulnerable to flooding. Lely died in 1929 having never witnessed the success of his master plan.

"Water is a double-edged sword," says architectural historian Marinke Steenhuis, a member of the Afsluitdijk quality team. "On the one hand, it is the black swan that can wreak havoc, displace people, and decimate cities. On the other, it is the basis of life and prosperity to every living creature on Earth."[13]

For the next thirty years, the Dutch continued to invest huge amounts of time, effort, and money into bolstering their defenses. More storm-warning systems, more dams, dikes, locks, sluices, and levees, and more storm surge barriers. In the Rotterdam port, the Oosterschelde and Maeslant barriers each have giant arms that cost $500 million to build and are nearly the size of the Eiffel Tower; these can close the mouth of the city's waterway in the event of a major storm. But building bigger, better, and higher, the Dutch eventually came to understand, could only get them so far.

Let the River Run

In the mid-1990s, Holland's Rhine, Meuse, and Scheldt Rivers swelled to unprecedented levels, forcing the evacuation of 250,000 people and a million farm animals – along with a dramatic change in the country's approach. "We were shocked," says Dutch water official Martin Hoenderkamp. "We thought we were prepared for everything. We thought evacuation was for other countries."[14]

By 2006, the Netherlands approved a multi-billion-dollar plan called Room for the River, which aimed to restore the natural floodplains of Holland's rivers and protect areas most likely to flood. "The Dutch had strayed too far away from the environment and over-engineered nature," says David Waggonner, a leading architect and founder of the Dutch Dialogues, a collaborative effort bringing together US and Dutch water experts. "They knew they needed a more sustainable solution based on landscape."[15] The government course-corrected, ultimately concluding that they could preempt future catastrophe by working with the rivers rather than trying to control them. "With the sea barriers, we sealed off our front door," one official said. "Alas, we've been caught off guard through the back entrance."[16]

With more frequent and intense rainfall and less area to actually work with, the country's rivers were discharging too much water, putting millions of lives at risk. Dutch officials put together a revolutionary proposal to expand the space for rivers at high water capacity. "It's no longer a war on water," says Tracy Metz, Holland-based journalist and author of *Sweet and Salt: Water and the Dutch*. "Holland's engineers and policymakers undertook the unprecedented, mind-bending exercise of lowering the dykes instead of raising them."[17]

In the end, the program implemented thirty-four measures that also included deepening floodplains, widening rivers, and moving two hundred families to drier areas. Work began in 2007 in thirty-nine separate locations and lasted for twelve years. Plans called for the demolition of farms and moving citizens out of harm's way. "The farmers were angry at first. We talked to them literally around their kitchen tables," says Hans Brouer, a senior rivers expert who worked on the project. "Half of them said they wanted to retire or try their luck in Australia. The other half wanted to build new farms for the future."[18] Ultimately, the government bought the land for market value.

As Henk Ovink explained to me, it's projects like Room for the River and the new Delta Committee that best encapsulate the success of the Dutch approach. The idea is that water mitigation is about much more than just engineering. It involves good governance, innovation, openness to new ideas and ways of thinking, and realizing that when the common good is in danger, individuals must rise to the occasion and put their interests aside.

Dealing with water intelligently isn't a luxury, but a matter of survival. And as the threat of climate-driven water catastrophe looms, the world will need to follow Holland's lead. Or hope for an email from Henk Ovink offering to help.

In March 2019, New York mayor Bill de Blasio announced a ten-billion-dollar climate-focused infrastructure project that implements many of Ovink's recommendations from the Hurricane Sandy Rebuilding Task Force. They include elevating parks, constructing a five-mile seawall around Staten Island and sand dunes around the Rockaways, and installing removable flood barriers around Lower Manhattan that could be used in the face of massive storms.

"It's a choice in the end, and our best opportunity," says Ovink, who is now a special envoy for the Kingdom of the Netherlands. "Nothing is too big to confront. We can close our eyes and do nothing and let the worst happen, see more assets and people lost, or think about that future as an opportunity. I choose the latter."[19]

CHAPTER 5

Hygiene: There Will Be Blood

Don't wait for a girl to become a woman to empower them. Empower a girl's life by giving them sanitary pads. With pads, we give them wings.

– Arunachalam Muruganantham,
aka the Pad Man, personal correspondence

Arunachalam Muruganantham (http://bharathbalasubramanian.com/, CC BY-SA 4.0)

Rag Time

Arunachalam Muruganantham was on the couch when he noticed his wife hiding something behind her back. On confronting her, he discovered it was a bloody rag. The twenty-nine-year-old Muruganantham had seen these rags in the family outhouse when he was growing up in a small village in southern India, but he had never known what they were for. His new wife Shanthi revealed that the rags were for menstrual blood, and that many women used rags, leaves, sand, sawdust, mud, or even ash to absorb the discharge.

Muruganantham asked his wife why she didn't just use pads that could be thrown out after each use. Shanthi explained that while she understood all the benefits of sanitary pads, using them would mean "we will have to cut down the family's milk budget."[1]

Muruganantham decided to go to the local pharmacy, where the cashier seemed horrified by the purchase. "He gave it to me as if it were contraband,"[2] recalls Muruganantham with a laugh. "I don't know why – I didn't ask for a condom."[3] It was 1998, and the 20 rupees (about 50 cents) the napkins cost him could have fed the family for several days.

When he inspected one of the pads at home, Muruganantham was shocked to discover that it was filled with nothing more than compressed cotton. Why were raw materials of such minimal value being sold for such an exorbitant amount of money? The son of a handloom weaver had a thought. He could make inexpensive pads of his own.[4]

Addicted to Sanitary Pads

Muruganantham knew his way around fabrics. The next day, he went looking for the softest cotton he could find at one of the textile mills in his village. But when he sewed it together for his wife to try out during her next menstrual cycle, she wasn't impressed. She told him his pad was no good and that she would stick to the rags, thank you very much.[5]

Muruganantham was undeterred. For the first time, he understood that women who couldn't afford sanitary pads were far more susceptible to disease and infection, and he was determined to change the situation. He started developing various pads based on different materials and gave them to his wife and sisters to test, but the cultural taboos against discussing anything connected to the subject were so strong that they'd never offer much feedback. Muruganantham briefly attempted to take on the testing himself, at one point tying a soccer-ball "uterus" filled with goat's blood to his hip, with a tube that ran down to his underwear. During the course of his day, walking or bicycling around town, he'd periodically press the ball to simulate squirting blood and discovered very quickly that his pads didn't work very well.[6] Muruganantham's undergarments were constantly wet, and the experience gave him a newfound respect for the strength of women. "Doing this for a week will make you sick and give you fever," he says. "Having wet genitals is really terrible."[7]

Muruganantham eventually got permission to expand his pool of test subjects with some female students at a local medical college. But because of deeply baked cultural norms, these women proved equally embarrassed when it came time to share their opinions. And then the rumors began to spread. "Under any circumstance, approaching female students for their used sanitary pads would be extraordinary," says Indian entrepreneur Vishal Gondal. "But in conservative rural India, this is simply unprecedented."[8]

While his wife Shanthi knew about her husband's ambitions to create a more cost-effective pad, Muruganantham never told her about his new test subjects. When she confronted him about it, he lied and told her he had gone to the local college to repair their main gate, suspecting she wouldn't understand. Shanthi decided to go back to her parents for a few days, then after a few weeks decided not to come home. "My wife thought I was a psychopath and decided to leave me after she saw me looking at sanitary pads," says Muruganantham. "Nobody understood me. People started to avoid me entirely."[9]

Soon after, a divorce noticed arrived. "My wife tried everything she could to make me stop doing research and bring me back to my senses," recalls Muruganantham. "But I was obsessed."[10]

Like Magic

Muruganantham was determined to figure out the exact material used in standard pads, which he surmised was some kind of unusual cotton, and he spent two and a half years calling American and European companies, hoping to get answers.[11] Calling from a local kiosk, often after midnight India time, he spent a small fortune on phone charges. He'd tell anyone who'd listen that he owned a textile mill and was considering entering the sanitary products business, and eventually he found a helpful female representative whom he convinced to send him a sample of the exact fabric they were using.[12]

When he opened the package a few weeks later, he was totally flabbergasted. He expected to find cotton, but instead he found something more like cardboard. Muruganantham had no idea what to do next. Eventually his dog scratched the material – and almost by happenstance, Murugantham discovered the fiber inside. Muruganantham determined that the cardboard was compacted cellulose manufactured from pine bark.[13]

It was immediately clear that one of his biggest challenges would be keeping costs in check – multinational companies mass produce sanitary pads with machines that cost millions of dollars. Muruganantham set about designing a low-cost machine that would allow people to make sanitary pads from the comfort of their homes or in small industrial centers.

When his wife left him, Muruganantham was forced out of his ancestral village after his neighbors decided that his quest to create a pad was keeping the protective goddess away from their community. He rented a room in downtown Coimbatore, sharing an apartment with five other tenants, and turned his living quarters into a laboratory workshop. His neighbors came to believe his spirit had been taken over by demons and that he had turned into a pervert. "We thought he had become a vampire who drank women's blood," recalls S. R. Ramesh, a fellow Coimbatorian and handloom weaver. "Many of the villagers wanted to chain him to a tamarind tree and hang him upside down."[14] Others thought he was doing "abnormal things," says Sashi Anand, vice president of the Coimbatore Women's Welfare Organization. "No one knew he was doing serious research. People thought he was doing black magic."[15]

When he wasn't working on his innovation, Muruganantham took roofing jobs to get by. But he was determined, and often staying up all night, Murugantham eventually created a pair of devices capable of producing sanitary pads at a much lower cost. The first, originally made of wood and iron, looks like a large meat grinder and is able to break up compacted cellulose. "It was beyond amazing to see the board transform itself into fluff," recalls Muruganantham, "I was so happy. I felt like a magician."[16] The second machine, which resembles a drill press, compacts the pulp into pads.

The whole process takes approximately two minutes and can be mastered in less than an hour of training. First the cellulose is ground into fluff, which is then placed into a metal compartment and pressed by plates into a hard, half-inch layer. The material is wrapped into a soft, gauzy cloth and sealed with a heating machine. At around $1,000, Muruganantham's machines are vastly less expensive than those used by multinational companies, which can run in excess of $1 million.[17]

With his new pad in hand, Muruganantham went back to the medical college in late 2004 to test his samples. "When I use your pad," said one of the volunteers, "I don't even think about my period."[18] After eight and a half years, Muruganantham knew he had finally done it.

His pad soon earned recognition and press coverage. In 2006, Muruganantham visited India's Institute of Technology in Madras to show them his invention, and they were so excited that they registered his pad for the National Innovation Foundation's Grassroots Technological Innovations Award.[19] Against 943 entries, he won first place for the Best Innovation for the Betterment of Society, an award granted personally by the president of India, Pratibha Patil. Suddenly Muruganantham was in the national spotlight.

Following a television interview, his ex-wife Shanthi phoned to express her congratulations.

"Do you remember my voice?" Muruganantham asked. "Why couldn't you understand what I was doing?"[20]

"I left you so you could follow your dreams,"[21] Shanthi replied.

He asked her to come back. Shanthi agreed, and they are still together to this day.

100 Percent Napkin-Using Countries

There are approximately 350 million women of menstruating age in India,[22] and when Muruganantham created his machine, just 12 percent of them were using sanitary pads. Today, that figure has climbed to about 57 percent, in large part because of Muruganantham's pad.[23]

Still, more than 20 percent of Indian girls drop out of school when they hit puberty because their menstruation cycles force them to miss so many days each month.[24] India's Ministry of Health estimates that because of unclean menstruation-management practices, 70 percent of women are at risk of severe infection. And in urban India, somewhere between 43 and 88 percent of women still use reusable cloth for their period.[25]

When Muruganantham finally figured out how to make his low-cost sanitary pad, he didn't set out to make money. "This has to be introduced to the world," he told Babu Yogeswaran, a close friend and filmmaker. "I must create a self-sustaining social revolution for those who need it most."[26] Muruganantham wanted to give the world's poorest communities the ability to make and distribute pads for themselves.

For about eighteen months starting in 2006, he built about five thousand machines and made thousands of trips to small villages across India,[27] in many of the poorest and least developed states, such as Bihar, Madhya Pradesh, Rajasthan, and Uttar Pradesh, sometimes called by the acronym BIMARU, which means "sick states." In places where women often walk for miles just to get water and where menstruation can cause further financial hardship, he sought out women who might be interested in making their own pads. In many of these villages, men resisted because of their traditional patriarchal traditions. But after months of Muruganantham's repeated visits and pleading, fathers and husbands eventually relented and allowed the women in their families to use his machine.

Over the next few years, female-run cooperatives and nonprofits around the country began adopting Muruganantham's model en masse. Women understood that using rags wasn't good for them, and once they had a low-cost way of getting pads, they were all too willing to use them. Many of these groups buy Muruganantham's machine, train over the course of one day, and immediately start producing about a thousand napkins daily.[28] And if they need financing, they often secure bank loans that can be paid off with the money earned from their sales.[29]

Murugantham owns the patent rights to his machine, but he gives away the right to produce pads with them free, so those who use the machines can profit off them – and enfranchise their ultimate customers. While Muruganantham's workshop also produces pads, the women around the country who purchase his machine to start pad-making centers produce their own versions of the pad, packaged under various labels. There are now more than twenty-one hundred local brands making his pads.[30]

Traditional pads costs around eight rupees – those made on Muruganthams machines sell for as little as three. Each machine provides jobs for ten to fifteen women, who are able to produce pads for thousands.[31] There are currently machines in over fifteen hundred Indian villages in all twenty-nine Indian states,[32] and the model has spread to at least thirty other countries around the world, including Bangladesh, Kenya, Mauritius, Nigeria, Nepal, Rwanda, the Philippines, and Zimbabwe, to name just a few.[33] His success will mean the transformation of an untold number of lives, and by extension of society at large.

Muruganantham has plans to expand the production of his machines around the world.[34] And Muruganantham turned down big multinational companies that approached him for help getting their expensive sanitary pads into local markets. He doesn't want to sell his innovation to big companies, preferring instead to empower women to produce sanitary pads themselves. "Muruganantham has succeeded where multinationals have failed," says Venkatesh Mahadevan, chief information officer of Dubai Investments. "He brought affordable sanitary napkins to those who need it most. He will go down in history as a true visionary and innovator."[35]

Muruganantham hopes to increase pad use among Indian women to 100 percent, while also giving those in the lowest socioeconomic strata an opportunity to earn a living selling pads produced at local centers – and facilitating a more effective feminine hygiene regime in communities that desperately need it. His priority is for sanitary pads to eventually be used everywhere around the world.

"The Pad Man is a god-sent angel," says Dr. Sujirtha Beena, an alternative medicine doctor in Coimbatore. "Above all, Muruganantham has succeeded in changing the culture of India's rural society. Young girls and women can now openly talk about and learn how to practice clean hygiene."[36]

His work has also encouraged a new generation of innovators. "Muruganantham's life story inspires millions of Indians like me," says Dr. Rengaraj Venkatesh, chief medical officer of Aravind Eye Care System, the largest primary eye care facilities in the world, "to reverse-engineer solutions to significant problems facing the entire developing world, and not just those facing the field of gynecological hygiene."[37]

In 2014, Muruganantham was named one of *Time*'s 100 Most Influential People in the World, and in 2016, he won the Padma Shri award, one of the highest civilian honors granted by the government of India. *Fortune* magazine listed him among the top fifty greatest leaders of 2019.

"On the one hand, I feel like a celebrity," says Muruganatham. "On the other, so many women are still using dry leaves, ashes, and sawdust powder during menstruation." He is keenly aware of how much more work is left to be done. "Many countries are trying to send rockets to the moon and Mars without empowering women," he says. "I love the idea of reaching the stars, but first let's make sure everyone on Earth who needs sanitary pads has them."[38]

CHAPTER 6

Medicine: Playing God – Genetic Editing

The unthinkable has become conveivable. We're on the cusp of a new era in human history.

– Nobel Prize laureate David Baltimore, personal correspondence

China's Dr. Frankenstein

It had been a little over a year since Chinese biomedical researcher He Jiankui made the announcement that stunned the world. It was

December 2019, and the first scientist in history to successfully produce a genetically edited baby – three of them, in all – now found himself standing in front of a judge in Shenzhen, China, accused of conducting illegal medical practices and deliberately violating Chinese national regulations on biomedical research and medical ethics.

The verdict from the Nanshan District People's Court was swift. Jiankui, known to his friends as JK, was guilty. He was sentenced to three years in prison and ordered to pay a 3 million RMB fine (roughly USD $430,000).[1] The judge also found that JK had forged ethical review documents and misled doctors into unknowingly implanting gene-edited embryos in two women. The ruling was harsh, but in one sense it was just a formality. JK had already been tried and convicted by his peers in the global scientific community.

In November 2018, when Jiankui first broke the news about the birth of Lulu and Nana, twin girls whose embryonic genes he had modified in secret using a new technology known as CRISPR (an acronym standing for clustered regularly interspaced short palindromic repeats), the initial astonishment quickly curdled into a collective gasp. Within a few weeks, as the potential ramifications of what Jiankui had done came into focus, he was fired from his job at Shenzhen's Southern University of Science and Technology. In short order, the experiment was castigated in the BBC as "monstrous," and academic journals from the *Lancet* to *Nature* piled on with scolding headlines like "Do Not Edit the Human Germ Line."[2] Revered bodies such as the National Institutes of Health condemned the work unequivocally, and hundreds of JK's fellow Chinese scientists issued a joint statement in resolute opposition, calling the study "crazy" and "a huge blow to the global reputation and development of Chinese science."[3] But the harshest rebukes came from renowned researchers in his own field, such as stem-cell biologist Kathy Niakan of the Francis Crick Institute, who denounced JK's endeavor as "highly irresponsible, unethical and dangerous," echoing the sentiments of so many of her colleagues.[4]

JK's crime was set in motion in June 2017, when he recruited a Chinese couple pseudonymously named Mark and Grace, who, along with six other couples he signed up, met the requirements of his experiment: a father infected with HIV attempting to have a baby with a healthy mother. JK offered the couples in vitro fertilization and gene editing so that their embryos – and offspring – would have innate resistance to HIV. The idea was that the technique could ultimately be used to reduce the high HIV/AIDS rate in places like Africa.

The hopeful parents all had similar fertility problems and signed informed consent forms. JK and his colleagues carried out the experiment in secret, taking the sperm and eggs from each pair and altering the genomes of the embryos using CRISPR-Cas9, a technology that enables geneticists to edit parts of a human genome by removing, adding, or altering sections of a cell's DNA sequence. In this case, however, unlike all the previous research done with CRISPR-Cas9, JK was editing the human germline – meaning whatever changes he made would be passed on to his subject's children, and their children's children, potentially forever altering the DNA of our species.

It was a red line no CRISPR scientist had yet dared to cross, and JK's gambit was deemed particularly premature and unjustified because of the exposure to risk in exchange for little, if any, upside. Critics of his editing method argued that Lulu and Nana, along with a third gene-edited baby later born to another set of recruits, could still be vulnerable to HIV with none of the purported benefits – and an even greater likelihood to develop cancer and heart disease, among other untold problems. Experts agree that there are other safer, more effective ways to prevent HIV in pregnancy.

What JK had done was let CRISPR out of the box – without

asking permission. And the world was left wondering if that genie could ever be contained again.

Opening the Genetic Floodgates

Humans have been modifying organisms for over thirty thousand years. While our ancient ancestors didn't have scientific laboratories, they manipulated DNA through what Darwin famously called selective breeding, or artificial selection. The process of choosing organisms with the most desired traits and mating them repeatedly over many generations yields significant genetic changes.

Dogs are generally considered the first organisms to undergo artificial selection.[5] As hunter-gatherers, early humans were a consistent source of food for East Asian wild wolves, who are thought to have first joined man as scavengers. Over time, they were domesticated, bred to increase docility and obedience, and eventually to manipulate specific traits such as size, hair length, and body shape. Today, they barely resemble their ancient wolf ancestors.

Humans have also been modifying food for nearly ten thousand years. Archeological sites in southwest Asia (in what is today modern Turkey) provide evidence that as far back as 7800 BCE, people were engaged in the selective breeding of corn, which before humans got to it had been a wild grass with small ears and few kernels.[6]

In 1973, scientists Herbert Boyer and Stanley Cohen produced the first modern genetically engineered organism by removing a gene that encodes antibiotic resistance from one bacteria and inserting it into another, conferring the same antibiotic fortitude upon the recipient. A year later, Rudolph Jaenisch and Beatrice Mintz became the first to introduce foreign DNA into mouse embryos. By 1974, scientists, governments, and global citizens were so concerned about the new technology that a universal moratorium was imposed for one year so that society could consider the potential ramifications. But in 1975, attendees of the Asilomar Conference established guidelines for genetic engineering that defined safety in order to mitigate risk, and with those standards in place, a new era of modern genetic research and development was ushered in.

The race to profit from this new industry was in full swing by 1980, when the US Supreme Court granted legal ownership rights over genetically modified organisms (GMOs) to their inventors. General Electric filed a patent on a bacteria used for spill mitigation that breaks down crude oil. The US Food and Drug Administration approved the first bacteria to synthesize human insulin, known as the drug Humulin, two years later. And in 1987, scientists successfully modified the DNA sequence of a Calgene Flavr Savr tomato to increase firmness and extend its shelf life.

The floodgates for genetically modified organisms have been open ever since, and over time, scientists have found ways to develop crops with increased nutritional value and bigger yields that are also drought resistant and easier and more efficient to cultivate. Critics have been challenging the safety of these organisms for decades, but the scientific community has by and large concluded that genetically engineered foods are just as safe as traditionally selected crops.

It was only a matter of time until humans began to gain control over our own DNA. But up until the last decade, gene editing was prohibitively labor intensive and inaccurate. Researchers would primarily load genes onto viruses that would take them to their target cells. "For a long time, people didn't really have a clue what these repeated DNA sequences in bacteria genomes really did,"[7] recalls Rodolphe Barrangou, a scientist who was working for the Danish food ingredient manufacturer Danisco in 2007 when he made a key discovery, thanks to your morning yogurt.

Barrangou was sequencing genomes from bacteria used as starter cultures and noticed that they contained DNA markers that played a role in infecting and destroying dairy cultures. These bacteria had repeating DNA sequences (CRISPRS) that confer a form of immunity to bacteria from invading viruses and ultimately keep dairy products from spoiling too quickly.[8] In nature, bacteria build up immunity from invaders by storing a small bit of a virus's DNA. And like a "most wanted" poster, the bacteria then relay that bad code to an extraordinary protein called Cas9, whose job it is to police the intracellular neighborhood for invading viruses. When Cas9 sees an exact match to the information on the poster, it cuts it out and kills it.

The discovery was extremely valuable to Danisco for manufacturing dairy cultures resistant to viruses. But once fully understood, the true value – to the future of the human race – was immeasurable.

Bigger Than Breakfast

In 2012, molecular biologists Jennifer Doudna and Emmanuelle Charpentier discovered that Streptococcus pyogenes, the bacteria that causes strep throat, had a kind of GPS that directed its Cas9 enzymes to a specific strand of DNA and sliced it in two. Cas9 could be sent anywhere in the genome in order to snip out mutated genes. A year later, in what scientists call the Great CRISPR Quake, Doudna and Charpentier announced that they had figured out a way to use the method on humans to target and edit human DNA.[9]

Published in *Nature*, their research, as Berkeley's Fyodor Urnov puts it, instantly became a "deservedly immortal science paper."[10] CRISPR-Cas9, or clustered regularly interspaced short palindromic repeats and the associated protein 9, "fundamentally allows us to change human evolution if we want to," Doudna explained. "It's that profound."[11]

Like a book that describes us, our genome is made up of three billion base pairs of DNA arranged in a variety of sequences, and their particular order dictates all human traits. Handed down from generation to generation, every cell in the body carries a copy of this code. By 2003, thanks to the Human Genome Project, computers had sequenced the entire genome, and for the first time were also able to identify the genetic mutations for diseases with no cure. But until CRISPR-Cas9, there wasn't a whole lot scientists could actually do with that information with any kind of efficiency.

"It's like having one typo in a book containing 3 billion letters," says Matthew Porteus, professor of pediatrics at Stanford and cofounder of CRISPR Therapeutics.[12] "[Before CRISPR,] we spent six years trying to repair one mutation." Now we have the genetic equivalent of Microsoft Word, allowing scientists to target, edit, and ultimately replace DNA as easily as cutting and pasting. CRISPR molecules have a pair of molecular scissors that can unzip DNA and snip off sections of the genome, and they can be programmed to find DNA that is either mutated or disease-prone. "It's often described as a kind of Swiss Army knife," says Emmanuelle Charpentier. "It's really a universal tool."[13]

CRISPR systems, according to science journalist Megan Molteniare, are "the fastest, easiest, and cheapest methods scientists have ever had to manipulate the code of life in any organism on Earth, humans included. It is, simply, the first technology truly capable of changing the fundamental chemistry of who we are."[14]

What many now call the eighth wonder of the world will impact us deeply as a species, in both the near and far term. And over the course of the last decade, there has been a flood of research and development to try to push forward CRISPR's potential.

"With CRISPR, we can do genetic experiments that would have been unimaginable just a few years ago, not just on inherited disorders but also on genes that contribute to acquired diseases, including AIDS, cancer and heart diseases," says Mark Mercola of the Stanford Cardiovascular Institute. "It's no exaggeration to say that CRISPR has been revolutionary."[15]

CRISPR could lead to previously unimaginable medical treatments and is already being used to cut out HIV and eliminate Huntington's disease from the brain. According to Dr. Daniel Kraft, founder and chair of Exponential Medicine, at a biomedicine and tech conference, ten years from now we'll have CRISPR-based xenotransplants from humanized pigs. "If you need a heart, liver, or a kidney, you can get one from a pig," says Kraft. "Might not be kosher, but you'll take it."[16]

Today there are thousands of scientists working with CRISPR on every conceivable organism, including dogs, butterflies, horses, wheat, and corn. And almost every major industry is investing tremendous amounts of money into exploring the technology's promise, putting us on the brink of what has already become a multi-billion-dollar DNA hacking industry. Pharma is researching all kinds of breakthrough medical applications, from cures for bacterial infections to cancer. Agriculture companies are trying to create crops capable of reversing climate change. Energy corporations are experimenting with algae that turn into biofuel.

Almost every week, the media churns out headlines announcing that CRISPR-enhanced products are going to help achieve humanity's greatest aspirations and horrors: everything from the mundane, like the ability to grow hair on command, to the miraculous, like an end to aging and pain – or to children programmed with bigger muscles and super intelligence who only need to sleep four hours a night. CRISPR not only has the potential to completely eliminate microbes that cause disease – and feed the world with sustainable and healthy new foods – it may even be used to resurrect extinct species. Yes, scientists are already working on bringing the wooly mammoth back.

"What sounds like science fiction today," says Rodolphe Barrangou, the first scientist to provide experimental proof of CRISPR, "will just be science in the next ten years."[17]

Working the Bugs Out

Almost immediately after the discovery of CRISPR, a debate began on whether the DNA sequencing of mosquitoes should be changed to render them incapable of harboring the parasites that transmit deadly diseases like malaria, ZIKA, and yellow fever. The mosquito is mankind's hunter king, killing about two million people annually – more than any other species does, including humans themselves. (By way of comparison, snakes kill fifty thousand of us annually, and sharks just ten.)

The female mosquito uses two serrated mandible cutting blades to slice open human skin and insert her proboscis, which acts as a syringe to extract human blood, which she needs to grow her eggs. Her saliva contains an anti-coagulant that shortens the amount of time she needs to do her job, decreasing the chances humans will detect her deadly act. And if she's carrying the malaria parasite, it gets injected into the bloodstream, too, where it spawns and attaches to our red blood cells, leading to chills, extreme fever, vomiting, yellow skin, spleen and liver enlargement, and often death.

Most scientists believe that the creature's only purpose is to control the human population. "Contrary to popular belief, the mosquito does not even serve as an indispensable food source for any other animal," explains Timothy Winegard, author of *The Mosquito: A Human History of our Deadliest Predator.* "She has no purpose other than to propagate her species, and perhaps to kill humans."[18]

In order to fight this assassin, humans currently spend in excess of $11 billion per year on shields, sprays, and other forms of protection. In 2016, the Bill and Melinda Gates Foundation donated $75 million to facilitate CRISPR mosquito research. Since then, one of the options being vigorously pursued by researchers includes making genetic edits that would wipe out only those mosquito species that carry parasites (roughly one hundred of the thirty-five hundred known), often referred to as an "extinction drive." The approach would mean eliminating through sterilization species such as the *Aedes,* a transmitter of diseases like yellow fever, Zika, West Nile, chikungunya, and Mayaro dengue, which kill over 700,000 people annually and account for 17 percent of all infectious diseases globally. "Countries seriously affected by dengue and Zika have shown real interest in testing this technology," says World Health Organization scientist Florence Fouque, who also points to the great potential for suppressing mosquitoes that are developing resistance to insecticides.[19]

Alternatively, researchers have been pursuing what's known as a "gene drive" approach, aimed at altering the DNA of mosquito species that transmit disease to prevent them from doing so. Using CRISPR, scientists at the Universities of California in San Diego and Irvine have engineered transgenic *Anopheles stephensi* mosquitoes that carry an anti-malaria parasite effector gene. "We know the gene works," says Anthony James, a professor of molecular biology, genetics, and biochemistry at UC Irvine and one of the first scientists to carry out a mosquito gene drive. "We now need the political will to put this disease into the history books."[20]

Mosquito-borne diseases, scientists predict, are just the beginning, as CRISPR's real-world potential to eliminate or alter other disease-carrying insects appears limitless – and imminent. Notable examples currently being considered include sand flies that transmit leishmaniasis, ticks that spread Lyme disease, and rats that decimate native bird and turtle populations in the Galapagos, among many others.

"This notion of permanently altering the genetics of an entire species – it goes against everything I was trained to think," says Todd Kuiken, a researcher at the Genetic Engineering and Society Center at North Carolina State University who served on the United Nations committee for gene drives. "What's hard to accept is that, at this point, it might end up being our best option."[21]

Beyond Hitler

For all its mind-blowing promise, CRISPR has put human beings on the precipice of playing god, and it's far from clear what this mighty power will unleash, either in the halls of science or writ large across nature itself.

"At a very instinctive level, there's a sense that these are things humans are not supposed to be doing," says Alta Charo, a professor of law and bioethics at the University of Wisconsin, Madison. "You can make that new animal, or you can wipe out that whole species. And if doing that turns out to have been a bad idea, it means you're going to have to deal with the consequences."[22]

The concern is nearly universal in academic circles. "When it comes to experiments on animals, plants and microbes, two things worry me," says Stanford bioethicist Hank Greely. "One is the intentional misuse of CRISPR. The other is that people with good intentions will inadvertently cause harm."[23]

In the near term, even in cases where the reward would seem to far outweigh the risk – like driving the extinction of mosquito species that have killed hundreds of millions of people – the unanswered questions remain daunting. What other species would step in and take their place? How would it impact the ecosystem and the equilibrium created by nature over millions of years? Would it provide carte blanche in other areas of science to engage in genetic engineering, leading us down that slippery slope to designer humans? And, perhaps most profoundly, should we even be doing this at all?

He Jiankui took it upon himself to answer the most difficult question for the rest of us – and the ripple effects of that decision will be felt for years to come.

Upon learning of Jiankui's experiment in 2018, Jennifer Doudna and the rest of the CRISPR community were outraged. "[I am] horrified, honestly," she said. "And disgusted."[24] It was the realization of Doudna and Charpentier's very worst fears, which they had outlined at the first International Summit on Human Gene Editing just three years earlier: that someone in the community would go rogue and conduct CRISPR experiments on human babies.

In a statement, Emmanuelle Charpentier and specialists from seven countries called for a "moratorium on all clinical uses of human germline editing," being careful to draw the distinction between JK's research, which involves gene edits that are both permanent and passed on to future generations, and genome editing that is limited to somatic cells, which only affect certain tissues and are not passed on. "That is," they clarified, "changing heritable DNA (in sperm, eggs or embryos) to make genetically modified children."[25]

Jiankui's experiment has now forced governments and scientists to grapple with how to regulate this powerful tool. And as with all new technologies, finding a global consensus on enforcement is unlikely, which means JK's CRISPR babies were the first, but not the last.

"Germline DNA is a public trust, just like freedom, equality, and ideas," says Dr. Tshaka Cunningham, a former scientific program manager at the US Department of Veterans Affairs. "It's like the air we breathe. And we should not alter it unless we are sure what the consequences will be. The train has left the station. We are going to have a take a side."[26]

The World Health Organization has called on all researchers to cease work on genome editing "until its implications have been properly considered" and created a global registry to track anyone conducting such research.[27] The United States, China, and other international organizations have also taken action.

American enforcement authorities are investigating Rice University's Michael Deem, a scientist suspected of knowledge of Jiankui's baby project. It also appears that two US-based companies were involved in supplying JK with some of the technology he needed. Massachusetts-based Thermo Fisher Scientific is alleged to have given JK Cas9, and the Bay Area's Synthego (whose cofounders worked as rocket engineers at SpaceX) is said to have provided the synthetic guide RNA – in other words, one company provided the tools and the other the manual.

In May 2019, China began drafting regulations that would hold anyone manipulating the human genome through gene-editing techniques like CRISPR responsible for any related adverse consequences. By then, the government had already begun the investigation that would lead to Jiankui's conviction, calling his work a "shocking and unacceptable" breach of ethics and a brazen violation of Chinese law.[28]

But in many ways, the Chinese government is speaking out of both sides of its mouth. In a potential harbinger of nefarious outcomes ahead, China is turning the other cheek as its scientists genetically engineer a wide variety of creatures, including dogs to have more muscle and humans to become biologically enhanced super soldiers.[29]

CRISPR, Russian president Vladimir Putin ominously presaged, "may be more terrifying than a nuclear bomb."[30]

The primary fear is that JK's research put a recipe out into the world that can't be unwritten, a first domino in what ultimately leads to a dystopian future of eugenics. Originally developed by English scientist Francis Galton in the late nineteenth century, the term, which outlines a method to increase certain desired heritable characteristics in an effort to improve the human race, fell into disfavor after being adopted by the Nazis, who used it as a pretext for monstrous ends.

Even Jennifer Doudna, the godmother of the CRISPR movement, which won her and Charpentier a Nobel Prize in 2020, is kept up at night by the thought. Over the years, she's had a series of disturbing dreams in which she walks into a room to find the silhouette of a man sitting in a chair with his back to her. As he turns around, she realizes with horror that it's Adolf Hitler, who leans over and whispers, "So, tell me about how Cas9 works." Doudna says she wakes up shaking and thinks, "Oh, my gosh, what have I done?"[31]

Putting aside the worst-case scenario, there is the very real concern that genetic editing could lead to tremendous societal inequality – exponentially greater than anything we have faced previously. The strong possibility exists that the rich will become genetically superior by paying to live longer, healthier lives. "Once you start creating a society in which rich people's children get biological advantages over other children, basic notions of human equality go out the window," says David King, founder of Human Genetics Alert, an independent watchdog group that opposes certain outcomes of genetic engineering. "Instead, what you get is social inequality written into DNA."[32]

Still, for all those sounding the alarm, there are just as many who are unapologetically bullish about the idea of improving the human condition, hypothetical pitfalls be damned.

Stephen Hsu, founder of Genomic Prediction, a company that develops technology for advanced genetic testing, insists that equating CRISPR and genetic testing with Nazism is "not just stupid but actually insane." Hsu believes that fifty years from now, people will look at the way we currently conceive children as old-fashioned. "Sex is for recreation," says Hsu, "and science is for procreation."[33] He is pushing for a general population that is smarter, healthier, lives longer, and has a decreased number of people with genetic disorders such as Down syndrome.

Many scientists are calling for a global debate on whether we should go in this direction, essentially turning reproduction into production – and our babies into consumer products.

"My grandchildren will be embryo-screened, germline-edited," says Daniel MacArthur, director of the Centre for Population Genomics at the Garvan Institute. "[It] won't change what it means to be human. It will be like vaccination."[34]

Some in the field believe that the same goes for humanity's relationship with nature. "We have been messing with nature ever since we came out of the trees," says Stanford law professor Hank Greely, who likes to remind people that corn used to be a grass and that tomatoes were once bitter berries. "Geneticists changed that. [But] we called them farmers."[35]

The history of the medical establishment is replete with examples of advances that were initially rejected but later widely adopted, including procedures that seemed impossible and in hindsight became the standard of care. Up until the middle of the nineteenth century, physicians ridiculed Ignaz Semmelweis for advocating hand washing in order to reduce contagious diseases. His critics forced him into a Viennese insane asylum, where he was beaten to death. For the first few decades after newborn incubators were invented, they could only be found at amusement parks and sideshows. Even Gregar Mendel, the father of the laws of genetic inheritance, was marginalized for most of his life as a "simple" monk from a provincial town in what is today the Czech Republic – and only proven correct decades after his death.

Whether He Jiankui is ultimately judged a crackpot or a pioneer, or maybe something in between, depends on the telescope of time. Some prominent CRISPR scientists, including Harvard's George Church and George Daily, believe that if JK's babies are healthy, they might be viewed more like Louise Brown, the first infant born through in vitro fertilization, as opposed to Jesse Gelsinger, the first person to die in an early gene therapy experiment in 1999.

"The fact that the first instance of human germline editing came forward as a misstep should not let us stick our neck in the sand," says Dr. George Daley, dean of Harvard Medical School, who has called for an end to the debates about ethical permissibility so that a path toward clinical research can be cleared. "It is time to move forward."[36]

CHAPTER 7

Disaster Resilience: The Ring of Fire

You can run, but you can't hide.

– Joe Louis, world heavyweight boxing champion

Emergency Road Sign in Suginami, Japan, 2012 (Abasaa)

Paper Houses

In January 1995, a young architect named Shigeru Ban traveled to Kobe, Japan, to witness the aftermath of one of the most destructive earthquakes in history. The magnitude 7.2 Great Hanshin-Awaji Earthquake and its ensuing seventy-four aftershocks killed 6,433 people, destroyed nearly 400,000 buildings, numerous roads and rail bridges, and the majority of the quays in the port. About three hundred fires raged in the city, disrupting water, gas, and electric lines.

Residents of Kobe told Ban about the violence that rattled them out of their beds in the predawn hours. For twenty seconds, the world was filled with the sound of glass breaking, roof tiles falling, and the shrieks of old wood bending. When the living began to venture outside, what they found seemed to defy reality. Shattered windows. Crushed cars. Collapsed buildings. And dead bodies everywhere. Seismologists later determined that the focus of the quake was located on the east-west strike-slip fault line where the Eurasian and Philippine plates meet. In all, it caused more than $100 billion in damage.

Ban was among the staggering 1.2 million volunteers who poured into the Kobe region to provide relief and assistance. As an architect, he felt that too many of his colleagues in the profession hadn't been doing enough to help the thousands of people who had lost their homes in natural disasters, including himself. "We [were] not working for society," he recalls. "[We were] working for privileged people, rich people, government developers."[1]

Ban knew what he needed to do to improve the situation. After seeing one of the local Catholic churches completely destroyed, he approached the priests with a radical proposal.

"Why don't we rebuild the church out of paper tubes?" he asked.

The interaction was as short as it was awkward. "Oh, God," the skeptical priests replied. "Are you crazy? After a fire?" They turned him down flat.[2] Which person of sound mind would agree to a building made out of paper?

Undeterred, Ban recruited volunteers and forged ahead with his plan to build houses for those in Kobe who needed them most – using recycled paper tubes. As a construction material, they proved strong, cheap, and durable in the face of moisture and termites, and, perhaps most surprisingly, also water- and fireproof. His temporary units, called Kobe Paper Log Houses, provided warm, safe shelter for thousands of displaced refugees. The structures rested on foundations of donated beer crates loaded with sandbags, and for insulation, Ban employed waterproof sponge tape backed with adhesive, inserted between the paper tubes of the walls. Each unit cost less than $2,000.

After Ban built fifty homes, the priests finally relented. "As long as you collect money by yourself, bring your students to build, you can do it,"[3] they told him. So Ban spent five weeks rebuilding the Takatori church, which would later become internationally known as the Paper Church. Meant to last three years, it was universally beloved and remained in Kobe for more than a decade. Eventually it was dismantled and shipped to Taiwan, where it serves to this day as a permanent house of worship, primarily serving victims of the island's various earthquakes. Ban's easily constructed paper homes are now routinely used all over Japan – and around the world – when natural disasters strike.

X-Day: Apocalypse

Experts say it isn't a question of if, but when. Government officials call it X-Day: the moment Japan's 125 million inhabitants experience the unthinkable, a catastrophic earthquake that strikes at the populated heart of the country and wreaks untold havoc.

The Land of the Rising Sun is one of the most vulnerable places on Earth to natural disasters like earthquakes, typhoons, and tsunamis because of its location in what's known as the "Ring of Fire," a twenty-five-thousand-mile horseshoe-shaped area defined by a series of oceanic trenches, volcanic arcs, and belts – all perched on four precariously shifting tectonic plates. With over 450 volcanoes, the Ring hosts more than 75 percent of the world's active and dormant volcanoes, and 90 percent of all earthquakes.

Though as many as two thousand earthquakes hit Japan each year, for the most part, the country has remained relatively lucky in terms of *where* they have struck for about a century.

The last seismic mega tremor to hit Tokyo was in 1923. Described by residents and escapees as "hell on earth," it killed 143,000 people and destroyed 695,000 homes across Tokyo and Yokohama.[4] The Great Kanto Earthquake measured 7.9 on the moment scale, an updated version of the Richter scale that measures energy at the epicenter of an earthquake. Experts predict that there's a 70 percent chance that the next magnitude 7 earthquake will hit Tokyo before 2050. And there's an even greater chance of a magnitude 9 quake in the Nankai Trough off the Pacific coast, which could inundate cities like Tokyo – and the entire island of Kyushu – with sixty-five-foot-plus waves.

Tohoku University's Yuichi Ono, a professor of disaster science and former UN official, puts it bluntly. "Japan is a ticking time bomb."[5]

Kobe was in many ways a wake-up call. Within moments of the disaster, thousands were dead, "crushed by buildings that Japanese engineers were amazed – and ashamed – to see crumple like cards," writes David Pilling, former Tokyo bureau chief for the *Financial Times*.[6] The Hanshin Expressway collapsed, and thousands of buildings erected before 1981 toppled.

"Earthquakes don't kill people," Shigeru Ban observed. "The collapse of buildings kills people. And that is the responsibly of architects."[7]

In the decades since Kobe, bold Japanese architects and engineers like Ban have proven themselves equal to the challenge. In the quake's immediate aftermath, officials tightened regulations – and then continued to tighten them again and again until the country had one of the world's most advanced building codes. In an unprecedented marshalling of resources, talent, and public policy, it has put in place a system that integrates design thinking and innovation, which has already saved an untold number of lives. After investing billions of dollars in technology and infrastructure, Japan is now the most-prepared country on the planet to protect its citizens from earthquakes and tsunamis – and no doubt will be well into the future.

"Following every natural disaster, the Japanese study carefully what happened in order to improve building design so that above all, we are able to better protect people's lives," says Keio University professor Rajib Shaw. "That is the very minimum requirement."[8]

Japan's buildings must now be designed to withstand smaller earthquakes unharmed, without a single repair. For major quakes (defined as larger than the magnitude 7.9 quake that hit Tokyo in 1923), preserving the building is secondary to the protection of human life. Among other techniques, Japan uses reinforced walls and foundations of a certain thickness. In many cases, buildings rest on massive rubber shock absorbers or Teflon-coated pegs embedded into the foundation, which essentially allow buildings to float, decoupling them from the ground so they can move semi-independently. "When the structure can absorb all the energy from the earthquake, it will not collapse." says Jun Sato, a structural engineer and associate professor at the University of Tokyo. "Great nature is our best teacher."[9]

For high-rises and skyscrapers, architects use multiple sophisticated techniques that allow for strong yet flexible structures that sway, including hollow walls that hide metal plates and fluid-filled shock absorbers that deploy thick oil in the opposite direction of movement. "It's like a yardstick when you bend it – it snaps back without any damage," explains John W. van de Lindt, a civil engineering professor at Colorado State University.[10]

According to the University of Tokyo, 87 percent of the city's buildings are now either retrofitted or constructed according to anti-seismic standards.[11]

In the process, Japan has also pioneered an entirely new and innovative industry. "What Japanese architects have done in the last twenty years is the envy of the world and nothing short of astonishing," explains Tokyo Institute of Technology's Toru Takeuchi. "We have pioneered an entire industry that manages to marry seismic engineering requirements with functional and elegant design."[12]

Many of Japan's most iconic buildings employ high-level seismic design, including Tokyo's Mori Tower, a fifty-four-story mixed-use skyscraper reinforced with steel piping that rests on 192 oil-charged shock absorbers. Tokyo's Skytree radio tower, the second tallest building in the world, relies on a core vibration system, an idea that originated in Japan's rich architectural history of five-story traditional wooden pagodas. "It's an incredible energy dissipating system," says Sigrid Adriaenssens, a civil and environmental engineering professor at Princeton University, of the highly innovative feature. "It allows for independent movement of the tower itself, which counterbalances and suppresses the swaying caused by earthquakes."[13]

Still, for all the levitating buildings and paper houses, the Japanese are well aware that their future safety can never rest solely on the promise of miraculous feats of engineering and technology. On a societal level, their most impressive achievement might be the successful mobilization of an entire populace toward readiness.

"While Japan is renowned internationally for its resilient infrastructure and its seismic technologies," says Meri Joyce, a Tokyo-based specialist in disaster preparedness and response, "where the country really excels is its preparedness at the community and individual level."[14]

There's a Plan for That

Every day at five in the evening, hundreds of loudspeakers across Tokyo and other parts of Japan boom out a children's song called "Yuyake Koyake," encouraging children and workers to go home. The melody announces the end of the day, but more importantly, it also tests an emergency system designed to save lives when an imminent disaster is approaching. If an earthquake is about to strike one of the biggest and most important metropolises in the world, the loudspeakers blare the real-time early alarm, an eerie noise similar to the *Godzilla* theme song. TVs and radios blast it, too, and cell phones erupt in a series of long beeps.

"These warnings are terrifying," admits Dr. Satoru Nishikawa, a professor of disaster resilience at Nagoya University and former senior Japanese cabinet official. "It makes my blood freeze as I brace for safety."[15] And that, of course, is the intended effect.

For all the energy it has channeled toward engineering and design innovation, the Japanese government has sunk that much or more into meticulous planning and preparation. It has spent over a billion dollars to build a seismic alert system that can detect earthquakes eighty seconds before they arrive, giving residents enough time to turn off gas mains, step away from windows, and make their way down from high floors. If a tsunami follows, citizens are directed to the closest defense bunkers.

Were a magnitude 7 earthquake to strike the northern Tokyo Bay, the effects would be calamitous: an estimated ten thousand people killed; almost 150,000 injured; over three million in need of immediate evacuation; over five million stranded; and at least 300,000 buildings destroyed.[16] The government would aim to restore power within a week and the water supply within a month.

"We know from experience that critical infrastructure such as electricity, gas, and water gets hit hard as a result of disasters," says Takeshi Komino, a disaster management expert who has worked all over the world. "It is baked into Japanese culture to prepare ahead of time for the inevitable and do what it takes to get basic lifelines up and running as quickly as possible."[17]

Tokyo's 338-page *Disaster Preparedness Tokyo*, issued in multiple languages, outlines what people should expect when disaster strikes – with zero sugar coating. A comic strip depicts a forlorn figure navigating falling objects, derailed trains, crashed vehicles, collapsed buildings, and a lack of power and cell service. Advice includes "accepting death."[18]

"This is not a 'what if' story," the manual warns. "In the near future, this story is sure to become reality."[19]

Every September 1, drills take place in workplaces and schools across the country to inculcate the need to seek refuge in safe places. The government instructs residents to use L-shaped brackets to fasten furniture, and anti-slip pads to secure cabinets, chairs, and tables. Citizens are also encouraged to be vigilant about storing extra canned food, water, flashlights, medicine, and batteries.

Tokyo also has the world's largest fire brigade trained to deal with earthquake-sized natural disasters. "Japanese firefighters are often regarded like members of the Tokkotai," says Mikio Ishiwatari, a senior advisor at the Japan International Cooperation Agency, referring to the revered kamikaze pilots of WWII. "They are prepared to sacrifice everything in order to save lives in the face of tsunamis, earthquakes, and volcanic eruptions."[20]

In Tokyo, preparedness is so much a part of the culture, it's quite literally everywhere you look. Like an architect's plans, every last detail has been carefully thought through and anticipated, which is one of the reasons the *Economist* Intelligence Unit now ranks Tokyo as the safest city in the world.[21] Major roads marked with large blue catfish – the mythological Namazu, causer of earthquakes – are specifically designated for fire trucks and rescue vehicles. Over a dozen of Tokyo's parks and three thousand of its schools and community centers have been fully prepped as evacuation centers.

Manhole covers across the city are designed to be quickly removed and refitted with special seats and privacy tents, turning them into emergency toilets. Benches double as cooking stoves. Solar-powered charging stations are ubiquitous for electric bikes and smart phones. And city officials have built in enough underground water reservoirs and food stores to allow entire districts to survive the first seventy-two hours.[22]

A Global Shake-Up

In 2011, Tohuko was hit by a magnitude 9.0 earthquake, the fourth most powerful in the history of seismology, and Japan learned once again that no matter how precisely it plans, the country still has significant vulnerabilities.[23] The earthquake was so fearsome, it knocked the earth six and a half inches off its axis and moved Japan thirteen feet closer to America. And the tsunami that followed wreaked mayhem and unspeakable disaster. Over 130 feet high at their peak, the giant waves killed eighteen thousand people, and more than 120,000 buildings were destroyed. The devastation only mounted when three reactors of the Fukushima Daiichi Nuclear Power Plant melted down, spreading radiation in the worst nuclear accident since Chernobyl. Tohuko soon became the most expensive natural disaster of all time, causing more than $235 billion in damage.

But, as with so many of the rude geo-seismic awakenings that preceded it, Japan absorbed the lessons, adapted, and improved – and is now that much more prepared in the bargain. On the whole, the earthquakes that have hit Japan over the last two decades have only proven the effectiveness of its regulations and push toward technological innovation. Even though the epicenter of the quake was 230 miles from Tokyo, the city experienced extended shaking. But the buildings did exactly what they were supposed to do: sway like trees in the breeze. Not a single building fell in Tokyo, despite the record-breaking magnitude of the quake – a testament to the high level of engineering used in the construction of the city's structures.

At the time of the disaster, twenty-seven Shinkansen high-speed trains were on the move. As a result of advanced seismographic technology placed throughout the country, they detected the earth's tremors ahead of time. This allowed the train conductors to put on the emergency breaks a few precious moments before the most violent movements took place. No passengers lost their lives – or were even injured.[24]

Despite extensive damage to the track network and infrastructure, bullet train service resumed just forty-nine days later. "The speed of restoration was incredible," says Takeshi Fukayama, senior consultant and expert in railway development at Mitsubishi Research Institute.[25]

When Typhoon Lan reached landfall in 2017, Japan had even more proof that its redoubled efforts were paying off. Despite a storm surge of nearly ten feet, not a single home flooded. The ten-foot surge caused by 1949's Typhoon Kitty, by comparison, flooded nearly 140,000 homes.

The Land of the Rising Sun has indisputably set the bar for the rest of the world, and to varying degrees, countries such as Chile, China, Italy, Mexico, Peru, and Turkey are adopting seismic technologies innovated in Japan. Unfortunately, too many are lagging behind in this regard, including the United States, which has used the methods sparingly, with a few notable exceptions (such as Apple's new headquarters in Silicon Valley). Safety advocates argue that it's a huge missed opportunity to save billions in reconstruction costs after an inevitable earthquake hit.

"Short-term thinking is absolutely the biggest villain," says Evan Reis, a cofounder of the US Resiliency Council. "People are willing to roll the dice."[26]

The urgent need for long-term, low-cost systemic solutions has never been greater. It is only a matter of time before the next large earthquake or tsunami hits one of the world's major population centers on a fault line or near the ocean. "The next ten years will witness extreme levels of climate change and disaster risk," says Keio University professor Rajib Shaw. "Except these risks will no longer be 'extreme.' Extreme events will be the new normal."[27]

Though the cost of disaster mitigation has come down significantly in the last few decades, the reality for most countries is that the price is still too high. Engineering and planning at such a high level requires tremendous investment – Japan now spends $46 billion annually to maintain, repair, and manage water mains, ports, airports, and other large assets that make up its disaster resilience program.[28]

"If you need evidence of why earthquakes are political and economic, almost as much as natural," observed the *New Yorker*'s Amy Davidson after Tohoku's magnitude 9 monster in 2011, "look at the pictures from Japan. Then remember the ones from Haiti."[29]

The magnitude 7.0 quake that hit Haiti in 2010 completely decimated the country's ill-prepared infrastructure, flattened Port au Prince, and killed hundreds of thousands of people.

Once again, Shigeru Ban got on a plane to lend a hand. Together with local professors and university students, he built fifty shelters made of paper tubes and local materials that cost less than $300 each.

"These natural disasters will continue happening," said the Pritzker Prize–winning architect. "More problems are coming. I need to solve them one by one."[30]

CHAPTER 8

Energy: The Electric Slide

If we keep pulling death from the ground,
we will reap death from the skies.

– Van Jones, *The Green Collar Economy*

A Charge to Keep

When entrepreneurs Martin Eberhard and Marc Tarpenning sold their e-book business, NuvoMedia, for $187 million in 2000, the partners weren't sure what they wanted to do next. With backgrounds in computer science and electrical engineering, both had become obsessed with the pernicious effects of fossil fuel dependence. Tarpenning had spent his formative years after college working for the conglomerate Textron in Saudi Arabia. "If you want to see what oil does to a local economy, and where our treasure ends up going," Tarpenning says, "spending six years in Saudi Arabia is quite illuminating."[1]

Eberhard didn't need any convincing. "Somewhere along the way, I started thinking that this global warming thing was not a hoax," he recalls. "Maybe the wars we were having in the Middle East were at least somehow involved in oil, and I decided that we needed to do something different."[2] Eberhard also happened to be recently divorced. "I was thinking that I should do what every guy who gets divorced does and buy a sports car," he says. "But I couldn't bring myself to buy one that got eighteen miles to the gallon."[3]

The car he really wanted didn't exist, so Eberhard created a spreadsheet that modeled out the various power sources available, including gasoline, diesel, natural gas, solar, wind, fuel cells, and batteries. "The results were startling," he recalls. Eberhard discovered that hydrogen was just as bad as gasoline, and that even under the best circumstances, it was only about 20 percent efficient with the technology available. "Your hydrogen fuel-cell car will require three to four times as much of whatever your source fuel is per mile compared to electric vehicles," says Eberhard. Lithium batteries, on the other hand, are about 95 percent efficient. "We were surprised to learn that electric cars were head and shoulders above everything else," he says.[4]

At the time, GM had already sunk more than $1 billion into the development of the EV1, the first modern electric car from a major manufacturer, before infamously repossessing and crushing the entire line because the cars were unprofitable. When the two engineers started digging into why the EV1 failed, they concluded it was the battery type and business model. And before long, the pair had convinced themselves they could do better.

Early EV1 models used lead-acid batteries, which yield a range of up to a hundred miles. But throughout the 1990s, the battery industry was rapidly changing, and Eberhard and Tarpenning felt that lithium-ion batteries, which today yield over four hundred miles of range, presented a huge opportunity. The battery itself was a real disruptor in the market – in terms of both cost and performance, it improved by about 7 percent each year. “It was so much better,” says Eberhard, “that it was stunning to us that nobody else was doing it.”[5]

In 2003, Eberhard and Tarpenning decided that the most effective way to battle climate change and reduce the world’s reliance on oil was to start their own electric car company. But they didn’t jump in blindly. “We had no experience making cars,” recalls Eberhard, “and we had a lot to learn.”[6]

For months, they struggled to come up with the perfect name for their new venture. One night, Eberhard was eating at the iconic Disneyworld restaurant Blue Bayou, overlooking the Pirates of the Caribbean, when it struck him. “What do you think about the name Tesla Motors?”[7]

The very engine they intended to use in their car, known as the AC induction motor, had been patented by the legendary Serbian-American inventor Nikola Tesla in 1887. In July 2003, the pair incorporated the company and rented office space in Menlo Park, California, which still had the sign from the previous tenant: “Bushtracks African Expeditions.” They flipped the sign over and wrote “Tesla Motors” on the back.[8]

Zero Hour

Early on, to get a sense for the market, Eberhard drove around Palo Alto trying to figure out who their ideal customer might be. He snapped a lot of driveway pictures and noticed that he often saw an old BMW or Porsche parked next to a Prius. And that’s when it hit him: “Someone who likes a fun-to-drive hot car,” says Eberhard, “but also feels bad about the gas mileage and oil consumption.”[9]

Eberhard and Tarpenning started working on their go-to market plan. Up until then, Eberhard recalls, electric cars “had been given such a black eye in the marketplace. People thought about them as punishment cars, dork-mobiles, just ugly golf carty things."[10] Their plan was to innovate on a successful business model that was already common in the tech industry: first create a high-end product that commands high prices, then produce a mid-market product with higher volume, and eventually leverage the skills, economy of scale, and price reductions to secure the mass market.

The two engineers started tinkering with a model car. One afternoon, Eberhard called an old friend, Malcolm Smith, who was a Silicon Valley product designer. "I can't tell you what we're doing," Eberhard said, "but why don't you come check out this car I have."[11] When Smith arrived, Eberhard showed him their business plan, then said, "Let's go for a ride."[12]

They jumped into a small yellow two-seater with a decal on the side that read "tZero," a reference to the mathematical symbol that means the beginning of time within a system. As the two turned slowly onto Sand Hill Road, Eberhard asked Smith to try to touch the dashboard. Just as he reached out, Eberhard hit the accelerator, and Smith's hand never made it – the sheer force of the acceleration drove him straight back into his seat. The car had an AC propulsion motor with a lithium-ion battery that allowed the tZero to go from zero to sixty in less than four seconds.

"I get it," Smith thought. "This isn't a nice little science experiment."[13] He eventually became one of Tesla's first employees.

"The tZero will be remembered as the father of all modern electric vehicles," says Seth Weintraub, founder of the popular news site Electrek. "It changed the world and ultimately became the modern Tesla."[14]

In the fall of 2003, Eberhard and Tarpenning started speaking to potential investors in the Silicon Valley ecosystem. But at first, they only pitched people they knew would never invest, so they could hone their pitch and not burn their chances with serious investors.

At the time, they acknowledged that building a high-performance electric car "sounded impossible," recalls Tarpenning.[15]

They eventually decided that the only model that made fiscal sense would be to sell directly to consumers, as opposed to going through traditional car dealerships. They also concluded that they shouldn't try to build their own car from scratch. Instead, they'd build on top of an existing car they could customize, eventually settling on Lotus, a British auto manufacturer known for their sexy sports cars. Eberhard and Ian Wright, the company's first employee, both drove Lotus Europas.[16]

With the pitch honed and the business plan complete, they went back out to raise money in earnest. And though they got a bit of capital from friends and family, no one seemed willing to step up with a big investment – until they reached out to one man who was willing to give them a shot.

"We would love to talk to you about Tesla Motors," wrote Eberhard in a short email on March 31, 2004. "Particularly if you might be interested in investing in the company."

Elon Musk sent a note back that very night. "Sure," he wrote.[17]

Eberhard and Ian Wright flew down to the SpaceX office in Los Angeles. "He gave me half an hour to make a presentation, which turned into two hours," recalls Eberhard. "And by the end, we had a handshake deal that he would lead the investment in the company."[18]

Musk shared their passion for building a superior electric car that could change the world. At their second meeting in LA, Tarpenning joined Eberhard to make the final pitch. "He liked the idea and peppered us with questions," Tarpenning recalls. And just like that, "He said, 'I'm in, you convinced me.'"[19]

In April 2004, Musk invested $6 million and became the company's largest investor.

Off to the Races

By the summer of 2004, Eberhard had a rough idea of how he wanted his first car to look. The engineer turned businessman-carmaker knew he needed a sports car, not just to play against the stigma that electric vehicles (EVs) were for crunchy environmentalists, but also because sports cars have higher margins. Top-tier buyers are traditionally less price sensitive, which meant they'd need to sell fewer cars to become profitable.

According to Tarpenning, their initial call for design submissions came back "terrible."[20] Yet no matter how hard they tried, they couldn't communicate to designers what they wanted. All except for one.

In the fall, Eberhard turned to his long-time friend and next-door neighbor Bill Moggridge, one of the cofounders of the world-renowned design firm IDEO and a longstanding faculty member of the Stanford Design School. Over wine, Moggridge peppered him with questions.

"OK, let's consider this axis, from retro to futuristic," Moggridge said. "Where would you want your car to be on that axis?" Retro was Eberhard's preference.

"Here's another axis, masculine to feminine," Moggridge said. "Where do you imagine your car on that axis?" Eberhard wanted something in the middle.

"Curvaceous or boxy?" Moggridge asked. "You could look at the classic old Ferraris, which are very curvaceous, and the modern Lamborghinis, which are very boxy."

"Somewhere in between, but closer to the curvaceous end," Eberhard replied.[21]

Moggridge, Eberhard recalls, was able to take his preferences and turn them into "magic." "He spoke the language of designers," recalls Eberhard, "but knew me well enough as an engineer to understand my vision."[22] The engineer was then able to go back to his designers and explain exactly what they wanted for their first car.

Around Christmas, Eberhard hosted a small house party for Tesla's employees, their families, and company advisors. He put the design sketches up on the wall and asked each guest to vote with green or red Post-it Notes. At the end of the night, there was one sketch covered in green. It was the car designed by Barney Hatt from Lotus Design Studio: the Roadster.

Problem Child

On July 19, 2006, the Tesla Roadster was unveiled to an elite group of Hollywood who's whos at the Barker Hanger in the Santa Monica Airport. Guests were told that the first one hundred people to pre-order the $100,000 two-seater would get an engraved plaque inside their cars. Within two weeks, Tesla had sold 127 Roadsters, with a first batch of customers that included George Clooney, Google founders Sergey Brin and Larry Page, inventor Dean Kamen, Hyatt heir Nicholas Pritzker, and sportswear designer Julie Chaiken.[23]

In short order, the Roadster received rave reviews from the *New York Times*, *Fortune*, *Wired*, CNET, and the *Washington Post*, which declared, "This is not your father's electric car."[24]

Yet from the start, the Roadster was plagued with bugs that would significantly delay its delivery date. "We have a tremendous number of difficult problems to solve just to get the car into production," Eberhard wrote to Musk in November 2006. "I stay up at night worrying about simply getting the car into production sometime in 2007."[25]

Slowing the production was what's known as the "elegance creep" phenomenon. The company kept customizing their car, making it as beautiful as possible and enhancing the user experience. Tesla started with five subassemblies – and eventually had hundreds. Over time, the Roadster barely resembled the Lotus Elis it was meant to be based on, and the two cars shared only 7 percent of their common parts. The cost to produce the first Tesla skyrocketed. Both investors and employees worried it would choke the company to death.

Between the Roadster's launch and the date customers actually got their cars, Tesla's leadership underwent significant changes. By August 2007, Eberhard had resigned as CEO, though he remained president and a member of the company's board of directors. A short time later, Eberhard was forced out. In the year that followed, the company hired and fired two more CEOs – until Musk finally took over in October 2008. By that time, Musk had already invested more than $55 million, and many feared the company was on the fast track to bankruptcy.

"I've got so many chips on the table with Tesla," Musk explained. "It just made sense for me to have both hands on the wheel."[26]

In his first year as CEO, Musk made drastic changes. In addition to firing 25 percent of his employees, he sold 10 percent of the company to Daimler AG for $50 million, raised an additional $40 million of debt, and secured a $465 million loan from the US Department of Energy.[27]

In fall 2008, a full year after its promised summer 2007 delivery date, the Roadster finally started going out to buyers. "We had incredibly supportive customers who stayed with us, " recalls Tarpenning, "I don't think anyone cancelled their reservations – it was craziness."[28]

That same year, Tarpenning decided it was time to leave the company.

Bigger, Stronger, Faster

On June 29, 2010, Elon Musk and members of the Tesla team walked into the NASDAQ exchange and rang the opening bell. Tesla officially became the first American carmaker since Ford in 1956 to go public. Musk secured $226 million, giving the company the capital infusion it needed to grow. The problematic Roadster was discontinued in

2012 after the company sold approximately twenty-five hundred cars.

To capture a larger share of the total auto market, Musk knew he needed to pivot. To do that, Tesla would need to develop advanced electric drive components and produce low-cost, high-energy-density lithium-ion batteries. He focused his energy on the Model S, a full-sized luxury car the company had been developing since 2007, which went to market in June 2012 and very quickly won multiple accolades. The National Highway Traffic Safety Administration called it "the safest car ever tested," *Consumer Reports* deemed it the "best car ever tested," and *Road & Track* rated it "the most important car America has made in an entire lifetime."

In 2013, Tesla announced its plan to build its multi-billion-dollar Gigafactory with the aim of bringing the cost of lithium batteries down by at least 30 percent. Located near Reno, Nevada, the $5 billion structure became the largest building in the world by footprint, with 5.8 million square feet of space roughly equal to a hundred football fields. The factory is also powered entirely by green energy, including solar, wind, and geothermal sources.

A year later, the company achieved a major milestone. For the first time, Tesla released an autopilot car, which gave the Model S semi-autonomous driving and parking capabilities.

In 2015, Tesla released its falcon-winged luxury crossover SUV, the Model X, to near universal acclaim. And a year later, it launched the Model 3, a cost-effective sedan targeting the mass market. The entry-level sticker price was $35,000, which led to a staggering half million pre-orders. Over the previous six years, Tesla had only sold about 200,000 cars – a figure many car manufacturers sell in a month.

Today, Morgan Stanley calls Tesla "the world's most important car company." With a valuation of more than $460 billion, it's also the seventh most valuable company in America. In 2020, it was added to the S&P 500.

The Future of Sustainable Energy

"When something is important enough," says Elon Musk, "you do it even if the odds are not in your favor."[29]

Tesla's official mission statement is "To accelerate the world's transition to sustainable energy,"[30] and to achieve this end, the company has moved well beyond electric cars. For homeowners and businesses, Tesla has developed a range of products that allow customers to produce, manage, store, and consume their own renewable energy. The ecosystem includes solar panels and solar roofs that provide 100 percent sustainable energy, which can then be stored in a Powerwall that both makes electricity available during peak hours and provides power when the grid experiences outages.

"In the next few years, we'll reach a decisive point," says Lawrence Burns, author of *Autonomy* and former vice president at General Motors. "That's the amazing moment when the technology works, the people want it, the price is low enough to make it work, and the companies can make money selling it."[31]

When Tesla launched, there were no viable electric cars on the market. Today, there are about three million on the road globally, including over a million Teslas.[32] To date, Teslas alone have driven about ten billion miles, saving the planet more than four million metric tons of CO_2, the equivalent of the annual emissions of one million cars.[33]

"Elon has managed to take it further than I could have imagined," says Tarpenning. "To see so many on the road and people really loving them just feels great."[34]

In the last few years, Musk has also unveiled the Tesla Semi, an electric hauling truck. (He says driving diesel-powered trucks is "economic suicide.")[35] And in the near future, Tesla intends to design autonomous mini-buses for mass transit.

"The transition is just getting started and there is no turning back," says Daniel Sperling, founding director of the Institute of Transportation Studies at the University of California, Davis. "Both the automotive industry and international policy leaders are committed to EVs, which represent the single most important transportation strategy to slow climate change."[36]

Every major car manufacturer now produces electric vehicles, and there are a significant number of world leaders now calling for an eventual ban on cars with internal combustion engines. Industry analysts predict that in the coming years, the market for EVs will explode as their prices decrease at an inverse proportion to their increase in range, and more government policies support their mass adoption.

"Regulation drives innovation," says Roland Hwang of the National Resources Defense Council, who points to California's zero-emission vehicle credits program, which incentivizes manufacturers to produce a minimum number of electric vehicles per year, as one of the prime movers of Tesla's success. "For decades, automakers have repeated one line: 'It can't be done.' Tesla is proof positive they were wrong."[37]

In 2017, China adopted its own New Energy Vehicle Regulation, based on California's program, and the country is now the largest market for EVs in the world, with well over a million sold in 2019 alone.[38] China has traditionally battled terrible smog in its larger cities, but it has also grown increasingly concerned about importing more than 70 percent of its oil from abroad. In EVs, it sees a golden opportunity to "leapfrog other countries in the auto sector as an industrial policy," says Yunshi Wang, director of the China Center for Energy and Transportation at the UC Davis Institute of Transportation Studies. "China is engaged in self-enlightened interest and is poised to dominate the clean car economy for a very long time to come."[39]

In Norway, which provides huge tax incentives for electric cars – not to mention exemptions from bridge tolls and free public parking – nearly 60 percent of all new car sales are EVs.[40] The government has decided to completely phase out the sale of new internal combustion engine cars starting in 2025, and many other European countries have proposed similar phase-out plans.

"When the industry combines electric drive and autonomy, we will witness the most profound change since the industrial revolution," says Robbie Diamond, founder of Securing America's Future Energy (SAFE), an organization based in Washington, DC, committed to combating the economic and national security threats posed by dependence on oil. "And it will happen faster than people think."[41]

In the coming decades, both Martin Eberhard and Marc Tarpenning envision a green revolution around the world, driven by electric vehicles and solar energy, that will significantly reduce the harmful emissions we are putting into the atmosphere. Eberhard likes to cite the late Saudi oil minister Ahmed Zaki Yamani, who once said, "The Stone Age did not come to an end for lack of stones, and the oil age will not come to an end for lack of oil."[42] What it will take is vision. "Progress is not foreordained," says Tarpenning. "It is created by people."[43]

The founders of Tesla have proven it.

"If you get up in the morning and think the future is going to be better, it is a bright day," Elon Musk says. "Otherwise, it's not."[44]

CHAPTER 9

Prosperity: The King of Microfinance

As long as poverty, injustice and gross inequality persist in our world, none of us can truly rest.

– Nelson Mandela

Nobel Peace Prize winner Muhammad Yunus at a summit hosted by the University of Salford, May 2013 (University of Salford Press Office, CC BY 2.0)

From Loaner to Laureate

On a frigid evening in December 2006, Muhammad Yunus walked jubilantly up the red carpet into Oslo City Hall and waved to the cheering crowds. It was the 105th anniversary of the Nobel Prize, and hundreds of dignitaries, government officials, and royals had already made their entrance, along with nine women from the villages of Bangladesh.

The excitement was palpable as Yunus made his way to the dais, with their Majesties the King and Queen of Norway following shortly behind. Wearing a tan *kurta* and matching *salwar* trousers, he smiled as he looked out at the attendees, who were there to award him the prestigious Nobel Peace Prize for pioneering the concepts of microcredit and microfinance. In large part because of his lending program, Bangladesh's poverty rate had dropped by more than 10 percent since Yunus launched his venture.[1] All told, the World Bank estimates that millions of people are either direct or indirect beneficiaries of the economic platform he brought to life.[2] Yunus had proven that even the poorest of the poor, unable to qualify for traditional bank loans, have the ability to work to bring about their own development.

"I firmly believe that we can create a poverty-free world if we collectively believe in it," he told the crowd. "We can put poverty in the museums."[3] The auditorium rose to its feet.

"I can't believe that this has really happened," the laureate said as he made his way off the stage.[4]

If receiving such an honor seemed implausible to Yunus, it would have been entirely inconceivable to him as the young and disillusioned economics professor who had set out thirty years earlier to try something different to alleviate poverty. What's even more incredible is that the world-changing financial system he eventually sparked started out with a simple bamboo stool.

A Step Up

It was 1976 when Yunus first walked into Jobra, a multiethnic village in southern Bangladesh where Muslims, Hindus, and Buddhists live side by side. The country was still reeling from the impact of a famine that had begun two years earlier, and the head of Chittagong University's economics program was at a professional crossroads. Since his return from the United States, where he completed his PhD at Vanderbilt University, Yunus had seen too many dead bodies, and too many of his countrymen "simply lay down very quietly on our doorsteps, wait[ing] to die."[5]

Yunus had begun to dread his own lectures. "What good were all my complex theories," he asked himself, "when people were dying of starvation on the sidewalks and porches across from my lecture hall?"[6] Nothing he was teaching was making a difference. The poor kept getting poorer, and life was getting worse for the majority of Bangladeshis.

"I needed to run away from these theories and from my textbooks and discover the real-life economics of a poor person's existence," he recalls. "To become a fugitive from academic life."[7]

To create a rural economic assistance program that actually worked, Yunus knew he would have to become a student all over again. "The people of Jobra would be my professors," he decided.[8]

To better understand the root of the poverty cycle, the young academic began visiting the poorest households in Jobra and spending time with the villagers. One morning, he stopped at a house with crumbling mud walls, where he saw a woman squatting on the dirt floor, gripping a half-finished bamboo stool between her knees. The mother of three was twenty-one-year-old Sufiya Begum, whose fingers were black and callused from her daily work turning strands of cane into Bengali mudda stools.

Sufiya told him she survived almost entirely on credit. She would buy five taka (about twenty-two cents) worth of raw materials from a *paikar* middleman and repay the loan by giving him finished stools. Yunus was stunned to learn that her income worked out to fifty poisha a day, or about two cents.

To acquire materials, Sufiya and the other villagers also had the option of going to local moneylenders, who were more than happy to dole out usurious loans. They charged anywhere from 10 percent a week to 10 percent a day. In every scenario, the poor would eke out a meager subsistence and almost certainly remain impoverished for the rest of their lives. Neither Sufiya nor her children were likely to ever break the cycle. Her kids didn't go to school, and Yunus couldn't see how Sufiya would ever be able to shelter, clothe, or feed them properly. "It seemed hopeless to imagine that her babies would one day escape this misery," he recalls.[9]

"I was angry," he says. "Angry at myself, angry at the economics department and the thousands of intelligent professors who had not tried to address this problem and solve it."[10] The professor was convinced that Sufiya's children would be condemned to a life of hand-to-mouth survival and penury, and that her income would be kept so low that she'd never be able to save or expand her economic base. Moneylenders made sure that their borrowers paid a price that barely covered the cost of the materials, earning them just enough to stay alive. Sufiya was living as a bonded laborer, but she might as well have been a slave. "It seemed impossible to me, preposterous," he says.[11]

Yunus wondered whether he should just reach into his pocket and give Sufiya the small amount of money she needed. "That would be so simple, so easy," he says. "But I resisted the urge."[12] Giving her money wouldn't address the root of the problem, and Sufiya wasn't asking for charity.

The professor made a list of the forty-two villagers who were dependent on moneylenders and tallied up how much each of them needed to buy their raw materials. "My god, my god," he said, as he calculated the modest figure that would help them shake the yolk of exploitation. "All this misery in all these families for the lack of twenty-seven dollars."[13]

Sufiya and those like her weren't poor because they were lazy or stupid. They worked long days full of tremendous physical exertion. But traditional financial structures weren't set up to provide the credit they needed. If Yunus personally lent the twenty-seven dollars to the villagers of Jobra, they could sell their products to anyone and get the highest return for their labor. They wouldn't be suffocated by the usurious rates they were being charged, and they could repay the traders and keep the profits themselves. He distributed the loans and told the villagers to pay him back whenever they could, with no interest.

From Untouchable to Huggable

The next day Yunus got into his white Volkswagen Beetle and drove to a local branch of the government-owned Janata Bank, Bangladesh's largest financial institution. He knew that for his plan to work on a larger scale, he would need to charge interest, and a bank would ultimately need to act as his partner in lending money to the poor. As Yunus explained his vision, the bank's manager laughed and listed all the reasons it wouldn't work, including the fact that preparing the documentation alone would cost more than the loan itself. "The bank is not going to waste its time on such a pittance,"[14] the manager told him. Besides, the people would need to be literate in order to fill out the forms and sign them, and the bank would need collateral, which the most destitute would almost certainly not have.

Yunus wasn't just up against the Janaa Bank, but rather the entire banking system, which was set up to service only those with money. "Poverty is not created by the poor," Yunus says, "but by the institutions built and the policies adopted."[15] Yunus continued to push, and after six months of haggling with the bank's head office in Dhaka, the loan was finally formalized in January 1977. The bank agreed to lend his project three hundred dollars, as long as he acted as the guarantor. Yunus and his colleagues would distribute the loans to Jobra's "banking untouchables" personally, in order to spare them the indignity of dealing with the bank, which required Yunus to sign for each and every request.

Out of a strong desire to solve a real-world problem, Yunus had cast aside the most central premise of banking – collateral – and become a moneylender himself. He had no idea what he was getting himself into. But he knew he needed to take his untouchables and make them "touchable, even huggable."[16]

Over the next two years, he planted the seeds of what would become an entirely new economic system. "I tried a great number of things," Yunus recalls. "Some worked. Others did not."[17] To his great surprise, over 98 percent of his initial loans would come to be repaid, a rate that far exceeded those backed by secured assets. And, as Yunus puts it, what started out as an attempt "to relieve my guilt and satisfy my desire to be useful to a few starving human beings" soon became much more.[18] By 1979, he had established Grameen Bank, which means "rural" or "village" in Bangla – "the bank for the poor."

After studying the loan operations of other banks, Yunus made it his mission not to repeat their mistakes. Grameen does not require borrowers to put up collateral for a loan, nor does it ask them to sign legal documents, since it does not want to take anyone to court.[19] "We have no guarantee, no references, no legal instrument," says Yunus, "and still it works. It defies all the conventional wisdom."[20]

Yunus also did away with the lump-sum payment most banks require at the end of a loan period. He felt this was psychologically trying for borrowers, who often try to delay repayment as long as they can – as the loan grows bigger and bigger – leading a significant percentage to eventually give up on paying it back. It also leads borrowers and bankers to ignore problems as they arise, in the hopes that they will go away as the loan is due.

To solve this problem, Yunus implemented a weekly repayment

program. The loan payments were so small that borrowers would not miss the money. And to make things extremely simple, loans were paid back fully after one year. So if you borrowed 365 taka (about sixteen dollars), you repaid at a rate of one taka a day (about four cents).

Seeing how crucial support groups were to success, Yunus required each loan applicant to join a group of like-minded individuals with similar economic and social circumstances. The groups not only provided succor, they also made borrowers much more reliable – because of peer pressure. All members had to approve the loan requests of the others in the group, building in a collective moral responsibility.

"The underlying genius of Yunus's model is the premise that relationships matter and provide a critical form of collateral," says Brown University professor Andrew Foster. "You are far less likely to fleece your friends and family, which helps ensure borrowers pay back their loans."[21]

One of the reasons Grameen's system works is that it requires the groups to form – and manage – themselves. A prospective borrower must take initiative and explain the way the bank works to a second individual, often a friend or family member who might be terrified or skeptical, or forbidden by a husband to deal with money in the first place. Those two courageous people then work together to find a third, fourth, and fifth person to join them. The group of five presents itself to the bank, and loans are initially given to just two members. If these loans are paid regularly over the next six weeks, two more members get their loans. And only after all four of those loans perform as expected for another six weeks does the fifth member of the group get his or her loan.

It takes anywhere from a few days to several months for a group to be certified by Grameen, and only after all of its members undergo seven days of training on the bank's policies and pass an oral exam.

Applicants know that if they fail, they aren't just letting themselves down, but the entire group.

The same pressure applies when the group receives their loans. If a borrower can't or won't repay a loan, the whole group becomes ineligible for larger loans until repayment is made or a program is put in place to remediate the issue. While there is no joint liability – the repayment responsibility rests solely on the individual borrower – the group helps to ensure no one defaults.

Every group is also required to build up a special savings account they can access if they encounter hard times, or even in the event they see an additional revenue opportunity. Each borrower deposits 5 percent of his or her loan into this general fund, and any member of the group may take an interest-free loan from it, provided all members approve, and it doesn't exceed half the reserve's total amount. Year after year, these group funds are used for seasonal malnutrition, medical bills, school supplies, natural disaster relief, and dignified family burials.

But the group also has an additional fail-safe measure. Up to eight borrower groups form what's called a "center," a loose federation that meets once a week with a bank employee. It is during these meetings that all the loan repayments and deposits are made. A center chief is elected by all members to solve any problems the groups are unable to solve themselves. And by design, the meetings happen in the open, reducing corruption, mismanagement, and misunderstandings. They also keep Grameen workers, the center chief, and the group chairperson accountable to the borrowers.

The group structure is not without its critics, who argue that the process intimidates borrowers. But Yunus maintains the practice ensures that only those who are the most needy and serious step forward. "We want only courageous, ambitious pioneers in our microcredit program," he says. "Those are the ones who will succeed."[22]

A vast majority – 97 percent – of those pioneers happen to be female.[23] Prior to the establishment of Grameen, women represented less than 1 percent of all Bangladeshi borrowers, primarily because traditional banks excluded them entirely. Yunus saw that women – with the most insecure social standing in Bangladesh and around the world – experience poverty and hunger far more acutely than men. But he also found that they tend to adapt quicker, perform more consistently, pay more attention, and work harder to lift themselves out of poverty. Grameen operates on the premise that the more credit women receive, the faster society can change.

One Poisha at a Time

By 1982, just three years after Grameen launched, the bank had grown from five hundred initial borrowers to twenty-eight thousand and dispersed roughly $25 million. By 1999, that number jumped to $190 million, with a near-perfect repayment record. And by the time Yunus received his Nobel Prize in 2006, Grameen had dispersed close to $8 billion, of which $7.16 had been repaid, with $720 million in outstanding loans. Today, Grameen has distributed more than $20 billion to at least eight million people, most of whom are women, in over one hundred countries around the world. The repayment rate is still 98 percent.[24]

"The Nobel Committee summed it up well when they said that loans to poor people without any financial security had appeared to be an impossible idea," says economist Christian Ahlin. "Yunus trailblazed an unforeseen explosion in access to formal credit throughout the developing world."[25]

With more than twenty-five hundred branches serving around eighty-five thousand Bangladeshi villages, and a staff of almost twenty-five thousand people, Yunus's bank for the poor has been a driving force in making a meaningful dent in his county's poverty rate. More than 68 percent of Grameen borrowers and their families have crossed the poverty line.[26]

Around the world, at least 1.9 billion people lived in extreme poverty in 1990. By 2020, that number had shrunk to 705 million – meaning every single day for the last thirty years, at least 105,000 people crossed over the poverty line.[27] Many of them leveraged microfinance programs to build better lives for themselves and their families.[28] The sector as a whole now has a combined portfolio of at least $124 billion in loans.[29] And according to Bangladeshi World Bank economist Shahidur R. Khandker, more than two hundred million people are either direct or indirect beneficiaries of microfinance programs today.[30]

In terms of sheer impact, Brown University's Mark Pitt compares Yunus to Ghandi. "On the one hand full of wisdom, and on the other super practical. As an economist, he created a well-thought-out, brilliant program to meet head-on a failure in the credit markets for the most underserved segments of society."[31]

From across the political divide, the development industry has for the most part adopted Yunus's vision. Though critics like the University of San Francisco's Bruce Wydick argue that microfinance is "far from the poverty elixir its advocates touted it to be"[32] – not to mention that measuring the direct impact of such programs is difficult – most economists now conclude that they produce broad positive effects, raise wages, encourage entrepreneurialism, and promote human flourishing. And at the very least, but perhaps most importantly, they provide a desperate segment of the population with a liquidity tool to ride out periods of instability or a sudden shock to their income, which can often pose a greater threat than low income itself.

"There is no silver bullet to completely eradicate poverty," says Tufts economist Cynthia Kinnan. "But Yunus and Grameen proved that microcredit could successfully bring financial access to millions of poor households. And we now have evidence that, for talented but low-wealth entrepreneurs, short-term access to credit can indeed unlock their escape from a 'poverty trap,' where low incomes can be self-perpetuating."[33]

By necessity, Grameen is operated as a for-profit institution, and it has turned a profit in all but three of its thirty years in existence (the exceptions were 1983, 1991, and 1992). It now offers the same products traditional banks do, including dividends to its owners and interest rate loans – 20 percent for income-generating loans, 8 percent for housing loans, 5 percent for student loans, and interest-free loans for the impoverished – plus generous deposit rates of 8.5–12 percent. But Grameen pursues profit as a double bottom line: the social purpose is paramount, and the business efficiency is a means to an end, not an end unto itself.

Still, Grameen requires subsidies on a continuing basis and relies on the generosity of donors and socially conscious investors. Studies show that microfinance programs that target the poorest borrowers generate sufficient revenue to cover about 70 percent of their full costs.[34] Timothy Ogden, managing director of the Financial Access Initiative, a research center based at NYU, warns that the microcredit industry itself could be threatened unless investors and international organizations continue to subsidize it. But that just might be the cost of helping the poor. "The sacrifice that social investors make could take many forms, including lower or zero returns compared to alternative investments, higher risk, longer timeframes, or less liquidity," Ogden says.[35]

However microfinance evolves in the future, Yunus has built an enduring platform on the idea that getting credit is a human right, and that all human beings, including the poorest, are endowed with endless potential.[36] He's hoping to someday reach the estimated one billion potential micro-borrowers across the globe, but he's already improved the lives of millions.

When Yunus stepped off the stage that night in Oslo, he looked over at one of those people, Taslima Begun, a Bangladeshi housewife turned entrepreneur who was also awarded the Nobel Prize. She took her first loan in 1992 for fifteen hundred taka (about sixty-five dollars) to buy a goat, spun that into a *mishuk* (motorized rickshaw) business and a thriving watch store, and is now a member of Grameen's board.[37]

"My parents gave me birth," she said. "But Grameen Bank gave me a life."[38]

CHAPTER 10

Food: Green Meat

How many things, too, are looked upon as quite impossible, until they have been actually effected?

– Pliny the Elder, *Natural History*

Beyond Meat CEO *Ethan Brown, 2019 (Steve Jurvetson,* CC BY *2.0)*

Meat and Human Evolution

Twenty years ago, Tel Aviv University archeologists discovered a small cave in the Samarian foothills that completely upended our understanding of human evolution. Located near the Afek Industrial Park, where many modern-day Israelis work in tech, it contained the remains of Paleolithic-era *Homo sapiens* that lived in the area over 200,000 years ago. Geological shifts had sealed the underground limestone cavern shut, where the secrets of Qassem Cave remained lost to the world until construction workers widened a highway in 2001.

The inhabitants of the cave appeared to be the world's first industrial workers, producing thousands of knives and other tools for a mass-production line used for butchering and ultimately eating deer. Evidence suggests that these prehistoric humans stored bone marrow and consumed it as needed. Archeologists concluded that our human ancestors were engaged in nothing short of a robust food innovation business.

The story of human evolution has always been intimately tied to the consumption of meat. Using fire to cook animals gave humans more energy, which led to larger brains and bigger bodies. This shift produced increasingly sophisticated hominids and transformed our ancestors from scavengers into hunters. As a species, we are designed to love meat. Sociologists believe that when early humans came back from hunts with prey, they danced. This model worked well for millions of years and provided the necessary calories to grow the human race into the roughly 7.9 billion we number today. Over the last few centuries, as our population increased exponentially, we have leveraged science and technology to get more food output per acre and more meat per species of livestock. But feeding the modern world is a problem that is growing more complex – and dire – by the day.

Research now shows a direct correlation between animal protein consumption and epidemics of cancer and diabetes.[1] Medical evidence also suggests that eating a diet rich in red meat triples the levels of chemicals linked to heart disease.[2] But more urgently, meat production is currently responsible for 18 percent of the planet's greenhouse gas emissions – and takes up 70 percent of our arable lands. Each year, more than seventy billion chickens, pigs, cattle, and other livestock are slaughtered globally for human consumption. The toll that takes on our natural resources is devastating: it takes approximately fifteen hundred gallons of water to produce a single pound of meat. And as the population increases to approximately ten billion by 2050, the human demand for meat will double, along with the ramifications.

Google founder Sergei Brin, who has invested millions in both commercial and nonprofit ventures that empower social entrepreneurs to take on climate change, has suggested three possible outcomes for the human race when it comes to meat: we can do nothing and suffer continued environmental damage; we can all become vegetarians or use vegetable-based protein alternatives; or we can produce meat from cells rather than raise entire animals.[3]

Humans will not need to make a binary decision on whether to eat or abandon meat. But we will need to consider the origin of our meat. If we choose poultry, pigs, and cows, then our choices are limited. If, on the other hand, we choose to define meat by its composition and chemical structure – amino acids, lipids, trace minerals, vitamins, and water made to taste, feel, and look like meat – then we have unlimited innovative solutions.

For now, 75 percent of our food system is based on just twelve plants and five animals, which experts say is nothing short of a global crisis. "For the human race to survive," says Scott May, head of MISTA, an innovation platform for the food industry that was launched by Givaudan, "we must remove the large-scale animal practices while increasing biodiversity sooner rather than later."[4]

Ethan Brown, the founder of Beyond Meat, wants to do it with science.

"People love meat, and we've been consuming meat since before we were human," Brown says. "So I think the fix here is technological. We bypass the animal, agriculture's greatest bottleneck."[5]

High Steaks

There is nothing new about veggie burgers. The first commercial veggie burger was sold in a natural foods restaurant in London in 1982.[6] But these patties are infamous both for being tasteless and for having a strange texture.

Ethan Brown used to eat his favorite Double R Bar Burgers from Roy Rogers as often as he could get them, but around the time he turned thirteen, he started to lose his appetite for beef. His father owned and operated a dairy farm in rural western Maryland,[7] and Brown became attached to the livestock. "I can just remember thinking: 'This is like an orgy of animals I am eating,'" he says. "I started to feel different about it."[8]

Brown embarked on an intellectual and emotional journey that eventually led him to take up veganism and adopt Hinduism's deep respect for all living creatures. He came to believe that the human tendency to see ourselves as starkly different from the rest of life on the earth wasn't just convenient, but incredibly harmful. "You can't sort of say 'us-them' in a scientific way," he says. "The reason that you think that way is because you want to create something that is separate, because it's easier to exploit."[9]

Brown was deeply influenced by his father, Peter Brown, a geography and government professor who also owned a dairy farm as a hobby.[10] The elder Brown's primary area of focus is how humanity has contributed to the deterioration of the earth's life support capacity.[11] When Brown graduated from Connecticut College in 1994, his father asked him what he wanted to do with his life. Brown said he didn't really know. "Well, what's the biggest problem facing the world?" his father asked.

"Climate," Brown responded. "If we don't get climate right, nothing else matters."[12]

Brown enrolled at Columbia University to pursue a master's in business administration, then got a job at Ballard, a Canadian company that develops fuel cells, where he worked on developing a proton-exchange membrane that promised to help curb climate change.[13] But over time, Brown says he is becoming increasingly dumbfounded by the fact that the meat industry, not automobiles or power plants, is actually the single largest contributor to greenhouse gas emissions around the world.[14] "At these big conferences, MBAS and engineers would go on and on about fuel cells, and then go out and have a steak," he says.[15]

Brown invested in Worthington Foods, a company that had a significant ownership stake in Morningstar Farms, best known for its veggie burgers. Brown also started thinking about the kind of plant-based venture that would appeal to a mainstream fast-food consumer. "I wanted to create the whole thing," he says. "I wanted to prove that you could do it."[16]

Though his friends and family were skeptical, after eight years at Ballard, Brown quit to start a company that would create "meat" directly from plants. "Here was a chance to change the world," he says, by tackling the issues of climate change, natural-resource depletion, human health, and animal welfare with a single product. "It wasn't brilliant foresight," Brown likes to remind people. "It was being unable to tolerate not doing it."[17]

Prius for the Plate

Per pound, beef is far and away the largest single contributor to the world's greenhouse gases, producing seven times as much as the equivalent amount of chicken – and sixty times as much as peas.[18] Consuming less of it would mean less demand on land, feed crops, and water supplies. But the typical veggie burger was never going to appeal to beef's mass audience. The traditional stand-in for the meat-averse simply couldn't replicate the hamburger experience. With only 5 percent of the US population keeping a vegetarian diet, Brown understood that he'd need a different path to go after the trillion-dollar US meat and poultry market.[19] He identified three potential avenues: lab-grown meat, in vitro meat, or developing a process that used plant protein to make food taste and feel like real meat. Brown thought the latter would be the most palatable to consumers.[20]

"The thing that I think most people didn't capture that well is that meat is a completely knowable entity. You can unpack it. You can figure out what's in it. You can analyze it to death," says Brown. "And so, to do that and rebuild it was the approach."[21]

Brown started by investing in a vegan restaurant that served textured soy protein from Taiwan. The establishment did so well – even attracting a non-vegetarian consumer base – that he opened a business in Maryland importing the faux meat from Taiwan. He made one of his first deals with Whole Foods to distribute it throughout the mid-Atlantic region.[22]

But Brown knew that textured protein was never going to fool anyone into believing they were eating real meat. So he began looking for a substitute that had the same taste and consistency and came across researchers from the University of Missouri who were claiming amazing results. In 2009, he set up a visit and got a first glimpse at the technology that would eventually become the heart and soul of his company, Beyond Meat.

Food science professor Fu-Hung Hsieh and researcher Harold Huff had managed to transform powdery pea protein isolate and soy protein into something that tasted like fibrous protein. Using a sophisticated piece of equipment called an extruder, which is something like a pressure cooker crossed with a food processor, dry ingredients are mixed with water, kneaded and heated, then pushed through a specially designed cooling process that realigns the food's protein molecules in ways that give them a texture that closely resembles meat.[23] "It's never about taste," says Huff. "It always about fibrous structure. Mouthfeel is everything."[24]

Hsieh and Huff did everything they could think of to manipulate the moisture, pressure, and temperature of various ingredients to get the texture right. After making some significant progress, they began showing their product to a wide variety of individuals, including Ethan Brown. Luckily for Brown, he was the only entrepreneur who clearly saw the market potential and started the long process that eventually took Hsieh and Huff's food to local supermarkets.

The following year, Brown exhausted every resource he had to license the technology and launch Beyond Meat.[25] He started by sending emails to Silicon Valley venture capital firms, calling his pitch the "Prius for the Plate," but only got a few replies. "My inbox wasn't exploding," he jokes.[26]

There was no shortage of doubters and potential investors who thought the idea was "bizarre" or "gross." He'd respond by asking, "Have you ever seen how a piece of meat is made?"[27]

In the Wilderness

Eventually, he got a response from Raymond Lane, a partner at Kleiner Perkins, one of Silicon Valley's premiere venture capital firms. To satisfy his curiously, Lane got onto his private jet and flew out to Columbia, Missouri. When he tasted Brown's food, "It had a texture and a fiber that was meat-like," Lane recalls. "And then as I took the product and tore it apart it looked like chicken."[28]

Lane was sold, but he had to persuade his firm to invest. They had reservations about the food market and Brown's lack of experience. "There were all sorts of skeptics about doing this," Lane recalls.[29] But in April 2011, Kleiner Perkins invested $2 million.

Twitter cofounders Biz Stone and Evan Williams were also interested. "My first thought was, 'Oh boy, some kind of hippie guy who is going to preach about how mean eating animals is,'" says Stone, who's a vegetarian himself, of their first meeting with Brown.[30] After questioning Brown, they were impressed with his business experience and his strategy to get into supermarkets and restaurants. They decided to invest. "This was no novelty food thing that was going to go into the freezer section next to the silly products," says Stone. "This was a guy who from the beginning was saying, 'We are going to be in McDonald's.'"[31]

Eventually, other big players joined in, including Taiwan's Tsai family, Morgan Creek Capital, DNS Capital, and Honest Tea founder Seth Goldman. But at that point, money was still tight. Brown had already run through his 401K and savings accounts, and even sold one of his houses. "I was in the wilderness," Brown says. "For a couple of years, it was everything going into it. There is a lot of collateral damage that goes along with that."[32]

When Brown traveled to pitch Bill Gates, who eventually invested in Beyond Meat, he had a hard time paying for his hotel room, having maxed out his credit cards.

But that same year, Beyond Meat released its first product at Whole Foods: chicken strips. Bill Gates wrote a blog post about his experience eating a Beyond Meat chicken taco: "The 'meat' was made entirely of plants. And yet, I couldn't tell the difference. What I was experiencing was more than a clever meat substitute. It was a taste of the future of food."[33]

The company went on to introduce precooked frozen burgers and eventually their signature burger made from peas. And by 2016, Beyond had a giant breakthrough: Whole Foods decided to sell Beyond Burgers in the meat section. In short order, other major food retail stores followed suit. By now it was clear: Brown's products were not just speaking to vegans. Mainstream consumers were finally ready.

Is it Fake or Real?

Today, Beyond's products are available in over nineteen thousand retail stores and restaurants, to the tune of $250 million in annual revenue. The Beyond Burger can be found in the meat sections of more than five thousand grocery stores and on the menus of over four thousand hotels, college cafeterias, and restaurants, including TGI Fridays, Bareburger, Beggie Grill, and BurgerFi.[34] Brown wants to eventually rival JBS, one of the world's largest meat processors with $45 billion in revenue. "Our ambition is to literally be in every major continent changing the way people eat," he says.[35]

Every day, Beyond tests and develops products in its food lab in El Segundo California, called the Manhattan Beach Project Innovation Center. (Brown liked the atomic allusion.) The lab's scientists have made it their mission to take plants and transform them into burgers, sausages, and chicken that taste exactly like the real thing – with a significantly smaller environmental footprint and no cholesterol.

"Not only are food innovators trying to build a new kind of rocket ship to take us someplace that we have never been before," says culinary scientist Ali Bouzari, "but we are also trying to turn that vehicle into a Honda Civic and bring it to the mass market in a few years. It is the food industry's version of SpaceX."[36]

A recent study by the University of Michigan showed that the Beyond Burger generates 90 percent less greenhouse gas emissions and requires 46 percent less energy to produce than a beef burger.[37] They also need less water and land to produce.

In 2019, Beyond Meat went public, and shortly after, McDonald's announced that the company would be its Canadian plant-based burger supplier.[38] That same year, Dunkin' introduced its Beyond breakfast patty.[39] And as demand for meat alternatives increases across the industry, more and more food providers are either using Beyond products or adopting their own. Burger King has experienced record sales with one of Beyond Meat's largest competitors, Impossible Burger. Nestlé is trying to compete with its Awesome Burger.[40]

If Brown is right, in the not-too-distant future we will be able to buy plant-based burgers from most supermarkets and food establishments. "The consumer is turning so quickly," says Brown. "Once we break the code and get to the point where it's indistinguishable from animal protein, I think you will see that shift."[41]

"This is the first time in human history we are literally building food from the bottom up," says Nir Goldstein, managing director of Israel's Good Food Institute. "And this is just the beginning."[42]

For now, plant-based protein represents what Rusty Schwartz of food-innovation incubator KitchenTown calls "the tip of the spear for disrupting the meat industry."[43] And the industry has taken notice. In a tactic reminiscent of the milk industry's efforts to derail the exploding popularity of nut-based beverages, Big Meat is on the attack, lobbying for regulatory changes to ban the labeling of plant-based burgers as "meat." One meat industry–backed nonprofit called the Center for Consumer Freedom recently launched a national marketing campaign. "Fake meats grow on vines," reads one of its newspaper ads. "They're ultra-processed imitations that are assembled in industrial factories."[44]

Brown, for his part, is fighting back. "The automobile is not a fake horse-drawn carriage," he says. "It's just plain meat. We are building it from the same stuff, so we should be able to call it meat."[45] And to detractors who argue that burgers made by companies like Beyond Meat have larger carbon footprints than traditional veggie burgers, Brown has an easy answer. "How many people are eating black-bean burgers?" he asks. "That is like taking a Tesla, which does have a good environmental outcome relative to an internal combustion engine, and comparing it to a bike. It's just not relevant."[46]

If capitalism is a relentless march toward efficiency, the meat industry has plenty of cause for concern. "Plant-based protein and 'clean' meat is without a doubt the future," says vegan advocate and CEO of data. World Brett Hurt. "We can save our planet's most precious resources at one tenth the cost – without giving up taste."[47]

CHAPTER 11

Water: Every Last Drop

When the well is dry, they know the worth of water.

– Benjamin Franklin, *Poor Richard's Almanack*, 1746

Burst water main (Richard Webb, CC BY-SA 2.0)

No Magic

On a cloudless summer afternoon in 2014, UCLA students heading to class were met with the terrifying sound of rushing water. Just before 3:30 p.m., a ninety-year-old water pipe ruptured, creating a thirty-foot geyser that was disgorging thirty-five thousand gallons per minute onto Sunset Boulevard. Firefighters arrived at the scene within minutes and used inflatable boats to save at least five stranded people. The school's storied athletic facility, Pauley Pavilion, was flooded, and parts of campus were unrecognizable. Hundreds of cars were abandoned in underground parking lots as water cascaded down stairs, and some students resorted to getting around on body boards. Emergency responders spent hours looking for the right valves to shut off.

While they did, about ten million gallons of water were lost during the worst drought in the city's history.[1]

Jim McDaniel of the Los Angeles Department of Water and Power later explained that the city is on a three-hundred-year replacement cycle for water main lines, and that there's "no magic" technology to determine when new lines are needed. "Every city that has aging infrastructure has issues like this and we're no exception," he said.[2]

But halfway around the world in Israel, a man named Amir Peleg had already designed such a software system. Called TaKaDu, it employs algorithms and statistical analysis to identify problems like leaks and burst pipes in real time, using data to revolutionize the efficiency and maintenance of water utility infrastructure.

Passing the Water Buck

More than two billion people currently lack access to safe drinking water, and by 2025, at least half the world's population is expected to be living in water-stressed areas.[3] The entire Middle East is headed toward massive water shortages, which in some places will likely lead to disaster of biblical proportions. The United Nations predicts that Egypt will approach a state of "absolute water crisis" in just a few years;[4] Jordan is set to run out of potable water in a matter of decades, and Iranian government officials project that in fewer than twenty-five years, over half the country's population will need to be relocated as water refugees.[5] In Iraq, engineers warn that the Mosul Dam could collapse at any minute, killing 1.5 million people.

According to the World Resources Institute, a global nonprofit that focuses on sustainable development, the crisis is the result of a confluence of issues, from climate change and its effects on a region's aridity to water demand, depleted groundwater, poor water infrastructure, lack of healthy ecosystems, water waste management, and the rising price of water.[6]

Many will be surprised to learn that the trend also holds true for the United States. According to the US Government Accountability Office, over the next ten years, forty states will have at least one region with a water shortage due to the lack of fresh water in lakes, rivers, reservoirs, and aquifers. And in just four years, at least a third of US households will lack potable water.[7] America's water and wastewater bills have increased by more than 30 percent in the last ten years, rising faster than inflation,[8] in large part because rising energy and transportation costs have made repairing and renewing pipes much more expensive. And as the price of water increases, millions of Americans will be unable to pay their water bills.[9]

The United States has approximately 1.6 million miles of water and sewer pipes that are, on average, forty-five years old – with many installed more than a century ago, and at least ten million made of lead that contaminates our water.[10] There are now more than seven hundred water main breaks in the United States every day.[11]

Americans are generally unaware that most of their water utility companies either lose money or barely break even. Between government subsidies and household water bills, water providers collect just enough revenue to conduct their business and handle ongoing problems. According to a recent study by Michigan State's Elizabeth Mack, utility companies are now spending approximately 80 percent of their revenue to maintain infrastructure that was built primarily in the 1930s and 1940s.[12] Mack estimates that updating aging infrastructure will cost over $1 trillion over the next twenty-five years, and that water prices will increase to four times their current levels.

Around the world, there are approximately 250,000 water utility authorities locked in the very same battle with leaks and burst pipes. They often don't discover the problem for days or weeks – if they do at all – or until the leak causes a rupture, impacting homes and businesses and electrical and telecommunications equipment underground, compounding the cost of water even more. Water authorities use a variety of methods to detect leaks, including fiber-optic cables, radar, and physical inspection crews that use listening sticks and acoustic loggers. But these efforts are generally futile, and the highest costs are actually associated with round-the-clock staffing to ensure continuity of service.

"Ninety-nine times out of a hundred, water authorities have traditionally relied on their customers to tell them there is a problem," says Tony Kelly, former CEO of the Yara Valley Water authority in Melbourne, Australia. "And ultimately, it is the consumer that pays the price for this kind of inefficiency."[13]

When water does not reach its customers, it is referred to as non-revenue water, or NRW. Experts believe that 25 to 30 percent of the world's water supply – which could otherwise be used to drink, shower, and irrigate – is NRW lost to distribution, costing about $14 billion annually.[14] The status quo, according to Kelly, isn't just uneconomical – it is unacceptable.

The Mega Problem

In 2008, Israel's Amir Peleg decided to do something about it. A self-described serial entrepreneur with a passion for data analytics, Peleg had already launched and sold three successful companies by the age of forty-two, including Elbit Vision Systems, which specialized in quality-control technology for fabric manufacturing; Cash-U, which focused on the mobile gaming industry; and YaData, a behavior-based digital marketing firm that was acquired by Microsoft. As Israel's drought problems worsened, the Hebrew University of Jerusalem graduate had grown increasingly interested in clean tech. "Israel is scarce of water so we are taught to appreciate it," Peleg says. "I didn't find many companies dealing with innovative water technologies."[15]

Peleg didn't know much about the industry, so he began reading voraciously on the subject and following global corporations investing in the field. One of the companies he did find looking into IT solutions for water utilities was IBM. "If a company like IBM believes in this market," Peleg thought, "there must be something there."[16] He cold-called Peter Williams, then the chief technology officer of IBM's Big Green Innovations, which was set up to tackle environmental problems. "We got along right from the start," Williams recalls. "He is your typical hard-charging driven character and wanted to get things done."[17]

Peleg also formed a relationship with Oded Distel, then the director of the Israel NewTech program at the Ministry of Economy and Industry. "At the time, Peleg only had a vague notion of how he wanted to revolutionize the water industry," Distel says. "I knew I wanted to help him achieve his vision and help guide him to become the new poster boy of saving one of the most important resources on the planet."[18]

Peleg had identified what he saw as a "mega problem" in a very global market.[19] "Water authorities are collecting data, but they aren't doing anything with it," he realized.[20] These entities have sensors on their pipes that measure water flow, pressure, quality, and many other important pieces of data, yet they were still losing so much of their treated water. Over time, he knew that water authorities would only increase the number of sensors they used, making data acquisition faster, cheaper, and more reliable. "The minute I understood a problem existed, that data is collected but not used, and that I could find people to work on the problem, I felt I had the makings of a new company," he says.[21]

In 2009, Peleg hired a few scientists and developed the software they needed, calling the new company TaKaDu. But the first problem they encountered was getting real data to test their algorithms.

Most water authorities they approached were deeply skeptical that data could be properly analyzed to yield actionable results.

Thanks to a phone call from Oded Distel, one company in Jerusalem decided to give them a chance. Distel set up a meeting with Aharon Rosenberg, chief technology officer of Hagihon, Israel's largest water utility, which services around one million people with 750 miles of pipes – on the promise that his colleague's company had a series of tools that could potentially change the way the water authority operated.

"How much will it cost me?" Rosenberg asked Peleg at their first meeting.

"Nothing," replied Peleg. "All I need from you is a CD-ROM with your data from the last year or two."[22] Peleg told him he would use unique algorithms and statistical tools to detect his leaks.

"Let's do it," said Rosenberg, burning the CD for him on the spot.[23]

High-Tech Plumbing

What Peleg's system discovered was astounding. Hagihon was losing a significant amount of water to people tapping into their network and stealing water. And TaKaDu pinpointed the exact location where the thieves were stealing it. His team was also able to identify the exact times and the general location of major burst pipes.

Eventually, more water companies got on board. "These utilities essentially said, 'If you can solve this problem, we're in,'" says Peleg. "'Here's our past data.'"[24]

TaKaDu's data analytics soon discovered that the UK's largest water authority, Thames Water, was losing 25 to 40 percent of its water to leaking pipes before it reached any of its fifteen million customers. Thanks to Peleg's system, they were now able to locate broken pipes in half the time. With the UK facing water shortages that the head of its Environment Agency has called the "jaws of death," the ability to curb the problem is critical.[25]

In Melbourne, the Yarra Valley Water utility, which serves over 700,000 customers with six thousand miles of pipes, was aware of the fact that they were losing about 13 percent of their water to leaky pipes. The managing director, Tony Kelly, was dubious about TaKaDu's ability to make any significant impact, but Peleg convinced him to give the software a chance in 2011. Yarra's pilot program with TaKaDu was able to detect large leaks two weeks faster than it was able to without it.

Over the next few years, Peleg's company started getting more paid clients, including Chile's Aguas de Antofagasta, a utility company that serves 100,000 customers in one of the earth's driest areas.

"We are very proud to be early adopters of the TaKaDu tools," says Marco Kutulas Peet, general manager of Aguas de Antofagasta, which became the first utility in Latin America to use the technology.[26] The Chilean utility increased the efficiency of its repair crews by 50 percent and doubled the amount of water saved with repairs.

TaKaDu software analyzes data collected by sensors located in a variety of places, including pipes, valves, and pumping stations. The challenge has always been what to do with all this information. If they've analyzed the data at all, many water companies have attempted to do so manually or through static algorithms. "This is a lot of information for people to stare at," says Peleg, "and [it] is very labor intensive."[27]

Peleg's system ingests data every fifteen minutes, stores it in the cloud, analyzes it, and compares it to historical data. It is able to detect any unusual activity and immediately classify what is happening, like the crucial difference between a leak, a burst, and a faulty meter. It has also evolved to become a full-blown operational platform that can detect and manage any incidents in a utility's network, including data problems, meter issues, and water quality control. It knows, for example, if an uptick in water usage is due to a national holiday, or if one part of the system is receiving less water because a crew is working on pipes and turned off certain valves. With a map-based user interface, water authorities using TaKaDu can see exactly where the potential issues are and deploy a fix. It also works as an early warning system, predicting adverse events and pinpointing parts of the physical infrastructure that need updating. In effect, Peleg says, TaKaDu has become "the eyes and ears" of water utility companies around the world. "I see myself as a high-tech plumber," he says.[28]

"Utilizing big data has yet to be universally adopted in the water industry," says Keith Hays, managing director of Bluefield Research, an independent water advisory firm. "What is inspiring about TaKaDu is that it is helping water authorities around the world get their act together in order to move from the twentieth century into the twenty-first."[29]

In the last decade, Peleg has secured water authorities in thirteen countries, including Australia, Brazil, Chile, Colombia, Finland, Holland, Israel, Peru, Romania, Spain, the United States, and Vietnam. The utilities using TaKaDu have reported a 30 to 40 percent drop in water loss. The company has since been named a "Technology Pioneer" by the World Economic Forum, "Company of the Year" for Europe and Israel by Bloomberg, and one of the top one hundred companies in the world by Global Cleantech.

"TaKaDu is playing a significant role around the world in helping countries curb the loss of water," says George Theo, CEO of Unitywater in Queensland, Australia.[30]

Christopher Gasson, publisher of Global Water Intelligence, agrees. "Digital systems like TaKaDu are completely changing how we find and stop leaks," he says. "They are making the invisible visible."[31] TaKaDu's reach is helping solve a local issue that when taken in the aggregate represents a massive global challenge.

Thanks to TaKaDu, water utilities can now be the first to detect the problem and proactively initiate repairs, greatly minimizing customer and community impacts. "Water isn't just water," says Seth

M. Siegel, water expert and author of *Let There Be Water.* "In the case of Israel, it's also an inspiring example of how vision and leadership can change a nation and transform the world."[32]

That vision was on full display in Mountain View, California, in 2014 when Governor Jerry Brown and Israeli prime minister Benjamin Netanyahu met to sign a monumental agreement that would transfer Israeli water technology to the Golden State to help mitigate its water crisis. At the time, 74 percent of the state was dealing with extreme drought. Israel, though 60 percent desert, had just announced that it had achieved water independence through smart planning and innovative thinking. No longer reliant on the weather – or its neighbors – for its water needs, Israel became a water superpower by charging users the real cost of water, desalinizing sea water, reusing treated sewage for agriculture, implementing drip irrigation techniques, and mandating that authorities actually spend 100 percent of all water and sewage fees on water and maintaining water infrastructure. And, of course, by using Peleg's software, which warns authorities about leaks, cutting waste by as much as 40 percent.

"California, I hear, has a big water problem," said Netanyahu after the signing. "We in Israel don't have a water problem. We use technology to solve it."[33]

CHAPTER 12

Governance: The Digital Republic

I should have called the Estonians when we were setting up our health care website.

– US President Barak Obama

Toomas Hendrik Ilves, President of Estonia, March 2016 (Chatham House, CC BY 2.0)

Leapfrogging: Analog to Digital

It was evening when Estonia declared independence from Mother Russia on August 20, 1991. Soviet tanks rolled through the countryside in an attempt to quell the movement. Paratroopers were on their way to the capital, Tallinn, to cut off the newly formed country's TV tower, which provided a critical means of communication with the population. Civilians and members of the Estonian Defense League, a paramilitary force, rushed to protect various strategic buildings, including the parliament and public broadcasting facilities. The Soviet military action failed, and within a few weeks, Boris Yeltsin's government recognized Estonian independence.[1]

Out from under the shadow of the Iron Curtain, the country's policymakers now vigorously debated how to supercharge the future of their citizens. After fifty years of Soviet occupation, much of the country's public infrastructure dated back to the 1930s, and fewer than half the country's homes had physical phone lines. As a first step, Estonia's leadership knew they had to find a seamless way to communicate with the outside world. They still had a rotary telephone exchange. Before World War II, Estonia's GDP per capita was roughly the same size as Finland's, but in the decades since, Estonians could only watch as their northern neighbors lapped them. Finland was implementing a digital phone network in Helsinki and as a beau geste offered its old 1970s analog telephone exchange to Tallinn. It would have covered all of Estonia's needs at no cost. But free isn't always free.

"I yelled and screamed and fought tooth and nail," recalls Estonia's then-ambassador to the United States and later president Toomas Hendrik Ilves. "We would be stuck with legacy technology."[2]

In a move that would set the stage for Estonia's remarkable reinvention in the decades that followed, then prime minister Maart Laar ultimately declined, knowing that his country's path to success would be tied to rapid technological advancement.[3] Despite the expense, Estonia decided to go with a digital phone exchange similar to Finland's.

With a small population – 1.3 million, about half the size of Silicon Valley – and a landmass roughly the size of Switzerland, the tiny satellite of the Soviet Union hadn't developed an international reputation for anything much beyond its lush forests, high literacy rates, and garlic bread. But in 1993, with the release of the original internet browser, Mosaic, Toomas Hendrik Ilves had two epiphanies: "Everyone," he intuitively felt, "was on a level playing field."[4] The world was now flat, and Estonia could compete at a global level. Ilves also read Jeremy Rifkin's *End of Work*, a book that described how automation put twelve hundred people out of work in a Kentucky steel mill and replaced them with just a hundred employees. Computerization, Rifkin believed, would significantly change the type of work humans did in the future.

Ilves understood that economic success no longer required an economy of scale. If his country played its cards right, the future president thought, Estonia could compete with the superpowers – or at the very least, much larger countries – and reinvent itself as "E-Stonia," a term he later coined.[5]

"From Kentucky's perspective, this was of course terrible," recalls Ilves. "But for Estonia, it was intriguing, because our fundamental existential angst is tied to our smallness."[6]

Ilves was a US citizen until he renounced his citizenship to become the Estonian ambassador to the United States,[7] and growing up in New Jersey, he learned how to code in high school. This deeply influenced his perspective on the role of schools and how to mold young minds. Two years after reading Rifkin's book, Ilves got together with the Estonian minister of education, and together they drew up a plan to computerize the country's school system. "We need to really computerize in every possible way," said Ilves, "to massively increase our functional size."[8]

By 1998, 100 percent of Estonian schools were online.[9] This kicked off a trend in Estonia to look for IT solutions where other countries couldn't or wouldn't because they were locked into legacy technology. "We leapfrogged developed countries that were still depreciating their huge investments in mainframes and couldn't adopt the new technologies as easily as we could," says Estonian IT visionary Linnar Viik.[10] Two years later, the government passed a law mandating internet access as a basic human right, propelling a series of events that fundamentally changed Estonian society.

In 2000, citizens could file their tax returns digitally for the first time. "We also had the additional sweetener that if you filled out your tax return online, you got your money back in a week," says Ilves. "If you did it on paper, it might take maybe three months."[11] Today, 99 percent of Estonians file their returns electronically.[12]

Free wi-fi soon became commonplace, and Estonia installed a widespread fiber-optic cable that reportedly gives citizens the fastest internet connection speeds in the world.[13] The government gave 10 percent of the adult population free computer training, and internet use increased from 29 percent of the population in 2000 to 91 percent in 2016.

Most people would be shocked to learn that Estonia is now the most advanced digital society in the world: the first country to permit online voting; to ensure all tax preparation and payments can be carried out seamlessly in minutes online; to guarantee health records are accessible from anywhere, anytime; to make starting a company as easy as clicking a button; and to teach every child how to code – starting at five years old. Everything from authenticated signatures to paying for parking, school fees, homework, government services, and tax preparation is connected on one digitally linked platform in the cloud using public internet.

"Politically and economically," says Viljar Lubi, deputy secretary general for economic development in the Ministry of Economic Affairs, "Estonia's goal is to become another rich, boring Nordic country."[14]

X Marks the Spot

In Estonia, the most important piece of infrastructure isn't a bridge or a highway – it's a digital network called X-Road. The dynamic platform allows information to be safely stored and exchanged online. Developed by the country's Information System Authority in conjunction with Cybernetica, an Estonian tech company, it resides on local individual servers with end-to-end encryption. Doctors, banks, and schools, for example, each might hold an individual's relevant data. "When a user requests a piece of information," explains journalist Nathan Heller, "it is delivered like a boat crossing a canal via locks."[15]

To get online services, citizens insert their integrated national ID card into a computer, then go through a two-layer authentication process. In most parts of the world, this is now accepted as the gold standard. At the time Estonia adopted these security protocols – long before Google did – it was revolutionary.[16] The country was also a pioneer of the digital signature, which allows citizens to sign legal documents and authenticate themselves to others in both the physical and digital worlds. The government has since integrated all kinds of public and private services into X-Road, from banking and medical records to phone bills, along with thousands of other services. And when its citizens voted in national elections using the system in 2005, Estonia became the first country in the world to do so. Today, nearly half of the population votes online.[17]

If a doctor prescribes medication, he puts it into a user's electronic record. A user then goes to the pharmacy, inserts the ID card, and authenticates it with a personal code, and the pharmacist provides the medication. Medical records are now securely available twenty-four seven to both medical practitioners and patients. A patient who wants a second opinion authorizes a second doctor (or as many as desired) to look at the records and make a determination. Ilves says that Estonia has now "broken a two thousand year Hippocratic tradition in which the doctor is the priest and the patient is the supplicant."[18]

Estonian citizens have their entire lives on the system. Personal profiles include email addresses, phone numbers, educational degrees, current and previous employment, insurance in-force, tax payments due, real-estate ownership, and even pet ownership. But Estonia has also put in place a "once-only" principle, which mandates that the government not ask citizens for data already held by a national public body. Authorities seeking data must first attempt to acquire it from their in-government agency, and if necessary, get permission to do so from citizens. In Estonia, citizens own their data, not the government or private institutions.

"Information technology is the basis for how the country organizes itself," says Robert Krimmer, a professor of e-governance at Tallinn University of Technology, "and its fundamental outlook strives for a hassle-free interaction for its citizens with the State apparatus."[19]

Estonia's approach to data represents a radical departure from both the Chinese and US models. In China, citizens' data is centralized and owned by the government; in the United States, corporations such as Facebook or Google own it. Estonian citizens own their own data, and therefore can chart their own future.

The annual cost of the system is relatively cheap, especially when compared to other e-government systems; Estonia's costs 50–60 million euros annually (the entire government budget is $8 billion), versus Russia's at $90 million and Kazakhstan's at $150 million, neither of which is as good as X-Road.

X-Road reportedly saves Estonia about 2 percent of annual GDP in salaries and other expenses, which it can reinvest elsewhere. "This is roughly the amount Estonia pays to be a member of NATO and in return gets protection," says former president Ilves. "I often joke that Estonia gets its national security for free."[20]

"Estonia is fifty or sixty years ahead of what is going on in Silicon

Valley," says Ilves, "where the private sector is revolutionized and is constantly reinventing itself, but in the public sector, not much has changed. It's time for the public sector around the world to shift from a Nokia to an iPhone mindset."[21]

From Russia with Love

Of course, the move to a completely digital public sector is not without its security risks. Estonia has confronted this reality on more than a few occasions, most profoundly in 2007. It was an eerily quiet night when the country came under attack. As leading newspapers, banks, and government communications went down, Estonia's minister of defense, Jaak Aaviksoo, sent his aide outside to see if there were enemy jets overhead or warships on the shoreline. But this was an invisible enemy – assaulting hundreds of targets simultaneously. Fifty-eight websites went offline on the first day.

Rogue computer networks, known as botnets, were attacking. "This affected the majority of the Estonian population," says Aaviksoo. "This was the first time that a botnet threatened the national security of an *entire* nation."[22] Some analysts have gone so far as to call the incident Web War I.

The minister had trained his mind to think about how to counter missiles, bombs, and traditional military raids. But cyberattacks? He didn't have any relevant experience to draw on. Aaviksoo debated whether to invoke Article 5, which would obligate NATO to join Estonia in a counterattack. But in this war, it was almost impossible to determine who was leading the assault.

The attackers knew exactly what they were doing, evading filters and forcing servers to crash. Computer geeks tried to figure out where the onslaught was coming from – the largest numbers appeared to be coming out of Egypt, followed by Vietnam and Peru. But it was clear that whoever was attacking was just tunneling through these countries as proxies. As websites across Estonia were swamped by foreign visitors, the available bandwidth hit zero. Banks, newspapers, and government agencies were forced to shut down all international web requests. With a click, thousands of sites went dark, and as far as the world was concerned, Estonia had digitally disappeared.

The onslaught lasted three weeks, with over 120 servers affected. Evidence suggests – but could never prove beyond a shadow of a doubt – that Russia was the culprit.

During the attacks, the Estonian government was transparent with its citizens, which built up a tremendous amount of goodwill. And in the aftermath, policymakers took action, doubling down on cybersecurity and the policy of decentralization. Each database is now compartmentalized and therefore distinct from any other server. While this means less efficiency in the short term, it is infinitely more secure in the long term. Since each database or system has a dedicated secure server, the entire system isn't compromised in the event of a breach. "Hackers would need to instantly and successfully hack at least 150 databases," says Marten Kaevats, Estonia's national digital adviser, "to steal someone's identity and do something meaningful with it. That is close to impossible."[23]

Cyberattacks continue to be the norm in Estonia – the country faced more than ten thousand incidents in 2017 alone – and will be for the foreseeable future.[24] In order to deal with the very real possibility of a serious breach, the government has installed a radically new fail-safe system: the world's first digital embassy, housed in a high-security data center in Betzdorf, Luxembourg. In the event of a crisis, Estonians can continue to digitally authenticate themselves, and all public-private services continue to run seamlessly. "It is no longer possible to move back to a paper era," says Siim Sikkut, Estonia's current chief information officer. "The Luxembourg site stores copies of the most critical and confidential data."[25]

Estonia's digital embassy, which has the same protection and immunity afforded to traditional embassies, maintains and preserves ten priority databases, including the treasury information system, pensions insurance, business, land and population registers, and the identity documents database. All this ensures that the country can keep all systems operational under almost any scenario. There are currently plans to put in place a total of six data embassies. "Cut one head off," says Taavi Kotka, the country's previous information officer, "and like Hydra, three more will grow back."[26]

Given the sensitivity of the data, security concerns are addressed on an ongoing basis. "End-to-end verifiability is the holy grail," says Tarvi Martens, the National Electoral Committee's head of e-voting.[27]

In terms of personal data, Estonia has also put in place a policy of complete transparency. Citizens can actually view a log of who has accessed their information, and a user who believes data has been misused or viewed by anyone illegitimately can file a complaint with the country's data ombudsman, resulting in almost real-time remediation. Every time an official of any kind looks at the secure data of an individual online, whether it's a banker, police officer, or government agency, it is recorded. A civil servant must justify the intrusion – looking at someone's personal data without cause is a criminal offense.

"You have to trust the government a little," says Ilves. "Whom do you trust more – putting your credit card number online with a three-digit CVV code? Or is the government sort of the guarantor of your identity?"[28]

Some have likened the system to "Little Brother." Estonia's former chief information officer, Taavi Kotka, has a simple answer for that. "There is no government that knows more about you than Google or Facebook," he says.[29] Estonia has created what he calls a "Lockean contract," where the government cannot access data without a user's knowledge, but citizens no longer have anonymity either.[30] The Estonian government has consistently made the case that the twenty-first-century networked society is a justifiable transparency bargain. Computers can't be bribed, and that means less corruption. "Our goal," says Ilves, "is to make it impossible to do bad things."[31]

Borderless Country

In 2013, then chief information officer Kotka was given a challenge: expand the population of Estonia for economic gain. "Countries are like enterprises," he says. "They want to increase the wealth of their own people."[32] The program he started the following year, known as e-citizenship, had four thousand registrants within the first eighteen hours.[33]

"If everything is digital, and location-independent," says Kotka, "you can run a borderless country."[34]

By handing over €100, a photograph, and submitting fingerprints at one of thirty-eight Estonian embassies around the world, anyone can now become an e-resident of Estonia. In order to access the country's national system, newly minted residents are given an identity card, a cryptographic key, and a pin code. E-residency does not provide a path to citizenship or allow for legal residency, nor can it be used as a travel document. What it does do is provide an individual with a supranational digital identity issued by a nation-state. E-residency changes fundamentally how we think about citizenship, the role of government, and the services it offers, but at its most practical, it's a powerful business tool.

Estonian e-residency provides an ID card that allows its user to authenticate their identity online, digitally sign documents, register a company, conduct online bank transactions, and order prescriptions from anywhere, anytime. Registering a business through the program is particularly useful for internet entrepreneurs in emerging markets and for startups in countries with burdensome financial limitations.

Both Estonian citizens and e-residents can perform nearly every public and private sector transaction in digital form. "The adoption of non-resident ID cards," says Kotka, "is an additional argument in favor of investing in Estonia."[35]

The thinking here is clear: the hope is that e-resident status will attract foreigners of all kinds to Estonia, including investors, digital nomads, specialists, and international policymakers. Because of the country's advanced banking system, the government expects it will attract entrepreneurs and additional revenue.

In 2018, President Kersti Kaljulaid handed Pope Francis his own digital ID card, making His Holiness the 37,647th digital e-citizen of the tiny Baltic country. Japanese prime minister Shinzō Abe, German chancellor Angela Merkel, Luxembourgish prime minister Xavier Bettel, Microsoft founder Bill Gates, Asia's ultra-wealthy Mukesh Ambani, and TV personality Trevor Noah have all become e-residents, giving them access to the country's e-infrastructure and services so they can establish and manage their finances and businesses inside the EU from anywhere.[36]

Because men make up the vast majority of applicants, the United Nations has been encouraging female entrepreneurs from India to apply as well, to great effect.[37] As of 2020, there are more than seventy thousand e-residents from over 165 countries.[38] The program has been so wildly successful that the e-residency rate now exceeds the national birth rate. "We tried to make more babies," says Siim Sikkut, Estonia's chief information officer, "but it's not that easy."[39]

By 2025, Estonia aims to enroll ten million e-residents. "We had to set a goal that resonates," says Taavi Kotka. "Large enough for the society to believe in."[40]

If they get anywhere close to that target, it would be a financial windfall for the country. For every euro that Estonia has invested in developing the e-residency program, it brings in at least 43 euros to the economy, according to the head of the country's e-residency project, Kaspar Korjus. "I envision a time when Estonians will not actually have to pay taxes," Korjus says. "Instead we will offer telemedicine, online education, and pensions to people all over the world who will pay a minor fee which will cover all of our government's expenses. That is when we will achieve E-residency 2.0."[41]

To Infinity and Beyond

In 2015, Estonia took a bold step in making its X-Road technology open source and available to any individual, organization, or country to use as they please. The source code for the program has been published under an MIT license (a nonrestrictive software license developed by the Massachusetts Institute of Technology) and is available free. "This is the beginning of the erosion of the classic nation state hegemony," observes John Clippinger, a digital identity researcher at the Massachusetts Institute of Technology. "It's going to get whittled away from the margins."[42]

Both Finland and Iceland have begun to implement a partial version of X-Road as they begin their long journey toward digital societies. To date, countries such as Canada, Rwanda, New Zealand, Singapore, and Uruguay have made significant progress toward digitizing their e-government public services infrastructure. Many more countries have expressed interest by sending delegations to Tallinn to better understand the Estonian model.

"We are riding the crest of a very powerful wave," says Aaron Maniam, deputy secretary of Singapore's Ministry of Communications and Information, "that in the coming few years will revolutionize how citizens interface with government and their fellow countrymen. This type of experience is unparalleled in human history."[43]

Estonia's position as the global leader in e-government has not gone unnoticed. "You're so much further than we are,"[44] admitted German chancellor Angela Merkel at the European Digital Summit. Even President Obama joked that he "should have called the Estonians" when he was setting up his healthcare website.[45]

Estonian president Kersti Kaljulaid, who identifies herself as the "president to a digital society," sees a world in which international borders will become increasingly irrelevant. "Our citizens will be global soon," she says. "We have to fly like bees from flower to flower to gather those taxes from citizens working in the morning in France, in the evening in the UK, living half a year in Estonia, and then going to Australia."[46]

If she has her way, what is happening in this tiny corner of the Baltics will by no means end there. "We have to embrace digital society with our minds and our souls, without fear," Kaljulaid says. "It is the future."[47]

CHAPTER 13

Security: Google's X-Man

The one thing that unites all human beings, regardless of age, gender, religion, economic status, or ethnic background, is that, deep down inside, we all believe that we are above-average drivers.

– Dave Barry, *Dave Barry Turns Fifty*

Sebastian Thrun speaks onstage during TechCrunch Disrupt SF 2017 (TechCrunch, CC BY 2.0)

The First Self-Driving Cars

Sebastian Thrun was standing in the middle of the Mojave Desert waiting for Stanley. It was a hot, dry day in October 2005, and along with about a thousand other people who had gathered in the grandstands for the $2 million Grand Challenge race held by the Defense Advanced Research Projects Agency (DARPA), Thrun was staring intently out at the horizon when the loudspeaker crackled. "I see a dust trail in the vicinity!" boomed the announcer. "I have a visual on bot number three! Ladies and gentlemen, here comes Stanley!"[1]

Stanford University's self-driving car, a modified Volkswagen Touareg, was about five hundred feet from the finish line when it stopped. Thrun and his racing team held their breath, along with the rest of the crowd. And then Stanley lurched back to life and crossed the finish line, becoming the first autonomous vehicle ever to complete what had started to seem like an unachievable mission set by DARPA.

"The impossible has been achieved," declared Sebastian Thrun, Stanford's Artificial Intelligence Lab director.[2] He and his team had proved that machines could drive safely and speedily over rugged terrain without human assistance. Roughly seven hours after the 132-mile race began, the packed grandstand erupted.

In the months leading up to the event, Stanford graduate students spent countless hours programming Stanley with algorithms that would allow it to self-learn the unforgiving desert contours of DARPA's track. The bot was connected to sensors and customized software that made it smarter with each new ride. That technology, when extrapolated and shared across a connected network as envisioned in autonomous rideshare models, is now known within the automated vehicle industry as fleet learning, which allows cars to improve exponentially with each mile driven. Fifteen cars had entered the previous year's Grand Challenge, but not a single one completed the course. Most crashed, flipped, or rolled within minutes – and the one that traveled the furthest only made it eight miles before bursting into flames. After an eighteen-month development cycle, Stanley was now able to find the most efficient paths, overcome obstacles, and stay on course to navigate a treacherous desert journey the distance of New York to Wilmington – entirely on its own.

It was an astonishing feat that was not lost on Google cofounder Larry Page, who was watching incognito in the stands in a hat and sunglasses. That day, he set his sights on Sebastian Thrun, eventually luring him away from Stanford to Google. Thrun would go on to found and run the company's secretive think-tank research lab, Google X. Its first project? Self-driving cars.

Mr. Roboto

Born in Solingen, Germany, in 1967, Thrun was obsessed with building things since childhood. He spent hours with Legos, putting together trains, cars, and complex structures. And when he turned eleven, he got his first TI-57 programmable calculator, which allowed him to write programs up to fifty steps. "I got very enthusiastic about seeing just what you could do with that," recalls Thrun. "Could you program a game, could you program complex geometry, could you solve financial equations?"[3] Between the ages of twelve and fourteen, he spent every afternoon proving the answer was yes, documenting each successful program he created in a booklet. By sixteen, he was programming his own video games on a NorthStar Horizon computer.[4]

But it was only after the tragic traffic death of his best friend at the age of eighteen that Thrun realized what he really wanted to do with his preternatural talents. Traveling on an icy road, the vehicle's driver had lost control and collided head-on with a truck. "He didn't pay attention for a fraction of a second and that was it,"[5] says Thrun. He knew he couldn't bring his friend back, but he could do something for all the people who die needlessly in car crashes. "I decided I would dedicate my life to saving one million people every year."[6]

Thrun enrolled at the University of Hildesheim and later the University of Bonn for his college and graduate studies. He became interested in the brain and human intelligence and took classes in philosophy, medicine, and psychology, but found what he truly enjoyed was bringing those subjects together holistically through computer science and artificial intelligence. "You could actually build something from the ground up that would then manifest intelligence," says Thrun. "That fascinated me."[7]

Thrun eventually found his way to robotics, which he felt offered the best way to study intelligence. In 1994, at age twenty-seven, he programmed his first robot, called RHINO, which was able to clean kitchens. The state-of-the art functionality was unheard of at the time. He won second place in a US robotics competition. Three years later, he created a robotic tour guide for museums. Visitors to the Deutsches Museum in Bonn and the Smithsonian's National Museum of American History were greeted by Minerva, an autonomous guide that made all its own decisions. It found museum guests by itself, interacted with them, directed them to specific exhibits, and even explained what they were seeing. "It had a face, it could smile, it could frown," says Thrun. "And it was great fun."[8]

One night, he was programming Minerva for the next day when a human guide came by and started speaking to the robot. She looked her up and down, Thrun recalls, and said, "You are not going to replace me."[9]

As an adjunct professor at Carnegie Mellon in Pittsburgh, Thrun took on a variety of projects, including robots that could tend to the elderly and map out abandoned coalmines, as well as autonomous helicopters that could map terrain. In 2003, he moved to Stanford, where he ultimately took over as the director of its Artificial Intelligence Lab and won DARPA's Grand Challenge. When Thrun left Stanford to work as a full-time manager at Google in 2007, he thought it would only be a sabbatical. For his first three years, he developed Street View, a feature of the site's map program that has now systematically photographed more than ten million miles of the world's streets and houses.

By 2010, Thrun was appointed to run Google X, a research lab and innovation center with, as Thrun put it, the "authority to just do crazy things."[10] It's where they work on the moon shot problems, and the self-driving car was first on the agenda. But at the time, Thrun wasn't convinced it was even possible. And at one of the project's early meetings with Google founders Larry Page and Sergei Brin, he let them know as much.

"OK, Sebastian," Page challenged him. "Prove to me it's not possible."

The three sat down and carved out a thousand miles of California streets "that are really insane to drive," recalls Thrun. They included downtown San Francisco and the famously steep twists of Lombard Street, along with Highway 1 from the Bay Area to Los Angeles.[11]

"If you drive those well and completely autonomously," Page and Brin told him, "you'll make a bonus." Thrun insisted it couldn't be done, but Google's founders were adamant. "It can be done," they said. "Just do it."[12]

And that's just what Thrun did. Working in secret, he and a team of engineers took the technology developed for the DARPA race and applied it to the challenge laid down by Page and Brin. "We had a separate building that no one knew about," recalls Thrun. "At least for a year and a half, no one in Google had a clue we existed."[13]

By October 2010, with the help of fellow engineers such as Anthony Levandowski, who would later go on to run Uber's driverless car operation, and Chris Urmson, who founded Aurora, Thrun and his team had once again achieved the impossible. "We have developed technology for cars that can drive themselves," he announced in a statement on Google's blog. The company had officially succeeded in getting their self-driving cars from Google's

Mountain View campus to its office in Santa Monica – down Lombard Street and over the Golden Gate Bridge. In all, their cars had logged over 140,000 miles without a single accident.[14]

"We think this is a first for robotics research," wrote Thrun.[15]

In 2012, the Nevada Department of Motor Vehicles became the first state entity to "license" a driverless car, Google's modified Toyota Prius, after it took officials for a successful cruise down the Las Vegas Strip. California, Florida, and Michigan passed laws permitting autonomous vehicles a year later. After a series of subsequent prototypes, the project became officially known as Waymo, short for "A new way forward in mobility."[16]

In 2013, Google constructed testing grounds for its self-driving cars two hours south of San Francisco in Atwater, California. Originally an Air Force base where bomber pilots trained for war, the secretive ninety-one-acre compound known as the Castle, which is surrounded by an eight-foot perimeter fence, is now used by Google to perfect its self-driving car. Engineers, test drivers, and designers choreograph a universe of driving scenarios, structured tests, and drills that its cars must repeat until they get them right every time. The obstacle course includes curbs, complicated highway merges, busy intersections, roundabouts, driveways, stop signs, traffic lights, and even a railroad crossing. Cars grapple with bikers, malicious potholes, flying debris, orange cones – and reportedly even people leaping out of portable toilets.

The testing at the Castle compliments the billions of miles Waymo's vehicles "drive" in simulators and the millions of actual miles they have now driven on America's roads, in more than twenty-five separate cities. By 2018, the technology was far enough along for Google to launch its self-driving taxi service Waymo One, which allows users in the pilot city of Phoenix, Arizona, to hail driverless rides with their phones. In 2019, Waymo One became the first autonomous taxi service in the world to operate without any safety drivers in the car.

Thrun, who himself drives a Tesla, believes we have already arrived at a point where autonomous vehicles can actually drive better than the best humans.[17] He jokes that one person in particular – his wife – agrees with him: "When you are in the self-driving car," she chides him, "can you please let the car take over?"[18]

Joy Ride

Automobile accidents currently rank as the eighth leading cause of death worldwide, with nearly 1.3 million fatalities each year and an additional twenty to fifty million victims injured or disabled. An estimated 94 percent of those crashes are caused as a result of human decisions, including drunk driving, distractedness, and drowsiness.

"When you put these figures into context, the human tragedy couldn't be more clear," says Eric Danko, director of federal affairs for Cruise, a leading self-driving vehicle company. "We're literally talking about more than a hundred people dying every day on US roads, the equivalent of two 747s falling from the sky every week. AVs hold enormous potential to save lives."[19]

Thrun and others believe that cars will soon become almost entirely immune to fatal collisions by eliminating human intervention. Drunk driving will be all but eradicated. And the shift will represent a revolution for the many millions of people who are blind, have neurological diseases such as Alzheimer's or Parkinson's, or any other physical condition that impairs driving. They'll also give people who are too young or too old to drive a tremendous amount of freedom. And then there's the potential to profoundly reduce traffic – not to mention the hours of productive time we'll all have back when we're not wasting so much of it driving.

"AVs will fundamentally alter transport systems," says Kara Kockelman, professor of transportation engineering at the University of Texas at Austin. "They not only have the potential to avert deadly crashes, but they can also provide mobility to the elderly and disabled, increase road capacity, save fuel, and lower harmful emissions."[20]

But perhaps most fundamentally of all, Thrun also envisions a future society where we don't have a need for private car ownership – and the tremendous environmental impact that will bring beyond just reduced emissions. Cars would come to consumers on demand, meaning we wouldn't need our own garages or driveways, and workplaces wouldn't need as many parking spots.[21] Cities currently allocate approximately a third of their usable land for parking. Thrun and other industry experts hope to reinvent the curb for green spaces and pedestrian use.

"I'm really looking forward to a time when generations after us look back at us," Thrun says, "and say how ridiculous it was that humans were driving cars."[22]

Perfect: The Enemy of Good

Like all other revolutionary innovations, self-driving cars will force society to grapple with technical and moral questions, particularly when it comes to safety. Can autonomous vehicles ever be 100 percent safe? And do they actually need to be? In the United States, human-driven cars cause one death for every hundred million miles driven.[23] In a Rand Corporation study titled "The Enemy of Good," researchers concluded that deploying cars that are just 10 percent safer than the average human driver would save more lives over the long term than waiting until they are almost perfect. In other words, more lives could be lost while we wait for the unflawed autonomous vehicle.[24]

"In a perfect world, autonomous vehicles should be at least as good as human-driven conventional ones," says the director of Princeton's transportation program, Alain Kornhauser, "but for the foreseeable future, we cannot produce that because when humans pay attention, we are really good. The problem is we're rarely ever fully focused, so over the long haul, we're better off relying on technology that doesn't need sleep, argue with its spouse, play with a cellphone, or have one too many drinks."[25]

For now, the biggest obstacle in getting self-driving cars safely on the road appears to be human. According to *Axios*, from 2014 to 2018, people were responsible for eighty-one of the eighty-eight accidents involving AVs in the State of California. Of the sixty-two accidents that took place while the car was in complete autonomous mode, only one was caused by computer error.[26]

One accident in particular captivated the world precisely because it was technical, not human, error that caused it – leading many to question the wisdom of self-driving cars. In March 2018, an Uber autonomous vehicle killed a pedestrian for the first time in history. The accident took place at night in Tempe, Arizona, just as forty-nine-year-old Elaine Herzberg was crossing the road with her bicycle outside of the sanctioned crosswalk.

According to the police report, the self-driving Uber's safety employee behind the wheel was not watching the road in the moments leading up to the crash, but rather watching Hulu. In September 2020, the driver was charged with negligent homicide. In the aftermath of the accident, Uber temporarily suspended autonomous road testing in Tempe, Pittsburg, San Francisco, and Toronto. "It was not an indictment of the industry," says Uber senior safety program analyst Chan Lieu, "but it was a wakeup call that showed we needed to dial it up to eleven to ensure complete safety."[27]

While Google and Thrun have done amazing things to further the safety of autonomous vehicles, they are far from alone in the space. Beyond Waymo and Uber's Advanced Technologies Group, almost every automaker in the world is working feverishly to create the ultimate self-driving machine – and carve out a piece of what is expected to be an $8 trillion industry. There's the billion-dollar GM acquisition Cruise and the billion-dollar Ford acquisition Argo AI, as well as many more smaller competitors like Aurora, which was started by former Google and Tesla chiefs. And then there's Tesla itself, which – unlike Waymo and other companies developing autonomous vehicles – doesn't rely on an internal test fleet to collect data.

By 2020, there were 730,000 Teslas on the road that logged over three billion miles using its autopilot capabilities.[28] "More than any other car manufacturer, Tesla is racking up autopilot miles to ultimately ensure road safety," says Scott Hardman of the Institute of Transportation Studies. "This is the long tail of black swans that will train automated vehicles to deal with extremely rare events."[29]

In recent years, Waymo announced that its 360 degree LiDAR (light detection and ranging) system can now accurately provide a bird's-eye view of cars, cyclists, and pedestrians surrounding its vehicles. The technology enables Waymo cars to navigate the complexities of chaotic urban streets and spot hazards such as highway debris from hundreds of meters away. It makes Stanley's 2005 amble through the empty desert look like a baby's first steps.

When my youngest son, Yaniv, first saw one of these giant LiDAR cylinders spinning around on top of an autonomous vehicle near the US Capitol building, he pointed at it in excited confusion. I explained what the funny-looking thing did and why it meant that humans wouldn't be driving for much longer. "No way, Baba," he said. "You mean when I'm big, the car will drive itself?!"

Phantoms at the Wheel

Inventors have been dreaming of autonomous vehicles since shortly after the birth of the first motorcar around 1885. Forty years later, in 1925, Francis Houdina innovated a radio-controlled car that drove through the streets of New York City without a human physically steering. According to the *New York Times*, the vehicle was able to start its engine, shift gears, and sound its horn, "as if a phantom hand were at the wheel." And in the 1990s, Dean Pomerleau created the first bare-bones self-driving vehicle that was able to use raw images from the road to steer in real time. Pomerleau and Todd Jochem, both Carnegie Mellon researchers, drove their minivan 2,797 miles from Pittsburgh to San Diego in what they called "No Hands Across America," by only having to control speed and breaking.

But even then, the idea of being driven around by a series of zeros and ones still seemed insane. Two decades later, thanks to Thrun and scores of others in the industry, summoning robot taxis from wireless devices in our pockets is an actual reality. General Motors CEO Mary Barra predicts that the automotive industry will change more in the next five to ten years than it has in the last fifty. "I believe only 1 percent of interesting things have been invented yet," says Thrun.[30] By the end of this decade, fully autonomous vehicles are expected to be available at a car dealership near you, ready to drive themselves off the lot.

"You can't just sit back and say, no, it's not going to happen," says former GM vice president Lawrence Burns, whose book, *Autonomy*, explores the potential for driverless cars to completely reshape our world. "You can't hide from the amount of technology coming. We have passed the tipping point, it's inevitable."[31]

AFTERWORD

Humanity 2.0

The future is brighter than we think.

– Ray Kurzweil, CeBIT keynote

"Kilometer 104," the conductor barked over the loudspeaker. The train stopped, and my guide and I stepped into the darkness. Outside, there was no real station. Just a dirt road and a sign that read, "Welcome to the Inka Trail." I had come to Peru to see Machu Picchu, one of the Seven Wonders of the World.

As Andean condors hissed above us, my guide and I set off on our journey up the mountains, equipped with only what we could carry in our backpacks: snacks, water, bug spray, and coca leaf candy for altitude sickness.

Soon, the sun began peaking over the horizon, and I could see what the darkness had hidden – lush green gorges and ancient stone temples, majestic valleys, and llamas grazing in the grass. As we climbed high into the Andes, I took in the panoramic views, but by the time we reached the ten-thousand-foot summit, I was lightheaded. More than once, I had to stop and take a deep breath.

After hours of hiking, we reached the majestic Inti Punku, or the Sun Gate, which marks one of the main entrances to the ancient city. I tried to imagine the place before it was abandoned in the sixteenth century, before the mighty Incan empire fell to the Spanish conquistadors. I wondered: Was the collapse of this civilization inevitable? Was the collapse of *every* civilization inevitable?

In the relatively short arch of recorded human history, few advanced societies have lasted more than a few hundred years, and just a handful more than a thousand. The Incans came and went in just three hundred years. Could technology and innovation have saved them?

Archeological evidence suggests that most of the world's great civilizations, from ancient Rome to the Mayans, fell due to familiar forces such as population movement, disease, famine, climate, and war. As I stared at the Urubamba River thousands of feet below, I thought about the challenges we will face in the future, and how they'll dwarf anything that's come before.

For the most part, we only have ourselves to blame for most of the Earth-sized problems we now face. From food scarcity to water insecurity, pollution to global warming, we have been hastening the planet's destruction at an alarming rate. And if we don't change our behavior, we may drive ourselves to extinction. But we are also at a unique inflection point, because for the first time in our history, technology has given us the ability to change our destiny. We can consciously choose to turn the tide, to break the patterns that have undone us in the past. And for our civilization to survive – and even thrive – will require a generation that's not only inspired by the question of what's next, but hears it as a clarion call to action.

The visionaries I've had the privilege of writing about have already demonstrated that our future *can* be better – and has the potential to be more wonderful than anything we've yet experienced as a species.

Underneath it all, there's also a compelling economic argument for progress. Until the Industrial Revolution, the world was always a winner-takes-all kind of place. The majority of people worked on farms, and for thousands of years, nothing much changed: a neighbor's loss was your gain, so conquering, plundering, and stealing generally worked. The result was extreme economic inequality. But in the age of high tech, the more people become educated, the more they can produce. The more they can produce, the more money they make, and the more money they make, the more they are able to pay to find solutions to longstanding problems, whether they be cures for diseases or more efficient ways to travel. Theoretically, this creates a virtuous cycle of increased innovation, profit, and abundance – and not just for the people at the top; there's an economic incentive to reduce inequality. "In a positive-sum world," says economist Max Roser, "the more people are well-off, the better your own life is."[1]

The innovations and entrepreneurs presented in this book are taking Roser's theoretical idea and making it a reality. We all have the power to identify, solve, and implement long-lasting environmental and societal change for the benefit of humanity. "From government officials and corporations to do-it-yourself innovators and techno-philanthropists, we can all do our part to improve the world around us," says Gidi Grinstein, founder of Tikkun Olam Makers, a global network of communities creating solutions for people with disabilities.[2]

"It's not your way or my way," wrote *Seven Habits of Highly Effective People* author Frank Covey, "it's a better way."[3] But the path forward requires taking Covey's famous "win-win" solutions one step further. The innovators here have adopted a win-win-win outlook: good for me, good for you, and good for humanity. As John Mackey, the founder of Whole Foods, explains, "Every problem we have is a failure of imagination. Everything is solvable, it just requires innovation and thinking about win-win-win solutions for the world's greatest challenges."[4]

The future belongs to those who choose hope – and those who leverage technology to positively impact what's next for humanity. It's the answer to my son's question: a reality that can be better, brighter, and more life affirming than anything we've seen before. That's what's next, if we will it.

Acknowledgments

I have been blessed with a rich circle of friends and colleagues who have broadened my vision of what humanity can and must achieve in the decades ahead to identify the grand global challenges humanity will face and to find the creative solutions necessary to solve them. Our ability to positively impact the lives of billions of people has never been more imperative, nor more possible. I interviewed nearly 150 people for this book (often more than once), including innovators, scientists, geneticists, doctors, astronauts, economists, venture capitalists, sustainability experts, engineers, teachers, food and water specialists, CEOs, NGO executives, university professors, lawyers, and social impact makers, among others. With every conversation, I could literally feel my entire body's DNA stretching. Each of these individuals was committed to making the world a better place and embodied the great privilege and obligation to tackle the greatest challenges of the twenty-first century.

My children – Eiden, Oren, and Yaniv – were the ultimate champions of this book. I would often share with them stories from my interviews. They in turn would grill me about what my interviewees said, what questions I asked, or about their own vision of the future. The boys would often come up with their own prospective solutions for solving these challenges. They are the greatest gift the universe has given me, and I am grateful for them every day.

Aviv and Einat Ezra not only served as great cheerleaders, but as a constant source of love, friendship, and support as I weathered highs and lows. Our meetings in Washington, DC, Chicago, and Israel made all the difference in the world as major changes presented themselves.

Jonathan Kessler, the CEO of Heart of a Nation, served as a stalwart sounding board, ally, friend, colleague, and mentor. As the world hunkered down for the coronavirus pandemic, we would meet often virtually on Zoom and on his building's roof deck off Dupont Circle. A force of nature, he would constantly push me hard to be a nuanced thinker and find compassion for those around me.

I am deeply grateful to Yaakov Katz, the editor in chief of the *Jerusalem Post*, who would check in on me regularly to assess my progress on the manuscript and my state of sanity. Our meetings in Jerusalem at various cafés in Baqa and on park benches throughout Capitol Hill were a source of inspiration, joy, and humor.

I had the great fortune to work with two brilliant editors, Ross Schneiderman and Adam Laukhuf, both of whom worked tirelessly to ensure this book's stories were told in the most compelling fashion possible and read well. Working with them both was among the most gratifying parts of writing *Next*.

My sincerest thanks to Ilan Berman, senior vice president of the American Foreign Policy Council, where I serve as a senior fellow. Ilan both read the manuscript and served as my travel partner to some of the world's most interesting countries. When we weren't looking into our crystal ball trying to discern what it would take to bring about a brighter future for humankind, we would head out on ATVs, sit cross-legged on the floor for dinner, and ask tough questions of some of the world's most important policymakers.

Michael Potter and Carol Goldstein served as amazing guides on my exploration to the world of Singularity University and to the space community. They opened their rolodexes and their hearts as they went to bat securing for me interviews with some of the world's most forward-thinking luminaries. Between our meetings in New

York, DC, and Austin, they read portions of the manuscript and made invaluable corrections.

I had an ongoing dialogue with my good friend and colleague Gidi Grinstein, founder of Reut Institute, that always forced me to look at the social impact side of the innovators I interviewed. Daniel Kraft, the founder of Exponential Medicine and Singularity University's Chair of Medicine, was both a gentleman and a scholar – I greatly appreciate his generosity with his time and contacts. During our walks to the National Mall, Eric Danko introduced me with his no-nonsense attitude to the world of autonomous and electric vehicles. Sema Sgaier, on our walks to Roosevelt Island, broadened my vision on pandemics, science, and human psychology. And Stratos Costalas, on countless phone calls and dinners in New York and Washington, DC, always made sure I was progressing and in good spirits.

My friends and colleagues in YPO, and in particular my chapter mates at Mosaic, the staff at M2MX, and Alina Eskinazi went above and beyond in connecting me to my interviewees, but also served as sources of support during the entire writing process. I also thank Ken Kwartler, who shepherded me once again through the copyright process at the Library of Congress (as it so happens, as Washington, DC, experienced unprecedented historical events).

Once again, Georgetown law professor Brad Snyder took me under his wing and pushed me regularly to get out another book. Our meetings often took place as we watched our kids run around the neighborhood playgrounds or in his backyard sipping coffee and noshing on goodies. I am grateful for his gift of friendship, sense of humor, and cheerful personality. My business partner, Larry Glick, remains one of my closest friends, mentors, and cheerleaders. I am grateful to the universe every day for bringing our paths together.

I am beyond grateful to Ilan Greenfield, the owner of Gefen Publishing House. Ilan has become a good friend, colleague, and member of the extended family. Between book adventures in New

York, DC, and around Israel, I could not have asked for a more creative publisher. And Kezia Raffel Pride's tremendous editing skills and eye for detail were a pleasure to work with. I am also extremely grateful to Lisa Mendelow, who designed the beautiful cover for *Next*.

As I shuttled back and forth to Israel for the writing process, I was grateful for the support of my mother, Marlene Schieber, and my truly exceptional sister Simone Pinsky and her tribe. Shlomit Shusan always kept the door open to her home and to her lovely bustan garden. Irit Lerner offered unwavering support, as did Avi Lichter, who continues to serve as a close friend, colleague, and family member. And my sincerest appreciation to the extended Gordon family for their ongoing support and love.

As I look back and think about all the amazing people featured in this book, alongside all the human beings working on making the world a better place, I know more than ever that our biggest mistake in life is not dreaming, but rather not dreaming big enough. Humanity's best days are ahead.

Avi Jorisch
Washington, DC
September 2022

Endnotes

Preface – The Future Is Now

1. David Brock, "How Gordon Moore Made 'Moore's Law,'" *Wired*, April 16, 2015, https://www.wired.com/2015/04/how-gordon-moore-made-moores-law/.
2. Thomas Friedman, "Moore's Law Turns 50," *New York Times*, May 13, 2015, https://www.nytimes.com/2015/05/13/opinion/thomas-friedman-moores-law-turns-50.html. Moore witnessed exponential growth in his own career. In 1968, he and Robert Noyce founded NM Electronics, which eventually became the Intel Corporation, the world's second largest semiconductor chip manufacturer and supplier of processors for computer systems including Apple, Lenovo, Hewlett-Packard, and Dell, along with many other devices that focus on communication and computing.
3. Ray Kurzweil, "The Law of Accelerating Returns," Kurzweilai.net, March 7, 2001, https://www.kurzweilai.net/the-law-of-accelerating-returns.
4. "The Future Is Better Than You Think: Predictions on AI and Development from Ray Kurzweil," *ITUNews*, May 29, 2019, https://news.itu.int/the-future-is-better-than-you-think-predictions-on-ai-and-development-from-ray-kurzweil/.
5. Monica Hunter-Hart, "8 Staggering Predictions from Ray Kurzweil," *Inverse*, May 29, 2017, https://www.inverse.com/article/30913-8-staggering-predictions-ray-kurzweil.
6. Christianna Reedy, "Ray Kurzweil Claims Singularity Will Happen by 2045," *Futurism*, May 10, 2017, https://futurism.com/kurzweil-claims-that-the-singularity-will-happen-by-2045.
7. Nick Bostrom, "Existential Risks: Analyzing Human Extinction Scenar-

ios and Related Hazards," *Journal of Evolution and Technology* 9, no. 1 (2002), www.nickbostrom.com/existential/risks.html.

8. Edward Geist and Andrew J. Lohn, "How Might Artificial Intelligence Affect the Risk of Nuclear War?" Rand Corporation, 2018, www.rand.org/pubs/perspectives/PE296.html.

9. Catherine Clifford, "Elon Musk: 'Mark My Words – A.I. Is Far More Dangerous than Nukes,'" CNBC, March 14, 2018, www.cnbc.com/2018/03/13/elon-musk-at-sxsw-a-i-is-more-dangerous-than-nuclear-weapons.html.

10. Elon Musk (@elonmusk), "Competition for AI superiority at national level most likely cause of WW3 imo," Twitter, September 4, 2017, 12:33 p.m., https://twitter.com/elonmusk/status/904638455761612800.

11. Marc Goodman, "A Vision of Crimes in the Future," TEDGlobal 2012, June 2012, www.ted.com/talks/marc_goodman_a_vision_of_crimes_in_the_future?language=en.

12. Fiona Harvey, "Humanity under Threat from Perfect Storm of Crises – Study," *The Guardian*, February 6, 2020, https://www.theguardian.com/environment/2020/feb/06/humanity-under-threat-perfect-storm-crises-study-environment?fbclid=IwAR3roRN91V7VtZY8vHsEQv3ScicNTSRJCr8hyay4u_Sm24J6DnB7lmsoyxQ.

13. Larry Elliott, "Climate Crisis Fills Top 5 Places of World Economic Forum's Risk Report," *The Guardian*, January 15, 2020, https://www.theguardian.com/business/2020/jan/15/climate-crisis-environment-top-five-places-world-economic-forum-risks-report.

14. John F. Kennedy, "Text of President John Kennedy's Rice Stadium Moon Speech," NASA, September 12, 1962, https://er.jsc.nasa.gov/seh/ricetalk.htm.

15. Greg Ip, "The Moonshot Mind-Set Once Came from the Government. No Longer," *Wall Street Journal*, July 14, 2019, www.wsj.com/articles/the-moonshot-mind-set-once-came-from-the-government-no-longer-11563152820. See also Casey Dreier, "Reconstructing the Cost of the 1 Giant Leap," *The Planetary Society*, June 16, 2019, https://www.planetary.org/blogs/casey-dreier/2019/reconstructing-the-price-of-apollo.html.

16. Astro Teller and Megan Smith, "What Is Your X? Amplifying Technology Moonshots," Google Official Blog, February 6, 2012, https://googleblog.blogspot.com/2012/02/whats-your-x-amplifying-technology.html.

17. Cited in John Bell, "To the Moon: Google X's Astro Teller Encourages Creatives to Think Big," *The Guardian*, June 20, 2013, www.theguardian

.com/media-network/media-network-blog/2013/jun/20/google-x-astro-teller-creatives.

18. "Singularity University Addresses the World's Greatest Challenges at SU Global Summit," Singularity University, August 9, 2016, https://su.org/press-room/press-releases/singularity-university-addresses-the-worlds-greatest-challenges-at-su-global-summit/.

19. Afshin Molavi, author's phone interview, March 16, 2021.

20. Jun Sato, author's phone interview, May 4, 2020.

21. Pagan Kennedy, "William Gibson's Future Is Now," *New York Times*, January 13, 2012, https://www.nytimes.com/2012/01/15/books/review/distrust-that-particular-flavor-by-william-gibson-book-review.html.

Chapter 1 – Space: Print Me Up, Scotty!

1. Jason Dunn, author's phone interview, July 20, 2020; NASA's Ames Research Center, "Jason Dunn – The Future of Making Things in Space," January 31, 2017, www.youtube.com/watch?list=PLfnpkfDmrBqYgX1O3kRQnc5Sjzz_LMwDo&time_continue=3313&v=ZXYCKoTclCc.

2. Olivia Solon, "'It's about Expanding Earth': Could We Build Cities in Space?" *The Guardian*, April 21, 2018, https://www.theguardian.com/science/2018/apr/21/expanding-earth-could-we-build-cities-in-space.

3. Zachary T. Sampson, "First 3-D Printing in Space Had Bay Area Start," *Tampa Bay Times*, December 8, 2014, https://www.tampabay.com/archive/2014/12/08/first-3-d-printing-in-space-had-bay-area-start/.

4. Solon, "'It's about Expanding Earth.'"

5. Jason Dunn, author's phone interview, July 20, 2020.

6. Jason Dunn, author's phone interview, July 20, 2020. See also Solon, "'It's about Expanding Earth.'"

7. Aaron Kemmer, author's phone interview, July 3, 2020.

8. "ISS National Laboratory," NASA, https://www.nasa.gov/mission_pages/station/research/nlab/index.html.

9. Aaron Kemmer, author's phone interview, July 3, 2020.

10. Jason Dunn, author's phone interview, July 20, 2020. See also Solon, "'It's about Expanding Earth.'"

11. Jason Dunn, author's phone interview, July 20, 2020.

12. Aaron Kemmer, author's phone interview, July 3, 2020.

13. Aaron Kemmer, author's phone interview, July 3, 2020.

14. Yvonne Cagle, author's phone interview, July 21, 2020.

15. Jason Dunn, author's phone interview, July 20, 2020. See also Sampson, "First 3-D Printing in Space."

16. Jason Dunn, author's phone interview, July 20, 2020. See also Solon, "'It's about Expanding Earth.'"

17. Bernard Kutter, author's phone interview, May 20, 2020.

18. Ramin Khadem, author's phone interview, April 27, 2020.

19. Brian Weeden, author's phone interview, April 30, 2020.

20. Yonatan Winetraub, author's phone interview, April 24, 2020.

21. Dave Mosher, "Congress and Trump are Running out of Time to Fix a $100-Billion Investment in the Sky, NASA Auditor Says," *Business Insider*, May 18, 2018, www.businessinsider.com/space-station-trump-plan-problems-astronaut-risk-2018-5.

22. Jackie Wattles, "NASA Says Moon Rocket Could Cost as Much as $1.6 Billion Per Launch," CNN, December 9, 2019, www.cnn.com/2019/12/09/tech/nasa-sls-price-cost-artemis-moon-rocket-scn/index.html.

23. NASA's Ames Research Center, "Jason Dunn – The Future of Making Things in Space," January 31, 2017, 25:00, www.youtube.com/watch?list=PLfnpkfDmrBqYgX1O3kRQnc5Sjzz_LMwDo&time_continue=3313&v=ZXYCKoTclCc.

24. Chris Stott, author's phone interview, April 21, 2020.

25. Andrew Wade, "Made in Space: New Production Frontiers," *The Engineer*, February 8, 2016, https://www.theengineer.co.uk/made-in-space-new-production-frontiers/.

26. Michael Potter, author's phone interview, April 17, 2020.

27. Carol Goldstein, author's phone interview, April 19, 2020.

28. Rachel Layne, "Space Case: Why Reaching for the Stars Could Soon Be a $1 Trillion Industry," CBS News, July 16, 2019, https://www.cbsnews.com/news/space-is-a-more-than-400-billion-market-and-getting-bigger/.

29. Peter Garretson, author's phone interview, May 12, 2020.

30. "Made in Space, Inc. to Demonstrate Manufacture of Exotic Optical Fiber in Space," PR.com, July 20, 2016, https://www.pr.com/press-release/680087.

31. Andy Patrizio, "IDC: Expect 175 Zettabytes of Data Worldwide by 2025," NetworkWorld, December 3, 2018, www.networkworld.com/article/3325397/idc-expect-175-zettabytes-of-data-worldwide-by-2025.html.

32. Andrew Rush, CEO, Made in Space, testimony to Senate Commerce, Science and Transportation Subcommittee on Space, Science, and Compet-

itiveness Hearing, "Reopening the American Frontier: Reducing Regulatory Barriers and Expanding American Free Enterprise in Space," April 26, 2017, https://www.commerce.senate.gov/2017/4/reopening-the-american-frontier-reducing-regulatory-barriers-and-expanding-american-free-enterprise-in-space. See also "Made in Space, Inc. to Demonstrate Manufacture of Exotic Optical Fiber in Space."

33. Mike Wall, "In-Space Manufacturing Is about to Get a Big Test," Space.com, December 11, 2017, https://www.space.com/39039-made-in-space-off-earth-manufacturing-test.html.

34. Stephen Eisele, author's phone interview, April 28, 2020.

35. William Kramer, author's phone interview, April 28, 2020.

36. Aaron Kemmer, author's phone interview, July 3, 2020.

37. Christian Sallaberger, author's phone interview, May 7, 2020.

Chapter 2 – Learning: The Internet Academy

1. Clive Thompson, "How Khan Academy Is Changing the Rules of Education," *Wired*, July 15, 2011, www.wired.com/2011/07/ff_khan/. See also www.linkedin.com/in/khanacademy/. See also Bryant Urstadt, "Salman Khan: The Messiah of Math," *Bloomberg Businessweek*, May 19, 2011, www.bloomberg.com/news/articles/2011-05-19/salman-khan-the-messiah-of-math.

2. Claudia Dreifus, "It All Started with a 12-Year-Old Cousin," *New York Times*, January 28, 2014, http://www.nytimes.com/2014/01/28/science/salman-khan-turned-family-tutoring-into-khan-academy.html.

3. Michael Noer, "One Man, 1 Computer, 10 Million Students: How Khan Academy Is Reinventing Education," Forbes, November 19, 2012, www.forbes.com/sites/michaelnoer/2012/11/02/one-man-one-computer-10-million-students-how-khan-academy-is-reinventing-education/#7e17495444e0.

4. Sarah Benson, "Salman Khan," Sutori, www.sutori.com/story/salman-khan--Lvux5bFpvn8zTBS222BUgLRN.

5. Helena de Bertodano, "Khan Academy: The Man Who Wants to Teach the World," *The Guardian*, September 28, 2018, www.telegraph.co.uk/education/educationnews/9568850/Khan-Academy-The-man-who-wants-to-teach-the-world.html. See also Nikhil Menon, "How Sal Khan Is Exorcising the Math Spectre from the Minds of Millions of Students Worldwide," *The Economic Times*, August 10, 2012, https://economictimes

.indiatimes.com/how-sal-khan-is-exorcising-the-math-spectre-from-the-minds-of-millions-of-students-worldwide/articleshow/15425169.cms.

6. Salman Khan, *The 1 World School House* (New York: Twelve, 2012), 152–55.

7. Somini Sengupta, "Online Learning, Personalized," *New York Times*, December 4, 2011, https://www.nytimes.com/2011/12/05/technology/khan-academy-blends-its-youtube-approach-with-classrooms.html. See also "Grace King High School," *U.S. News and World Report*, www.usnews.com/education/best-high-schools/louisiana/districts/jefferson-parish-public-school-system/grace-king-high-school-8639.

8. "Sal Khan's Academy Lights the Way," *Financial Express* (Bangladesh), December 26, 2010.

9. Par Parekh, "The Math of Khan: Salman Khan: Founder of the Kahn Academy," Bidoun, www.bidoun.org/articles/the-math-of-khan.

10. Menon, "How Sal Khan Is Exorcising the Math Spectre."

11. Thompson, "How Khan Academy Is Changing the Rules of Education."

12. Bruce Love, "Sal Khan: An Educational Entrepreneur," CampdenFB, May 7, 2015, https://www.campdenfb.com/article/sal-khan-educational-entrepreneur.

13. Guy Raz, "Sal Khan: Can Technology Help Create a Global Classroom," *NPR TED Radio Hour*, August 11, 2017.

14. Dreifus, "It All Started with a 12-Year-Old Cousin."

15. Dreifus, "It All Started with a 12-Year-Old Cousin." See also Raz, "Sal Khan."

16. Khan, *The 1 World School House*, 25.

17. "How Khan Academy Is Changing Education with Videos Made in a Closet," Mixergy, June 28, 2010, https://mixergy.com/interviews/salman-khan-academy-interview/.

18. Dreifus, "It All Started with a 12-Year Old Cousin."

19. Thompson, "How Khan Academy Is Changing the Rules of Education."

20. Khan, *The 1 World School House*, 26–27.

21. "Next Knowledge Factbook 2010," Kellogg Northwestern, 2010, www.kellogg.northwestern.edu/faculty/jones-ben/htm/NextKnowledgeFactbook2010.pdf.

22. Noer, "One Man, 1 Computer, 10 Million Students."

23. Kasra Koushan, "Sal Khan on Creating a Better Future," The Signal, July 20, 2017, https://thesign.al/sal-khan/.

24. Sowmya Nandakumar, "Salman Khan: World's Most Influential Online

Educator," *Indo-American News*, May 17, 2012, page 6, https://issuu.com/indoamericannews/docs/may18_pages_1-36.

25. Khan, *The 1 World School House*, 219–21.

26. Nicole Stott, author's phone interview, May 11, 2020.

27. Aleta Margolis, author's phone conversation, May 26, 2020.

28. Robert Murphy, author's telephone interview, May 27, 2020.

29. Noer, "One Man, 1 Computer, 10 Million Students."

30. Surojit Gupta, " Salman's India Dream: Coaching 450m in 10 years," *Times of India*, December 5, 2015, https://timesofindia.indiatimes.com/india/salmans-india-dream-coaching-450m-in-10-years/articleshow/50036337.cms.

31. De Bertodano, "Khan Academy."

32. Khan, *The 1 World School House*, 155–56.

33. Noer, "One Man, 1 Computer, 10 Million Students." See also Menon, "How Sal Khan Is Exorcising the Math Spectre." See also de Bertodano, "Khan Academy." See also David A. Kaplan, "Bill Gates' Favorite Teacher," *Fortune*, August 24, 2010, https://archive.fortune.com/2010/08/23/technology/sal_khan_academy.fortune/index.htm. See also Jeffrey R. Young, "College 2.0: A Self-Appointed Teacher Runs a One-Man 'Academy' on YouTube," *Chronicle of Higher Education*, June 6, 2010, www.chronicle.com/article/College-20-A-Self-Appointed/65793.

34. Khan, *The 1 World School House*, 157.

35. Thompson, "How Khan Academy Is Changing the Rules of Education."

36. Khan Academy, "Bill Gates Talks about the Khan Academy at Aspen Ideas Festival 2010," July 19, 2010, www.youtube.com/watch?v=6A07Pj71TUA. See also Chidanand Rajghatta, "Modi to Meet Salman Khan, the Educator not the Movie Star," *Times of India*, September 28, 2015, https://timesofindia.indiatimes.com/india/modi-to-meet-salman-khan-the-educator-not-the-movie-star/articleshow/49113900.cms.

37. De Bertodano, "Khan Academy."

38. Rajghatta, "Modi to Meet Salman Khan." See also Gupta, "Khan Academy CEO Salman Khan's India Dream." See also "Businessman Carlos Slim Will Invest More Than 317 US Dollar Millions in Education," European Press Agency (EPA), January 19, 2013. See also "Tata Trusts to Support Khan Academy for Free Online Education," *Hindustan Times*, December 7, 2015, www.hindustantimes.com/education/tata-trusts-to-support-khan

-academy-for-free-online-education/story-kpQuo1Ph4QGi9HDBrT3SVP .html.

39. "Sal Khan on Why Khan Academy Is a Non-Profit," *Overheard with Evan Smith*, season 5, episode 18, aired April 20, 2015, www.youtube.com /watch?v=LI7UOuX9Qjs.

40. Gupta, "Khan Academy CEO Salman Khan's India Dream." See also Alex Wagner, "Can Sal Khan Reform Education in America?" *Huffington Post*, June 4, 2011, www.huffingtonpost.com/2011/04/04/the-khan-academy -and-educ_n_844390.html.

41. Thompson, "How Khan Academy Is Changing the Rules of Education."

42. Gupta, "Khan Academy CEO Salman Khan's India Dream." See also Arie Passwaters, "President's Lecture Series to Present Educational Innovator Salman Khan," Rice University Office of Public Affairs, September 24, 2018, https://news.rice.edu/2018/09/24/presidents-lecture-series-to-present -educational-innovator-salman-khan/.

43. "What Languages Is Khan Academy Available In," Khan Academy, May 1, 2020, https://support.khanacademy.org/hc/en-us/articles/226457308 -Is-Khan-Academy-available-in-other-languages-.

44. Supriya Tripathi, author's phone interview, May 23, 2020.

45. David Arnold, author's phone interview, May 28, 2020.

46. Kaplan, "Bill Gates' Favorite Teacher."

47. Gary Stager, author's phone interview, May 22, 2020. See also Thompson, "How Khan Academy Is Changing the Rules of Education."

48. Karim Ani, author's phone interview, May 19, 2020.

49. Sylvia Martinez, author's phone interview, May 19, 2020. See also Thompson, "How Khan Academy Is Changing the Rules of Education."

50. Valerie Strauss, "Khan Academy: The Revolution that Isn't," *Washington Post*, July 23, 2012, www.washingtonpost.com/blogs/answer-sheet/post /khan-academy-the-hype-and-the-reality/2012/07/23/gJQAuw4J3W _blog.html?utm_term=.36e13b742375.

51. Thompson, "How Khan Academy Is Changing the Rules of Education."

52. Dana Goldstein, "Research Shows Students Falling Months Behind during Virus Disruptions," *New York Times*, June 5, 2020, www.nytimes .com/2020/06/05/us/coronavirus-education-lost-learning.html.

53. "Khan Academy Launches Free LSAT Prep for All," Law School Admission Council, June 1, 2018, https://www.lsac.org/about/news/khan -academy-launches-free-lsat-test-prep-all.

54. Kent D. Lollis, "Living, Working, and Achieving While Black," Law School Admission Council, June 16, 2020, www.lsac.org/blog/living-working-and-achieving-while-black.

55. Nadia Rahman, LinkedIn profile, retrieved May 23, 2020, www.linkedin.com/in/nadia-rahman-62203182/.

56. Helen Walters, "Two Giants of Online Learning Discuss the Future of Education," Ideas.Ted.com, January 28, 2014, https://ideas.ted.com/in-conversation-salman-khan-sebastian-thrun-talk-online-education/.

Chapter 3 – Shelter: Let There Be Light

1. Netta Achituv, "The Israeli Who Is Bringing Electricity, Agriculture and Medicine Cheaply to the Poorest Countries in the World" [in Hebrew], *The Marker*, July 6, 2016, https://www.themarker.com/themarker-women/MAGAZINE-1.2998657.

2. Sivan Yaari, author's interview, Innovation: Africa Office, Herzliya, August 8, 2019.

3. Sivan Yaari, "Africa: Is Good Good Enough?" TEDx Tel Aviv University, February 15, 2016, https://www.youtube.com/watch?v=tfgAexR5Z2w.

4. Ian Zelaya, "Lighting a Continent," *Washington Jewish Week*, November 13, 2013, http://washingtonjewishweek.com/7152/lighting-a-continent/news/world-news/. See also "New Perspectives: Influencing Circumstances through Human Connection– Spotlight: Sivan Yaari '02," *Altitude*, Winter 2020, https://www.alumni.pace.edu/s/1655/bp20/interior.aspx?sid=1655&gid=2&pgid=2383.

5. Sivan Yaari, author's interview, Innovation: Africa Office, Herzliya, August 8, 2019.

6. Shlomit Lan, "Positive Energy" [in Hebrew], *Globes*, May 17, 2010, https://www.globes.co.il/news/article.aspx?did=1000560436.

7. "Philip LaRocco: Adjunct Professor of International and Public Affairs," Columbia University School of International and Public Affairs, https://sipa.columbia.edu/faculty-research/faculty-directory/philip-larocco.

8. Phil LaRocco, author's phone interview, March 25, 2020.

9. Yaari, "Africa: Is Good Good Enough?"

10. Sivan Yaari, author's interview, Innovation: Africa Office, Herzliya, August 8, 2019.

11. Phil LaRocco, author's phone interview, March 25, 2020.

12. Yaari, "Africa: Is Good Good Enough?"

13. Yaari, "Africa: Is Good Good Enough?"
14. Sivan Yaari, author's interview, Innovation: Africa Office, Herzliya, August 8, 2019.
15. Stephen Oryszczuk, "Africa Gains from Israel's Solar System," *Jewish News*, April 12, 2018, https://jewishnews.timesofisrael.com/africa-gains-from-israels-solar-system/.
16. Yaari, "Africa: Is Good Good Enough?"
17. Yaari, "Africa: Is Good Good Enough?"
18. Rebeca Kuropatwa, "Helping 1 Village at a Time," *Jewish Independent*, June 16, 2017, https://www.jewishindependent.ca/tag/sivan-yaari/.
19. Gil Haskel, author's phone interview, March 10, 2020.
20. Sivan Yaari, author's interview, Innovation: Africa Office, Herzliya, August 8, 2019.
21. Natie Kirsh, author's phone interview, March 25, 2020.
22. Stephen Koseff, author's phone interview, March 19, 2020.
23. David Arison, author's phone interview, March 30, 2020.
24. Sivan Yaari, author's interview, Innovation: Africa Office, Herzliya, August 8, 2019.
25. Clive Cookson, "Africa to Propel World's Population towards 10bn by 2050," *Financial Times*, June 17, 2019, https://www.ft.com/content/868e20d0-90ec-11e9-b7ea-60e35ef678d2.
26. Sivan Yaari, author's interview, Innovation: Africa Office, Herzliya, August 8, 2019.

Chapter 4 – Environment: Water World

1. Henk Ovink, author's phone interview, April 1, 2020. See also Russell Shorto, "How to Think Like the Dutch in a Post-Sandy World," *New York Times Magazine*, April 9, 2014, https://www.nytimes.com/2014/04/13/magazine/how-to-think-like-the-dutch-in-a-post-sandy-world.html.
2. David Wolman, "Before the Levees Break: A Plan to Save the Netherlands," *Wired*, December 22, 2008, https://www.wired.com/2008/12/ff-dutch-delta/.
3. Denise Lu and Christopher Flavelle, "Rising Seas Will Erase More Cities by 2050, New Research Shows," *New York Times*, October 29, 2019, https://www.nytimes.com/interactive/2019/10/29/climate/coastal-cities-underwater.html.
4. Stefan Rahmstorf, author's phone interview, March 26, 2020.

5. "Dutch Flood Expertise Is Big Export Business," *Telegram & Gazette*, November 13, 2017.

6. Henk Ovink, author's phone interview, April 1, 2020. See also Shorto, "How to Think Like the Dutch in a Post-Sandy World."

7. Arun Gupta, "Disaster Capitalism Hits New York," *In These Times*, February 2013, https://inthesetimes.com/article/14430/disaster_capitalism_hits_new_york.

8. Jos Dijkman, author's phone interview, April 11, 2020.

9. Alexandra Wynne, "Special Report: Plugging the Floods Will Mean Going Dutch," *New Civil Engineer*, June 3, 2014, https://www.newcivilengineer.com/archive/special-report-plugging-the-floods-will-mean-going-dutch-03-06-2014/.

10. Mike Corder, "Water Wizards: Dutch Flood Expertise Is Big Export Business," Associated Press, November 12, 2017, https://apnews.com/article/1bc1f137cb134fdca67afdff08c242bc.

11. "Seven Wonders of the Modern World as Named by ASCE," Herff College of Engineering, University of Memphis, http://www.ce.memphis.edu/1101/interesting_stuff/7wonders.html.

12. Bas Jonkman, author's phone interview, April 14, 2020.

13. Marinke Steenhuis, author's phone interview, April 14, 2020.

14. Wynne, "Special Report: Plugging the Floods Will Mean Going Dutch."

15. David Waggonner, author's phone interview, April 13, 2020.

16. Marlise Simons, "Dutch Floods Came in the Back Door," *New York Times*, February 5, 1995, https://www.nytimes.com/1995/02/05/world/dutch-floods-came-in-the-back-door.html.

17. Tracy Metz, author's phone interview, April 6, 2020.

18. Will Storr, "Flooded Britain: How Can Holland Help?" *Daily Telegraph* (UK), April 26, 2014, https://www.telegraph.co.uk/culture/art/architecture/10769974/Flooded-Britain-how-can-Holland-help.html.

19. Henk Ovink, author's phone interview, April 1, 2020.

Chapter 5 – Hygiene: There Will Be Blood

1. Arunachalam Muruganantham, author's phone interview, February 21, 2020. See also "From Rags to Riches," *The Nation* (Nigeria), September 20, 2014, https://thenationonlineng.net/from-rags-to-riches/. See also Stephanie Nolen, "The Absorbing Tale of 1 Man's Quest for Better Feminine Hygiene," *Globe and Mail* (Canada), October 4, 2012.

2. Arunachalam Muruganantham, author's phone interview, February 21, 2020. See also Arunachalam Muruganantham, "How I Started a Sanitary Napkin Revolution," TED@Bangalore, May 2012, www.ted.com/talks/arunachalam_muruganantham_how_i_started_a_sanitary_napkin_revolution?language=en#t-70857.

3. Arunachalam Muruganantham, author's phone interview, February 21, 2020. See also Muruganantham, "How I Started a Sanitary Napkin Revolution."

4. Kristie Lu Stout, " India's 'Pad Man' Becomes Bollywood Blockbuster," CNN, February 8, 2018, http://edition.cnn.com/TRANSCRIPTS/1802/08/nwsm.01.html.

5. Arunachalam Muruganantham, author's phone interview, February 21, 2020; April 20, 2021.

6. Arunachalam Muruganantham, author's phone interview, February 21, 2020; April 20, 2021. See also "India's 'Menstruation Man' Reinvents Sanitary Pads," *Pakistan Today*, May 15, 2016, https://archive.pakistantoday.com.pk/2016/05/15/indias-menstruation-man-reinvents-sanitary-pads/. See also Nolen, "The Absorbing Tale."

7. Arunachalam Muruganantham, author's phone interview, February 21, 2020.

8. Vishal Gondal, author's phone interview, April 8, 2020.

9. Arunachalam Muruganantham, author's phone interview, February 21, 2020; April 20, 2021. See also Manjusha Radhakrishnan, "Meet the Real 'Pad Man' Arunachalam Muruganatham," *Gulf News*, February 7, 2018.

10. Arunachalam Muruganantham, author's phone interview, February 21, 2020; April 20, 2021.

11. Arunachalam Muruganantham, author's phone interview, February 21, 2020.

12. Arunachalam Muruganantham, author's phone interview, February 21, 2020.

13. Arunachalam Muruganantham, author's phone interview, February 21, 2020.

14. S. R. Ramesh, author's phone interview, March 28, 2020.

15. Sashi Anand, author's phone interview, March 30, 2020.

16. Arunachalam Muruganantham, author's phone interview, February 21, 2020.

17. Arunachalam Muruganantham, author's phone interview, February 21, 2020.

18. Arunachalam Muruganantham, author's phone interview, February 21, 2020.

19. Arunachalam Muruganantham, author's phone interview, February 21, 2020. See also "Who Is Arunachalam Muruganantham?" *Financial Express* (Bangladesh), February 8, 2018, www.financialexpress.com/entertainment/who-is-arunachalam-muruganantham-the-real-padman-who-was-the-inspiration-for-akshay-kumar-movie/1059638/.

20. Arunachalam Muruganantham, author's phone interview, February 21, 2020; April 20, 2021.

21. Arunachalam Muruganantham, author's phone interview, April 20, 2021. Yudhijit Bhattacharjee, "Launch Pad," *New York Times*, November 10, 2016, https://www.nytimes.com/interactive/2016/11/13/magazine/design-issue-sanitary-pads-india.html.

22. "Environment-Friendly Sanitary Pads: Work in Progress," *Free Press Journal* (India), December 6, 2018, https://www.freepressjournal.in/cmcm/environment-friendly-sanitary-pads-work-in-progress. See also Priyanka Golikeri, "An Enterprise Aimed at Boosting Women's Health," DNA, June 9, 2011, https://www.dnaindia.com/business/report-an-enterprise-aimed-at-boosting-women-s-health-1552766. See also Abigail Jones, "The Silence That Still Surrounds Periods," *The Independent* (UK), April 25, 2016, https://www.independent.co.uk/life-style/health-and-families/silence-around-menstruation-a7000051.html.

23. "National Family Health Survey (NFHS-4)," *India Fact Sheet, Ministry of Health and Family Welfare*, 2015-2016, http://rchiips.org/NFHS/pdf/NFHS4/India.pdf.

24. Arunachalam Muruganantham, author's phone interview, February 21, 2020. *Pakistan Today*, "India's 'Menstruation Man' Reinvents Sanitary Pads." See also Radhakrishnan, "Meet the Real 'Pad Man.'" See also Adeline Dorcas, "Pad Man Makes Nationwide Campaign for Use of Sanitary Napkins," *Medindia*, May 11, 2018, https://www.medindia.net/news/healthwatch/pad-man-makes-nationwide-campaign-for-use-of-sanitary-napkins-179368-1.htm. See also Jones, "The Silence that Still Surrounds Periods."

25. Radhakrishnan, "Meet the Real 'Pad Man.'"

26. Babu Yogeswaran, author's phone interview, March 26, 2020.

27. Arunachalam Muruganantham, author's phone interview, February 21, 2020

28. Arunachalam Muruganantham, author's phone interview, February 21, 2020

29. Arunachalam Muruganantham, author's phone interview, February 21, 2020. See also Betsy Teutsch, "How Sanitary Pads Can Help Women Improve their Health and Education," *The Atlantic*, April 22, 2014, https://www.theatlantic.com/business/archive/2014/04/sanitary-napkin-business/360297/.

30. Arunachalam Muruganantham, author's phone interview, February 21, 2020. See also Gilson, "India's Menstruation Man."

31. Arunachalam Muruganantham, author's phone interview, February 21, 2020

32. Arunachalam Muruganantham, author's phone interview, February 21, 2020. See also Brian Wang, "Cheap Sanitary Pads Are Better Than Apple iPads for Poor Women in India by Improving Lives, Improving Health and Providing Thousands of Jobs," *Next Big Future*, March 5, 2014, https://www.nextbigfuture.com/2014/03/cheap-sanitary-pads-are-better-than.html. See also Gilson, "India's Menstruation Man."

33. Arunachalam Muruganantham, author's phone interview, February 21, 2020. See also Gilson, "India's Menstruation Man." See also "This Man Is a Sanitary Pad Revolutionary. Watch Him Take On the Age-Old Taboo" [in Hindi], *Daily Bhaskar*, March 3, 2016, https://daily.bhaskar.com/news/TOP-sanitary-pad-revolutionary-arunachalam-muruganantham-menstruation-man-5265202-PHO.html. See also Wang, "Cheap Sanitary Pads Are Better Than Apple iPads."

34. Arunachalam Muruganantham, author's phone interview, February 21, 2020.

35. Venkatesh Mahadevan, author's phone interview, April 2, 2020.

36. Dr. Sujithra Beena, author's phone interview, March 28, 2020.

37. Dr. Rengaraj Venkatesh, author's phone interview, March 18, 2020.

38. Arunachalam Muruganantham, author's phone interview, February 21, 2020; April 20, 2021.

Chapter 6 – Medicine: Playing God – Genetic Editing

1. "The First Trial of 'Gene Editing Baby' Sentenced He Jiankui and 3 Other Defendants to Be Held Criminally Responsible" [in Chinese], *Xin-*

huanet, December 30, 2019, http://www.xinhuanet.com/legal/2019-12/30/c_1125403802.htm.

2. "He Jiankui Defends 'World's First Gene-Edited Babies,'" BBC, November 28, 2018, https://www.bbc.com/news/world-asia-china-46368731. See also Edward Lanphier, Fyodor Urnov, Sarah Ehlen Haecker, et al., "Don't Edit the Human Germ Line," *Nature*, March 12, 2015, www.nature.com/news/don-t-edit-the-human-germ-line-1.17111.

3. Akshat Rathi and Echo Huang, "More Than 100 Chinese Scientists Have Condemned the CRISPR Baby Experiment as 'Crazy,'" *Quartz*, November 26, 2018, https://qz.com/1474530/chinese-scientists-condemn-crispr-baby-experiment-as-crazy/. See also Jon Cohen, "The Untold Story of the 'Circle of Trust' behind the World's First Gene-Edited Babies," *Science*, August 1, 2019, www.sciencemag.org/news/2019/08/untold-story-circle-trust-behind-world-s-first-gene-edited-babies.

4. Alexandra Harney and Kate Kelland, "China Orders Investigation after Scientist Claims First Gene-Edited Babies," Reuters, November 26, 2018, https://www.reuters.com/article/us-health-china-babies-genes/china-orders-investigation-after-scientist-claims-first-gene-edited-babies-idUSKCN1NV19T.

5. Carl Zimmer, "From Fearsome Predator to Man's Best Friend," *New York Times*, May 16, 2013, www.nytimes.com/2013/05/16/science/dogs-from-fearsome-predator-to-mans-best-friend.html.

6. "The Evolution of Corn," Genetics Learning Center, University of Utah, July 2015, http://learn.genetics.utah.edu/content/selection/corn/. See also Hatice Bilgic, Erdogan Hakki, Anamika Pandey, Mohd Kamran Khan, and Mahinur Akkaya, "Ancient DNA from 8400 Year-Old Çatalhöyük Wheat: Implications for the Origin of Neolithic Agriculture," *PloS 1* 11, no. 3 (2016), https://www.ncbi.nlm.nih.gov/pmc/articles/PMC4801371/.

7. Rodolphe Barrangou, author's interview, October 12, 2020.

8. Prashant Nair, "QnAs with Rodolphe Barrangou," *PNAS*, July 11, 2017, www.pnas.org/content/114/28/7183. In 2011, Dupont purchased Danisco for USD $6.3 billion and began enhancing all of its yogurts and cheeses. This food giant owns a significant portion of the global dairy market, and so a significant number of people have consumed products that were manufactured using CRISPR cultures.

9. Martin Jinek, Krzysztof Chrlinski, et al., "A Programmable Dual-RNA-

Guided DNA Endonuclease in Adaptive Bacterial Immunity," *Science,* August 17, 2012, https://science.sciencemag.org/content/337/6096/816.

10. *Human Nature,* directed by Adam Bolt, Greenwich Entertainment, March 13, 2020.

11. *Human Nature,* directed by Adam Bolt, Greenwich Entertainment, March 13, 2020. See also Le Cong, F. Ann Ran, David Cox, Shuailiang Lin, et al., "Multiplex Genome Engineering Using CRISPR/Cas Systems," *Nature,* February 15, 2013, http://science.sciencemag.org/content/339/6121/819.long.

12. Mark Shwartz, "Target, Delete, Repair," *Stanford Medicine,* Winter 2018, https://stanmed.stanford.edu/2018winter/CRISPR-for-gene-editing-is-revolutionary-but-it-comes-with-risks.html.

13. *Human Nature,* directed by Adam Bolt, Greenwich Entertainment, March 13, 2020.

14. Megan Molteni, "The Wired Guide to CRISPR," *Wired,* March 12, 2019, https://www.wired.com/story/wired-guide-to-crispr/.

15. Shwartz, "Target, Delete, Repair."

16. Daniel Kraft, author's phone interview, February 19, 2020. See also Stephan Spencer, "Exponential Medicine in a COVID-19 World with Dr. Daniel Kraft," *Get Yourself Optimized,* no. 246, May 7, 2020, www.getyourselfoptimized.com/exponential-medicine-in-a-covid-19-world-with-dr-daniel-kraft/.

17. Rodolphe Barrangou, author's phone interview, June 16, 2020.

18. Timothy Winegard, "People v Mosquitos: What to Do about Our Biggest Killer," *The Guardian,* September 20, 2019, https://www.theguardian.com/environment/2019/sep/20/man-v-mosquito-biggest-killer-malaria-crispr.

19. World Health Organization, "Mosquito Sterilization Offers New Opportunity to Control Chikungunya, Dengue, and Zika," November 14, 2019, https://www.who.int/news-room/detail/14-11-2019-mosquito-sterilization-offers-new-opportunity-to-control-chikungunya-dengue-and-zika.

20. Anthony James, author's phone interview, June 12, 2020.

21. Jennifer Khan, "The Gene Drive Dilemma: We Can Alter Entire Species, but Should We?" *New York Times,* January 8, 2020, https://www.nytimes.com/2020/01/08/magazine/gene-drive-mosquitoes.html.

22. Alta Charo, author's phone interview, September 1, 2020. Khan, "The Gene Drive Dilemma."

23. Shwartz, "Target, Delete, Repair."

24. @BloombergQuickTake, "Co-Inventor of CRISPR says she's 'horrified,'" *Twitter*, November 28, 2018, https://twitter.com/QuickTake/status/1067772480511766528.

25. Eric Lander, Françoise Baylis, Feng Zhang, Emmanuelle Charpentier, Paul Berg, et al., "Adopt a Moratorium on Heritable Genome Editing," *Nature*, March 13, 2019, https://www.nature.com/articles/d41586-019-00726-5.

26. Tshaka Cunningham, author's phone interview, October 8, 2020.

27. Megan Molteni, "The World Health Organization Says No More Gene-Edited Babies," *Wired*, July 30, 2019, www.wired.com/story/the-world-health-organization-says-no-more-gene-edited-babies/.

28. Sui-Lee Wee, "Chinese Scientist Who Genetically Edited Babies Gets 3 Years in Prison," *New York Times*, December 30, 2019, https://www.nytimes.com/2019/12/30/business/china-scientist-genetic-baby-prison.html.

29. Ken Dilanian, "China Has Done Human Testing to Create Biologically Enhanced Super Soldiers, Says Top U.S. Official," NBC News, December 4, 2020, https://www.nbcnews.com/politics/national-security/china-has-done-human-testing-create-biologically-enhanced-super-soldiers-n1249914. See also "China Conducts Biological Test on Its Army to Create Super-Soldiers, Says Top US Official," *Business Today* (India), December 9, 2020, https://www.businesstoday.in/current/world/china-conducts-biological-test-on-its-army-to-create-super-soldiers-says-top-us-official/story/424341.html. See also Antonio Regalado, "First Gene-Edited Dogs Reported in China," *MIT Technology Review*, October 19, 2015, https://www.geneticsandsociety.org/article/first-gene-edited-dogs-reported-china.

30. Ruptly, "Russia: Putin Warns a GM Human Could Be 'Worse Than Nuclear Bomb,'" October 21, 2017, www.youtube.com/watch?v=E6pD96RRIUQ.

31. *Human Nature*, directed by Adam Bolt, Greenwich Entertainment, March 13, 2020.

32. Claudia Geib, "Genetic Engineering Can Treat Rare Diseases, but Could Also Give Wealthy Children a Biological Advantage," *Futurism*, August 9, 2017, https://futurism.com/neoscope/expert-argues-that-gene-editing-will-widen-economic-class-gap.

33. *Human Nature*, directed by Adam Bolt, Greenwich Entertainment, March 13, 2020.
34. Daniel MacArthur (@dgmacarthur), "Prediction: my grandchildren will be embryo-screened, germline-edited," Twitter, November 30, 2015, 4:31 p.m., https://twitter.com/dgmacarthur/status/671335629687406592?lang=en.
35. *Human Nature*, directed by Adam Bolt, Greenwich Entertainment, March 13, 2020.
36. Jenny Straiton, "Genetically Modified Humans: The X-Men of Scientific Research," *BioTechniques* 66, no. 6 (June 2019), https://www.future-science.com/doi/pdf/10.2144/btn-2019-0056.

Chapter 7 – Disaster Resilience: The Ring of Fire

1. Shigeru Ban, "Emergency Shelters Made from Paper," TEDxTokyo, August 13, 2013, https://www.ted.com/talks/shigeru_ban_emergency_shelters_made_from_paper/details.
2. Ban, "Emergency Shelters Made from Paper."
3. Ban, "Emergency Shelters Made from Paper."
4. Mark Schreiber, "Hell on Earth in '23," *Japan Times*, August 26, 2001, www.japantimes.co.jp/community/2001/08/26/general/hell-on-earth-in-23/#.Xkww7S2ZORu.
5. Yuichi Ono, author's phone interview, April 16, 2020.
6. David Pilling, "Lessons Learnt from Kobe Quake," *Financial Times*, May 11, 2011, https://www.ft.com/content/5290f95e-4bc4-11e0-9705-00144feab49a.
7. Ban, "Emergency Shelters Made from Paper."
8. Rajib Shaw, author's phone interview, April 20, 2020.
9. Jun Sato, author's phone interview, May 4, 2020.
10. John W. van de Lindt, author's phone interview, April 23, 2020.
11. "Tokyo, the Earthquake-Proof City," *We Build Value*, June 27, 2018, www.webuildvalue.com/en/reportage/tokyo-the-earthquake-proof-city.html.
12. Toro Takeuchi, author's phone interview, April 17, 2020.
13. Sigrid Adriaenssens, author's phone interview, April 16, 2020.
14. Joyce Meri, author's phone interview, April 7, 2020.
15. Satoru Nishikawa, author's phone interview, April 6, 2020.
16. "Disaster Prevention Information," *Disaster Preparedness Tokyo*, 2019,

https://www.bousai.metro.tokyo.lg.jp/book/pdf/en/01_Simulation_of_a_Major_Earthquake.pdf.

17. Takeshi Komino, author's phone interview, April 13, 2020.

18. "Disaster Prevention Information," *Disaster Preparedness Tokyo*, 2019.

19. "Manga Comic: 'Tokyo "X" Day,'" *Disaster Preparedness Tokyo*, 2019, https://www.bousai.metro.tokyo.lg.jp/book/pdf/en/comic.pdf.

20. Mikio Ishiwatari, author's interview, April 19, 2020.

21. "Safe Cities Index 2019," *The Economist*, https://safecities.economist.com/safe-cities-index-2019/.

22. Cameron Allan Mckean, "Tokyo's Disaster Parks," *The Guardian*, August 19, 2014, www.theguardian.com/cities/2014/aug/19/tokyo-disaster-parks-hi-tech-survival-bunkers-hidden-green-spaces-earthquake.

23. Thu Ngo Nha, "A Glimpse of Japan's Deadly Earthquakes," ArcGIS StoryMaps, August 18, 2019, https://storymaps.arcgis.com/stories/da0acfccaece4fc481353b1a01589259.

24. *We Build Value*, "Tokyo, the Earthquake-Proof City."

25. "Earthquake-Proof Tokyo Named Safest City in the World," *Architecture*, April 11, 2019, www.architecturelab.net/earthquake-safest-city-in-the-world/.

26. Evan Reis, author's phone interview, April 23, 2020.

27. Rajib Shaw, author's phone interview, April 20, 2020.

28. "Osaka Quake Exposes Japan's Aging Infrastructure," *Nikkei Asian Review*, June 19, 2018, https://asia.nikkei.com/Economy/Osaka-quake-exposes-Japan-s-aging-infrastructure.

29. Amy Davidson Sorkin, "Japan: Earthquake, Water, Fire," *New Yorker*, March 11, 2011, https://www.newyorker.com/news/amy-davidson/japan-earthquake-water-fire.

30. Nikil Saval, "Shigeru Ban," *New York Times*, October 18, 2019, www.nytimes.com/interactive/2019/10/15/t-magazine/shigeru-ban.html.

Chapter 8 – Energy: The Electric Slide

1. Marc Tarpenning, author's phone interview, November 17, 2020. See also Stanford Center for Professional Development, "Stanford Seminar: Electric Vehicles and Startups," January 5, 2017, https://www.youtube.com/watch?v=s47Cy8OUNcM.

2. Stanford Center for Professional Development, "Stanford Seminar: Electric Vehicles and Startups."

3. Martin Eberhard, author's phone interview, November 23, 2020. See also Stanford Center for Professional Development, "Stanford Seminar: Electric Vehicles and Startups." See also Drake Baer, "The Making of Tesla: Invention, Betrayal, and the Birth of the Roadster," *Business Insider*, November 11, 2014, www.businessinsider.com/tesla-the-origin-story-2014-10.
4. Martin Eberhard, author's phone interview, November 23, 2020. See also Baer, "The Making of Tesla."
5. Stanford Center for Professional Development, "Stanford Seminar: Electric Vehicles and Startups."
6. Baer, "The Making of Tesla."
7. Martin Eberhard, author's phone interview, November 23, 2020. See also Baer, "The Making of Tesla."
8. Marc Tarpenning, author's phone interview, November 17, 2020. See also Marc Tarpenning and Shernaz Daver, "The Story of Building Tesla," Startup Grind, June 3, 2018, https://www.youtube.com/watch?v=pGf1tyPXBpA.
9. Martin Eberhard, author's phone interview, November 23, 2020. See also Stanford Center for Professional Development, "Stanford Seminar: Electric Vehicles and Startups."
10. Chris Nuttall, "Tale of a Supercharged Start-Up Entrepreneurship," *Financial Times*, April 12, 2007, https://www.ft.com/content/ab94e730-e83c-11db-b2c3-000b5df10621.
11. Baer, "The Making of Tesla."
12. Baer, "The Making of Tesla."
13. Baer, "The Making of Tesla."
14. Seth Weintraub, author's phone interview, May 12, 2020.
15. Marc Tarpenning, author's phone interview, November 17, 2020. See also Baer, "The Making of Tesla."
16. Marc Tarpenning, author's phone interview, November 17, 2020. See also Baer, "The Making of Tesla."
17. Baer, "The Making of Tesla."
18. Martin Eberhard, author's phone interview, November 23, 2020. See also Nuttall, "Tale of a Supercharged Start-Up Entrepreneurship."
19. Marc Tarpenning, author's phone interview, November 17, 2020. See also John Reed, "Elon Musk's Groundbreaking Electric Car," *Financial Times*, July 24, 2009, https://www.ft.com/content/e117987e-74eb-11de-9ed5-00144feabdc0. See also Tarpenning and Daver, "The Story of Building Tesla."
20. Marc Tarpenning, author's phone interview, November 17, 2020. See

also MIT Club of Northern California, "Pioneers in Clean Technology – Marc Tarpenning – Tesla Motors," March 6, 2014, https://www.youtube.com/watch?v=EDCYoAQmmAA.

21. Martin Eberhard, author's phone interview, November 23, 2020. See also Baer, "The Making of Tesla."

22. Martin Eberhard, author's phone interview, November 23, 2020.

23. Michael Shnayerson, "Quiet Thunder," *Vanity Fair*, May 2007, https://www.vanityfair.com/news/2007/05/tesla200705.

24. Mike Musgrove, "An Electric Car with Juice," *Washington Post*, July 22, 2006, http://www.washingtonpost.com/wp-dyn/content/article/2006/07/21/AR2006072101515.html.

25. Baer, "The Making of Tesla."

26. Claire Cain Miller, "Musk Unplugged: Tesla CEO Discusses Car Troubles," *New York Times*, October 24, 2008, https://bits.blogs.nytimes.com/2008/10/24/musk-unplugged-tesla-ceo-discusses-car-troubles/.

27. Martin Eberhard, author's phone interview, November 23, 2020.

28. Marc Tarpenning, author's phone interview, November 17, 2020. See also Tarpenning and Daver, "The Story of Building Tesla."

29. Scott Pelley, "U.S., China, Russia, Elon Musk: Entrepreneur's 'Insane' Vision Becomes Reality," CBS News, May 22, 2012, https://www.cbsnews.com/news/us-china-russia-elon-musk-entrepreneurs-insane-vision-becomes-reality/.

30. "About Tesla," https://www.tesla.com/about.

31. Lawrence Burns, author's phone interview, April 27, 2020.

32. "The Global Electric Vehicle Market in 2020: Statistics & Forecasts," Virta, https://www.virta.global/global-electric-vehicle-market. See also Fred Lambert, "Tesla Produces Its 1 Millionth Electric Car," *Electrek*, March 9, 2020, https://electrek.co/2020/03/09/tesla-produces-1000000th-electric-car/.

33. "Shell Quest Carbon Capture and Storage Project Reaches Milestone of 4M Tonnes," *Canada Press*, May 23, 2019, https://www.cbc.ca/news/canada/calgary/carbon-storage-cnrl-shell-quest-1.5146857. See also Tesla, "Impact Report," https://www.tesla.com/ns_videos/tesla-impact-report-2019.pdf.

34. Marc Tarpenning, author's phone interview, November 17, 2020.

35. Julia Carrie Wong, "Elon Musk Unveils Tesla Electric Truck – and a Surprise New Sports Car," *The Guardian*, November 17, 2017, https://www

.theguardian.com/technology/2017/nov/17/elon-musk-tesla-electric-truck-sports-car-surprise.

36. Daniel Sperling, author's phone interview, May 11, 2020.

37. Roland Hwang, author's phone interview, May 15, 2020.

38. Mark Kane, "Close to 1.18 Million Plug-In Electric Cars Were Sold in China in 2019," *InsideEVs*, January 22, 2020, https://insideevs.com/news/394229/plugin-electric-car-sales-china-2019/.

39. Yunshi Wang, author's phone interview, May 12, 2020. Peter Marsters, "Electric Cars: The Drive for a Sustainable Solution in China," *Wilson Center*, August 2009, https://www.wilsoncenter.org/publication/electric-cars-the-drive-for-sustainable-solution-china.

40. David Nikel, "Electric Cars: Why Little Norway Leads the World In EV Usage," Forbes, June 18, 2019, https://www.forbes.com/sites/davidnikel/2019/06/18/electric-cars-why-little-norway-leads-the-world-in-ev-usage/.

41. Robbie Diamond, author's phone interview, April 17, 2020.

42. Martin Eberhard, author's phone interview, November 23, 2020. See also "Martin Eberhard Calls Tesla His 'Baby', Talks about Being Ousted from Company's Board," *Economic Times*, March 1, 2019, https://economictimes.indiatimes.com/magazines/panache/martin-eberhard-calls-tesla-his-baby-talks-about-being-ousted-from-companys-board/articleshow/68211761.cms.

43. Marc Tarpenning, author's phone interview, November 17, 2020.

44. Elon Musk (@ElonMuskNewsOrg), "If you get up in the morning…," Twitter, January 2, 2017, 3:30 a.m., https://twitter.com/elonmusknewsorg/status/815731862601666560?lang=en.

Chapter 9 – Prosperity: The King of Microfinance

1. "Bangladesh Continues to Reduce Poverty but at a Slower Pace," World Bank, October 24, 2017, www.worldbank.org/en/news/feature/2017/10/24/bangladesh-continues-to-reduce-poverty-but-at-slower-pace. See also Randeep Ramesh, "Giving a Hand Up, Not a Handout," *The Guardian*, December 27, 2006, https://www.theguardian.com/business/2006/dec/27/money.

2. Julia Kagen, "Microfinance," *Investopedia*, April 21, 2020, www.investopedia.com/terms/m/microfinance.asp.

3. "Muhammad Yunus – Nobel Lecture," Nobel Prize, December 10, 2006,

www.nobelprize.org/prizes/peace/2006/yunus/26090-muhammad-yunus-nobel-lecture-2006-2/.

4. "Yunus, Grameen Bank win Nobel Peace Prize for Loans to the Poor," *Salt Lake Tribune*, October 13, 2006, https://archive.sltrib.com/article.php?id=4486938&itype=NGPSID.

5. Muhammad Yunus, *Banker to the Poor: Micro-Lending and the Battle against World Poverty* (New York: PublicAffairs, 1999), vii.

6. Yunus, *Banker to the Poor*, viii.

7. Yunus, *Banker to the Poor*, viii.

8. Yunus, *Banker to the Poor*, ix.

9. Yunus, *Banker to the Poor*, 47.

10. Yunus, *Banker to the Poor*, 48.

11. Yunus, *Banker to the Poor*, 48.

12. Yunus, *Banker to the Poor*, 48.

13. Yunus, *Banker to the Poor*, 50. See also Doug Mellgren, "Bangladeshi Economist Wins Nobel Peace Prize," *Seattle Times*, October 13, 2006, https://www.seattletimes.com/nation-world/bangladeshi-economist-wins-nobel-peace-prize/.

14. Yunus, *Banker to the Poor*, 52.

15. "United Nations Dignitaries Hail Nobel Peace Prize Winners in Special Tribute to Muhammad Yunus, Grameen Bank," United Nations, November 17, 2006, https://www.un.org/press/en/2006/dev2610.doc.htm.

16. Yunus, *Banker to the Poor*, 55.

17. Yunus, *Banker to the Poor*, ix.

18. Yunus, *Banker to the Poor*, ix.

19. Muhammad Yunus, "Grameen Bank at a Glance," Yunus Centre, March 12, 2008, https://www.muhammadyunus.org/post/370/grameen-bank-at-a-glance.

20. Anand Giridharadas and Keith Bradsher, "Microloan Pioneer and His Bank Win Nobel Peace Prize," *New York Times*, October 13, 2006, https://www.nytimes.com/2006/10/13/business/14nobelcnd.html.

21. Andrew Foster, author's phone interview, May 18, 2020.

22. Yunus, *Banker to the Poor*, 64.

23. Yunus, "Grameen Bank at a Glance."

24. Yunus, "Grameen Bank at a Glance."

25. Christian Ahlin, author's phone interview, May 29, 2020.

26. M. Nurul Alam and Dr. Mike Getubig, "Guidelines for Establishing and

Operating Grameen-Style Microcredit Programs," Grameen Foundation, 2010, https://grameenfoundation.org/documents/GrameenGuidelines.pdf.

27. World Data Lab, "World Poverty Clock," September 17, 2020, https://worldpoverty.io.

28. Max Roser and Esteban Ortiz-Ospina, "Global Extreme Poverty," Our World in Data, 2013, https://ourworldindata.org/extreme-poverty.

29. "CGAP Launches Global Microfinance Survey to Inform COVID-19 Response," Consultative Group to Assist the Poor, June 1, 2020, www.cgap.org/news/cgap-launches-global-microfinance-survey-inform-covid-19-response.

30. Shahidur R. Khandker, author's phone interview, June 1, 2020. See also Shahidur R. Khandker, M. A. Baqui Khalily, and Hussain A. Samad, *Beyond Ending Poverty: The Dynamics of Microfinance in Bangladesh*, World Bank Group, 2016, http://documents.worldbank.org/curated/en/366431468508832455/pdf/106672-PUB.pdf.

31. Mark Pitt, author's phone interview, May 16, 2020.

32. Bruce Wydick, author's phone interview, May 14, 2020.

33. Cynthia Kinnan, author's phone interview, May 21, 2020.

34. Jonathan Morduch, author's phone interview, May 27, 2020. See also Jonathan Morduch, "The Role of Subsidies in Microfinance: Evidence from the Grameen Bank," *Journal of Development Economics* 60 (1999): 229–48, https://s18798.pcdn.co/jmorduch/wp-content/uploads/sites/5267/2016/12/Morduch-Grameen-Subsidies-JDE-1999.pdf.

35. Tim Ogden, author's phone interview, May 26, 2020.

36. "Is Grameen Bank Different from Convential Banks," Grameen Bank, March 12, 2011, https://www.grameen-info.org/is-grameen-bank-different-from-conventional-banks/.

37. Britt F. Hagstrom, "The Man behind the Nobel Prize," *Daily Star*, December 10, 2006, https://archive.thedailystar.net/suppliments/2006/december/bangladeshstandingtall/index.htm.

38. Syed Zain al-Mahmood, "Muhammad Yunus and Bangladesh Government Battle over Grameen," *The Guardian*, November 21, 2012, https://www.theguardian.com/global-development/2012/nov/21/muhammad-yunus-bangladesh-grameen-bank.

Chapter 10 – Food: Green Meat

1. Frederico Alisson-Silva, Kunio Kawanishi, and Ajit Varki, "Human Risk of Diseases Associated with Red Meat Intake: Analysis of Current Theories and Proposed Role for Metabolic Incorporation of a Non-Human Sialic Acid," *Molecular Aspects of Medicine* 51 (October 2016): 16–30, https://www.ncbi.nlm.nih.gov/pmc/articles/PMC5035214/.
2. National Institutes of Health, "Eating Red Meat Daily Triples Heart Disease-Related Chemical," *NIH Research Matters*, January 15, 2019.
3. Mosa Meat, "Why We Need to Change How We Make Meat," July 5, 2018, https://www.youtube.com/watch?v=oSl9yZuPrdo.
4. Scott May, author's phone interview, June 8, 2020.
5. Stephen J. Bronner, "With $72 Million in Funding, the Entrepreneur behind Beyond Meat Pursues Innovation over Profit," *Entrepreneur*, January 22, 2018, https://www.entrepreneur.com/article/307715.
6. Colin Finkle, "Beyond Meat Brand Profile," Brand Marketing Blog, July 21, 2019, https://brandmarketingblog.com/articles/good-branding/beyond-meat-brand-profile/.
7. Anna Starostinetskaya, "Beyond Meat Is about to Go Public; Founder Ethan Brown Speaks about the Evolution of Meat," VegNews, April 4, 2019, https://vegnews.com/2019/4/beyond-meat-is-about-to-go-public-founder-ethan-brown-speaks-about-the-evolution-of-meat. See also Alice Park, "Why We Don't Need Animals to Keep Enjoying Meat," *Time*, June 6, 2019, https://time.com/5601980/beyond-meat-ceo-ethan-brown-interview/.
8. Laurence Darmiento, "Ethan Brown Went Vegan but Missed Fast Food. So He Started a Revolution," *Los Angeles Times*, January 20, 2020, https://www.latimes.com/business/story/2020-01-08/beyond-meat-founder-ethan-brown.
9. Darmiento, "Ethan Brown Went Vegan but Missed Fast Food."
10. Shira Zakai, author's phone interview, January 27, 2020.
11. "Peter G. Brown," Faculty of Agricultural and Environmental Sciences, McGill University, www.mcgill.ca/nrs/academic-0/brown.
12. Tom Foster, "Friends Thought Beyond Meat's Founder Was Crazy. His Billion-Dollar IPO Proved Them Wrong," *Inc.* August 2019, https://www.inc.com/magazine/201908/tom-foster/ethan-brown-beyond-meat-alternative-protein-plant-burger-ipo-public-whole-foods.html.
13. Bronner, "With $72 Million in Funding."

14. “Livestock and Climate Change,” World Watch, November/December 2009, https://awellfedworld.org/wp-content/uploads/Livestock-Climate-Change-Anhang-Goodland.pdf.

15. Foster, “Friends Thought Beyond Meat's Founder Was Crazy.”

16. Darmiento, “Ethan Brown Went Vegan but Missed Fast Food.”

17. Foster, “Friends Thought Beyond Meat's Founder Was Crazy.”

18. “Meat: The Future Series; Alternative Proteins,” World Economic Forum White Paper, 2019, http://www3.weforum.org/docs/WEF_White_Paper_Alternative_Proteins.pdf. See also “Carbon Footprint Factsheet,” University of Michigan, Center for Sustainable Systems, Pub. No. CSS09-05, 2020, http://css.umich.edu/factsheets/carbon-footprint-factsheet. See Hannah Ritchie, “You Want to Reduce the Carbon Footprint of Your Food? Focus on What You Eat, Not Whether Your Food Is Local,” Our World in Data, January 24, 2020, https://ourworldindata.org/food-choice-vs-eating-local.

19. Zach Hrynowski, “What Percentage of Americans are Vegetarian?” Gallup, September 27, 2019, https://news.gallup.com/poll/267074/percentage-americans-vegetarian.aspx.

20. Bronner, “With $72 Million in Funding.”

21. Darmiento, “Ethan Brown Went Vegan but Missed Fast Food.”

22. “Live Episode! Beyond Meat: Ethan Brown,” NPR Radio, February 16, 2017, www.npr.org/2017/02/16/515420148/live-episode-beyond-meat-ethan-brown.

23. Fu-Hung Hsieh, author's written correspondence, May 30, 2020.

24. Mark Barna, “Bock Soy,” *Mizzou*, November 7, 2012, https://mizzoumag.missouri.edu/2012/11/bock-soy/index.html.

25. Bronner, “With $72 Million in Funding.”

26. Darmiento, “Ethan Brown Went Vegan but Missed Fast Food.”

27. Foster, “Friends Thought Beyond Meat's Founder Was Crazy.”

28. Darmiento, “Ethan Brown Went Vegan but Missed Fast Food.”

29. Darmiento, “Ethan Brown Went Vegan but Missed Fast Food.”

30. Darmiento, “Ethan Brown Went Vegan but Missed Fast Food.”

31. Darmiento, “Ethan Brown Went Vegan but Missed Fast Food.”

32. Darmiento, “Ethan Brown Went Vegan but Missed Fast Food.”

33. Bill Gates, “Future of Food,” *GatesNotes*, March 18, 2013, www.gatesnotes.com/About-Bill-Gates/Future-of-Food.

34. Bronner, “With $72 Million in Funding.”

35. Darmiento, "Ethan Brown Went Vegan but Missed Fast Food."
36. Ali Bouzari, author's phone interview, May 28, 2020.
37. Martin C. Heller and Gregory A. Keoleian, "Beyond Meat's Beyond Burger Life Cycle Assessment: A Detailed Comparison between a Plant-Based and an Animal-Based Protein Source," Center for Sustainable Systems Report, University of Michigan, September 14, 2018, http://css.umich.edu/publication/beyond-meats-beyond-burger-life-cycle-assessment-detailed-comparison-between-plant-based.
38. Marthe Fourcade and Corinne Gretler, "McDonald's Picks Beyond for Canada Trial, U.S. Prize Remains," Bloomberg, September 26, 2019, www.bloomberg.com/news/articles/2019-09-26/mcdonald-s-to-test-veggie-burger-from-beyond-meat-in-canada.
39. Michelle King, "Dunkin' and Beyond Meat Accelerate Nationwide Launch of Beyond Sausage Sandwich," Dunkin' News Release, October 21, 2019, https://news.dunkindonuts.com/news/releases-20191021.
40. Paul R. La Monica, "Impossible Whoppers Are a Huge Hit at Burger King, Fueling Its Best Quarter in 4 Years," CNN Business, October 28, 2019, www.cnn.com/2019/10/28/investing/restaurant-brands-earnings-burger-king-popeyes/index.html. See also "Sweet Earth Debuts New Plant-Based Culinary Innovations, Awesome Burger & Awesome Grounds," Cision PRNewswire, September 25, 2019, www.prnewswire.com/news-releases/sweet-earth-debuts-new-plant-based-culinary-innovations-awesome-burger--awesome-grounds-300924874.html.
41. Bronner, "With $72 Million in Funding."
42. Nir Goldstein, author's phone interview, June 25, 2020.
43. Rusty Schwartz, author's phone interview, June 2, 2020.
44. "Ad: Fake Meat Grows in Factories, Not on Vines," Center for Consumer Freedom, August 13, 2019, www.consumerfreedom.com/2019/08/ad-fake-meat-grows-in-factories-not-on-vines/.
45. Darmiento, "Ethan Brown Went Vegan but Missed Fast Food."
46. Darmiento, "Ethan Brown Went Vegan but Missed Fast Food."
47. Brett Hurt, author's phone interview, June 4, 2020.

Chapter 11 – Water: Every Last Drop

1. Associated Press, "A Broken Water Pipe Floods UCLA," *New York Post*, July 29, 2014, https://nypost.com/2014/07/29/a-broken-water-pipe-is-flooding-ucla/.

2. Associated Press, "A Broken Water Pipe Floods UCLA."
3. Simona Shemer, "On World Water Day, a Look at the Innovative Tech That Makes Israel a Water Superpower," No Camels, March 22, 2018, http://nocamels.com/2018/03/on-world-water-day-a-look-at-the-innovative-tech-that-makes-israel-a-water-superpower/. See also World Health Organization, "Drinking Water," June 14, 2009, http://www.who.int/news-room/fact-sheets/detail/drinking-water.
4. Mada Masr, "'We Woke Up in a Desert' – The Water Crisis Taking Hold across Egypt," *The Guardian*, August 4, 2015, www.theguardian.com/world/2015/aug/04/egypt-water-crisis-intensifies-scarcity.
5. Arash Karami, "Iran Official Warns Water Crisis Could Lead to Mass Migration," *Al-Monitor*, April 28, 2015, www.al-monitor.com/pulse/originals/2015/04/iran-water-crisis-mass-migration.html.
6. Leah Schleifer, "7 Reasons We're Facing a Global Water Crisis," World Resources Institute, August 24, 2017, https://www.wri.org/blog/2017/08/7-reasons-were-facing-global-water-crisis.
7. US Government Accountability Office, "Supply Concerns Continue, and Uncertainties Complicate Planning," May 20, 2014, www.gao.gov/products/GAO-14-430.
8. Rachel Layne, "Water Costs Are Rising across the US – Here's Why," CBS, August 27, 2019, https://www.cbsnews.com/news/water-bills-rising-cost-of-water-creating-big-utility-bills-for-americans/.
9. Sarah Frostenson, "America Has a Water Crisis No 1 Is Talking About," *Vox*, March 22, 2018, www.vox.com/science-and-health/2017/5/9/15183330/america-water-crisis-affordability-millions.
10. Hiroko Tabuchi, "$300 Billon War beneath the Street: Fighting to Replace America's Water Pipes," *New York Times*, November 10, 2017, https://www.nytimes.com/2017/11/10/climate/water-pipes-plastic-lead.html.
11. Allison Kosik, "Experts: U.S. Water Infrastructure in Trouble," CNN, January 21, 2011, http://edition.cnn.com/2011/US/01/20/water.main.infrastructure/.
12. Elizabeth Mac, "A Burgeoning Crisis? A Nationwide Assessment of the Geography of Water Affordability in the United States," *PloS 1* 12, no. 4 (January 11, 2017), https://journals.plos.org/plosone/article?id=10.1371/journal.pone.0169488.
13. Tony Kelly, author's phone interview, April 1, 2020.
14. Bill Kingdom, Roland Liemberger, and Philippe Marin, "The Challenge

of Reducing Non-Revenue Water (NRW) in Developing Countries – How the Private Sector Can Help: A Look at Performance-Based Service Contracting," World Bank, *Water Supply and Sanitation Sector Board Discussion Paper Series*, no. 8, December 2006, https://openknowledge.worldbank.org/handle/10986/17238.

15. Elie Ofek and Matthew Preble, "Takadu," Harvard Business School Case 514-011, July 2013 (revised August 2017).

16. Ofek and Preble, "Takadu."

17. Peter Williams, author's phone interview, March 11, 2020.

18. Oded Distel, author's phone interview, March 12, 2020.

19. Amir Peleg, author's interview, Washington, DC, April 2, 2019.

20. Amir Peleg, author's interview, Washington, DC, April 2, 2019.

21. Ofek and Preble, "Takadu."

22. Amir Peleg, author's interview, Washington, DC, April 2, 2019.

23. Amir Peleg, author's interview, Washington, DC, April 2, 2019.

24. Ofek and Preble, "Takadu."

25. Damian Carrington, "England Could Run Short of Water within 25 years," *The Guardian*, March 18, 2019, https://www.theguardian.com/environment/2019/mar/18/england-to-run-short-of-water-within-25-years-environment-agency.

26. Elise Ackerman, "Israeli Startup Mines Data to Solve Global Water Crisis," *Forbes*, December 31, 2013, https://www.forbes.com/sites/eliseackerman/2013/12/31/israeli-startup-mines-data-to-solve-global-water-crisis/?sh=4ff87d7d2f46.

27. Ofek and Preble, "Takadu."

28. Ofek and Preble, "Takadu."

29. Keith Hays, author's phone interview, April 4, 2020.

30. George Theo, author's phone interview, March 30, 2020.

31. Christopher Gasson, author's phone interview, March 30, 2020.

32. Seth M. Siegel, author's correspondence, October 29, 2016.

33. Jonathan Ferziger, "Netanyahu Offers to Help Brown Manage California Drought," Bloomberg, March 5, 2014, www.bloomberg.com/news/articles/2014-03-05/netanyahu-offers-to-help-brown-manage-california-drought.

Chapter 12 – Governance: The Digital Republic

1. "Estonia Celebrates the Day of Restoration of Independence," Estonian World, August 20, 2017, https://estonianworld.com/life/estonia-celebrates-the-day-of-restoration-of-independence/.

2. Toomas Hendrik Ilves, author's phone interview, March 27, 2020. See also Ripon Society, "The Estonian Example – Q&A with Toomas Hendrick Ilves," *The Ripon Forum* 47, no. 1 (Winter 2013), https://www.riponsociety.org/article/the-estonian-example-qa-with-toomas-hendrik-ilves/.

3. Wen Hoe, "E-Stonia: 1 Small Country's Digital Government Is Having a Big Impact," Harvard Government Innovators Network, June 7, 2017, https://www.innovations.harvard.edu/blog/estonia-one-small-country-digital-government-having-big-impact-x-road. See also e-Governance Academy, "Toomas Hendrik Ilves – Estonia as a Think Tank of e-Government. The eGA/GISI Initative," May 16, 2013, https://www.youtube.com/watch?v=BBcUORzU3CA.

4. A.A.K., "How Estonia Became a Leader in Technology," *The Economist*, July 31, 2013, https://www.economist.com/the-economist-explains/2013/07/30/how-did-estonia-become-a-leader-in-technology.

5. Toomas Hendrik Ilves, "e-Estonia and the New Europe," Välisministeerium, October 25, 2005, https://vm.ee/et/node/42610.

6. Toomas Hendrik Ilves, author's phone interview, March 27, 2020. See also Ripon Society, "The Estonian Example." See also Hoe, "E-Stonia." See also See also e-Governance Academy, "Toomas Hendrik Ilves."

7. Toomas Hendrik Ilves, author's phone interview, March 27, 2020. See also Jamie Kitman, "President Ilves: The Man Who Made E-Stonia," *The Guardian*, November 3, 2011, www.theguardian.com/world/2011/nov/03/president-ilves-made-estonia.

8. Toomas Hendrik Ilves, author's phone interview, March 27, 2020. See also Tim Mansel, "How Estonia Became E-Stonia," BBC, May 16, 2013, www.bbc.com/news/business-22317297.

9. e-Governance Academy, "Toomas Hendrik Ilves." See also Hoe, "E-Stonia."

10. Linnar Viik, author's phone interview, August 20, 2019.

11. Ripon Society, "The Estonian Example." See also e-Governance Academy, "Toomas Hendrik Ilves."

12. Toomas Hendrik Ilves, author's phone interview, March 27, 2020. See also Ott Ummelas, "World's Most High-Tech Voting System to Get Hacking

Defenses," Bloomberg, July 18, 2017, https://www.bloomberg.com/news/articles/2017-07-17/world-s-most-high-tech-voting-system-to-get-new-hacking-defenses.

13. Nina Kolyako, "Estonia Ranked First Worldwide in Terms of Broadband Internet Speeds," *Baltic Course*, January 27, 2012, http://www.baltic-course.com/eng/good_for_business/?doc=52217.

14. Viljar Lubi, author's phone interview, April 4, 2020.

15. Nathan Heller, "Estonia, The Digital Republic," *New Yorker*, December 11, 2017, https://www.newyorker.com/magazine/2017/12/18/estonia-the-digital-republic.

16. Ripon Society, "The Estonian Example."

17. "Statistics about Internet Voting in Estonia," Valimised, www.valimised.ee/en/archive/statistics-about-internet-voting-estonia.

18. Ripon Society, "The Estonian Example."

19. Dr. Robert Krimmer, author's phone interview, April 8, 2020.

20. Toomas Hendrik Ilves, author's phone interview, March 27, 2020.

21. Toomas Hendrik Ilves, author's phone interview, March 27, 2020. See also Georgeta Gheorghe, "Estonia's Toomas Hendrik Ilves Talks Digitalisation, from X-Road to Digital Embassies and E-Residency," *Business Review*, August 11, 2017, http://business-review.eu/featured/estonias-toomas-hendrik-ilves-talks-digitalisation-between-x-road-digital-embassies-e-resdiency-152141.

22. Joshua Davis, "Hackers Take Down the Most Wired Country in Europe," *Wired*, August 21, 2007, https://www.wired.com/2007/08/ff-estonia/.

23. Marten Kaevats, author's phone interview, April 6, 2020.

24. Elizabeth Schultze, "How a Tiny Country Bordering Russia Became 1 of the Most Tech-Savvy Societies in the World," CNBC, February 8, 2019, https://www.cnbc.com/2019/02/08/how-estonia-became-a-digital-society.html.

25. "Estonia to Open World's First Data Embassy in Luxembourg," e-estonia, June 2017, https://e-estonia.com/estonia-to-open-the-worlds-first-data-embassy-in-luxembourg/.

26. Taavi Kotka, author's phone interview, August 1, 2019.

27. Tarvi Martens, author's phone interview, July 30, 2019.

28. Ripon Society, "The Estonian Example."

29. Taavi Kotka, author's phone interview, August 1, 2019.

30. Andrew Keen, "E-Stonia: The Country Using Tech to Rebrand Itself

as Anti-Russia," *The Guardian*, April 21, 2016. https://www.theguardian.com/world/2016/apr/21/e-stonia-country-using-technology-to-rebrand-itself-as-the-anti-russia.

31. Toomas Hendrik Ilves, author's phone interview, March 27, 2020. See also Keen, "E-Stonia."

32. Taavi Kotka, author's phone interview, August 1, 2019.

33. "Ministry Hopes Great Success for Estonian e-Residency Service," LETA (Latvian Information Agency), October 7, 2014.

34. Taavi Kotka, author's phone interview, August 1, 2019.

35. Taavi Kotka, author's phone interview, August 1, 2019.

36. Xolo, "10 Notable People Who Are E-Residents of Estonia," October 11, 2016, https://blog.leapin.eu/10-notable-people-who-are-e-residents-of-estonia-c3b7dcd3c7d6. See also Kaspar Korjus, "Welcome to Our Digital Nation, Bill Gates," *Medium*, October 18, 2018, https://medium.com/e-residency-blog/welcome-to-our-digital-nation-bill-gates-a8c33edd79a8.

37. Heller, "Estonia, The Digital Republic."

38. Danica Kirka, "No More Paperwork: Estonia Edges toward Digital Government," Associated Press, December 26, 2018, https://apnews.com/article/da5e0dd3bf364bbda436fc16690f842c.

39. Heller, "Estonia, the Digital Republic."

40. Taavi Kotka, author's phone interview, August 1, 2019.

41. Kaspar Korjus, author's phone interview, April 29, 2020.

42. Hal Hodson, "E-Citizens Unite," *NewScientist* 224, no. 2991 (October 18, 2014): 24.

43. Aaron Maniam, author's phone interview, April 15, 2020.

44. Kersti Kaljulaid, "Tallinn Digital Summit – Opening Address by Estonia's President Kersti Kaljulaid," September 29, 2017, https://www.youtube.com/watch?v=h1I4LGaP6Io.

45. CNN, "Estonia: Obama on Obamacare Website," July 21, 2016, www.youtube.com/watch?v=v6vF_ARdlBA.

46. Kaljulaid, "Tallinn Digital Summit."

47. Kersti Kaljulaid, author's interview, Washington, DC, March 3, 2020.

Chapter 13 – Security: Google's X-Man

1. Bjorn Carey and Elizabeth Svoboda, "Live Reports from the 2005 Darpa Grand Challenge," *Popular Science*, October 8, 2005, https://www.popsci.com/scitech/article/2005-10/live-reports-2005-darpa-grand-challenge/.

2. CNBC International TV, "Sebastian Thrun, Founder of Udacity," *The Brave Ones*, May 15, 2017, https://www.youtube.com/watch?v=DtO3pcNnEuA.
3. Gideon Rose, "Google's Original X-Man: A Conversation with Sebastian Thrun," *Foreign Affairs*, November/December 2013, www.foreignaffairs.com/interviews/2013-10-15/googles-original-x-man.
4. CNBC International TV, "Sebastian Thrun, Founder of Udacity."
5. Rohan Abraham, "A Personal Loss, a $2-mm Win and a Soaring Ambition Led Sebastian Thrun on the Road to Innovation," *Economic Times*, April 2, 2019, https://economictimes.indiatimes.com/magazines/panache/a-personal-loss-a-2-mn-win-a-soaring-ambition-led-sebastian-thrun-on-the-road-to-innovation/articleshow/68680571.cms.
6. Sebastian Thrun, "Leave the Driving to the Car, and Reap Benefits in Safety and Mobility," *New York Times*, December 5, 2011, https://www.nytimes.com/2011/12/06/science/sebastian-thrun-self-driving-cars-can-save-lives-and-parking-spaces.html. See also Sebastian Thrun, "Google's Driverless Car," Ted2011, March 2011, https://www.ted.com/talks/sebastian_thrun_google_s_driverless_car?referrer=playlist-planes_trains_and_automobiles.
7. Rose, "Google's Original X-Man."
8. Rose, "Google's Original X-Man."
9. Rose, "Google's Original X-Man."
10. CNBC International TV, "Sebastian Thrun, Founder of Udacity."
11. CNBC International TV, "Sebastian Thrun, Founder of Udacity."
12. CNBC International TV, "Sebastian Thrun, Founder of Udacity."
13. Lucy Handley, "The Education of Sebastian Thrun," CNBC, January 1, 2017, https://www.cnbc.com/2017/06/01/sebastian-thrun-udacity-googlex.html.
14. Sebastian Thrun, "What We're Driving At," *Google Official Blog*, October 9, 2010, https://googleblog.blogspot.com/2010/10/what-were-driving-at.html.
15. Thrun, "What We're Driving At."
16. Darrel Etherington and Lora Kolodny, "Google's Self-Driving Car Unit Becomes Waymo," December 13, 2016, https://techcrunch.com/2016/12/13/googles-self-driving-car-unit-spins-out-as-waymo/.
17. Shernaz Daver, author's phone interview, November 16, 2020.
18. Rose, "Google's Original X-Man."
19. Eric Danko, author's phone interview, June 11, 2020.

20. Kara Kockelman, author's phone interview, April 27, 2020.
21. Thrun, "Leave the Driving to the Car."
22. Thrun, "Google's Driverless Car."
23. Insurance Institute for Highway Safety, Highway Loss Data Institute, "Fatality Facts 2018, State by State," https://www.iihs.org/topics/fatality-statistics/detail/state-by-state.
24. Nidhi Kalra and David Groves, "The Enemy of Good: Estimating the Cost of Waiting for Nearly Perfect Autonomous Vehicles," Rand Corporation, 2017, https://www.rand.org/pubs/research_reports/RR2150.html.
25. Alain Kornhauser, author's phone interview, April 29, 2020.
26. Kristin Houser, "Many Self-Driving Car Accidents Have Been Caused by Humans," *Business Insider*, September 5, 2018, https://www.businessinsider.com/self-driving-car-accidents-caused-by-humans-2018-9.
27. Chan Lieu, author's phone interview, April 23, 2020.
28. Johnna Crider, "Over 730,000 Tesla Vehicles with Autopilot 2 & 3 on the Road," *Cleantechnica*, January 5, 2020, https://cleantechnica.com/2020/01/05/over-730000-tesla-vehicles-with-autopilot-2-3-on-the-road/. See also Fred Lambert, "Tesla Drops a Bunch of New Autopilot Data, 3 Billion Miles and More," *Electrek*, April 22, 2020, https://electrek.co/2020/04/22/tesla-autopilot-data-3-billion-miles/.
29. Scott Hardman, author's phone interview, May 7, 2020.
30. Sebastian Thrun and Chris Anderson, "What AI Is – and Isn't," Ted2017, April 2017, https://www.ted.com/talks/sebastian_thrun_and_chris_anderson_what_ai_is_and_isn_t?language=en.
31. Lawrence Burns, author's phone interview, April 27, 2020. See also David Dorsey, "Expert Says Driverless Cars Will Disrupt Transportation Industry within 5 Years," *Fort Myers News-Press*, January 31, 2020, https://www.news-press.com/story/news/local/2020/01/31/driverless-cars-larry-burns-predicts-autonomous-cars-go-mainstream-within-five-years-disrupt-transpo/4625273002/.

Afterword – Humanity 2.0

1. Kurzgesagt – In a Nutshell, "A Selfish Argument for Making the World a Better Place – Egoistic Altruism," March 18, 2018, https://www.youtube.com/watch?v=rvskMHnosqQ&vl=en.
2. Gidi Grinstein, author's phone interview, December 17, 2020.

3. Frank Covey, *The 7 Habits of Highly Effective People* (New York: Free Press, 1989), 207.
4. John Mackey, author's phone interview, June 15, 2020.

Bibliography

Books

Burns, Lawrence. *Autonomy: The Quest to Build the Driverless Car – And How It Will Reshape Our World*. New York: Ecco, 2018.

Covey, Frank. *The Seven Habits of Highly Effective People*. New York: Free Press, 1989.

Diamandis, Peter, and Steven Kotler. *Abundance: The Future Is Better Than You Think*. New York: Free Press, 2012.

———. *Bold: How to Go Big, Bank, and Better the World*. New York: Simon and Schuster, 2015.

———. *The Future Is Faster Than You Think*. New York: Simon and Schuster, 2020.

Goodman, Marc. *Future Crimes: Everything Is Connected, Everyone Is Vulnerable and What We Can Do about It*. New York: Doubleday, 2015.

Khan, Salman. *The One World School House*. New York: Twelve, 2012.

Kurzweil, Ray. *The Age of Spiritual Machines: When Computers Exceed Human Intelligence*. New York: Viking, 1999.

———. *The Singularity Is Near*. New York: Viking, 2005.

Martinez, Sylvia Libow, and Gary S. Stager. *Invent to Learn: Making, Tinkering, and Engineering in the Classroom*. Torrance, CA: Constructing Modern Knowledge Press, 2013.

Siegel, Seth M. *Let There Be Water*. New York: Thomas Dunne Books, 2015.

Winegard, Timothy. *The Mosquito: A Human History of our Deadliest Predator.* London: Allen Lane, 2019.

Yunus, Muhammad. *Banker to the Poor: Micro-Lending and the Battle against World Poverty.* New York: Public Affairs, 1999.

Articles and Lectures

A.A.K. "How Estonia Became a Leader in Technology." *The Economist,* July 31, 2013. https://www.economist.com/the-economist-explains/2013/07/30/how-did-estonia-become-a-leader-in-technology.

Abraham, Rohan. "A Personal Loss, a $2-mm Win and a Soaring Ambition Led Sebastian Thrun on the Road to Innovation." *Economic Times,* April 2, 2019. https://economictimes.indiatimes.com/magazines/panache/a-personal-loss-a-2-mn-win-a-soaring-ambition-led-sebastian-thrun-on-the-road-to-innovation/articleshow/68680571.cms.

Achituv, Netta. "The Israeli Who Is Bringing Electricity, Agriculture and Medicine Cheaply to the Poorest Countries in the World." [In Hebrew.] *The Marker,* July 6, 2016. https://www.themarker.com/themarker-women/MAGAZINE-1.2998657.

Ackerman, Elise. "Israeli Startup Mines Data to Solve Global Water Crisis." *Forbes,* December 31, 2013. https://www.forbes.com/sites/eliseackerman/2013/12/31/israeli-startup-mines-data-to-solve-global-water-crisis/?sh=4ff87d7d2f46.

al-Mahmood, Syed Zain. "Muhammad Yunus and Bangladesh Government Battle over Grameen." *The Guardian,* November 21, 2012. https://www.theguardian.com/global-development/2012/nov/21/muhammad-yunus-bangladesh-grameen-bank.

Alam, M. Nurul, and Dr. Mike Getubig. "Guidelines for Establishing and Operating Grameen-Style Microcredit Programs." Grameen Foundation, 2010. https://grameenfoundation.org/documents/GrameenGuidelines.pdf.

Alisson-Silva, Frederico, Kunio Kawanishi, and Ajit Varki. "Human

Risk of Diseases Associated with Red Meat Intake: Analysis of Current Theories and Proposed Role for Metabolic Incorporation of a Non-Human Sialic Acid." *Molecular Aspects of Medicine* 51 (October 2016). https://www.ncbi.nlm.nih.gov/pmc/articles/PMC5035214/.

Architecture. "Earthquake-Proof Tokyo Named Safest City in the World." April 11, 2019. www.architecturelab.net/earthquake-safest-city-in-the-world/.

Associated Press. "A Broken Water Pipe Floods UCLA." *New York Post*, July 29, 2014. https://nypost.com/2014/07/29/a-broken-water-pipe-is-flooding-ucla/.

Baer, Drake. "The Making of Tesla: Invention, Betrayal, and the Birth of the Roadster." *Business Insider*, November 11, 2014. www.businessinsider.com/tesla-the-origin-story-2014-10.

Ban, Shigeru. "Emergency Shelters Made from Paper." TEDxTokyo, August 13, 2013. https://www.ted.com/talks/shigeru_ban_emergency_shelters_made_from_paper/details.

Barna, Mark. "Bock Soy." *Mizzou*, November 7, 2012. https://mizzoumag.missouri.edu/2012/11/bock-soy/index.html.

BBC. "He Jiankui Defends 'World's First Gene-Edited Babies.'" November 28, 2018. https://www.bbc.com/news/world-asia-china-46368731.

Bell, John. "To the Moon: Google X's Astro Teller Encourages Creatives to Think Big." *The Guardian*, June 20, 2013. www.theguardian.com/media-network/media-network-blog/2013/jun/20/google-x-astro-teller-creatives.

Benson, Sarah. "Salman Khan." Sutori. www.sutori.com/story/salman-khan--Lvux5bFpvn8zTBS222BUgLRN.

Bhattacharjee, Yudhijit. "Launch Pad." *New York Times*, November 10, 2016. https://www.nytimes.com/interactive/2016/11/13/magazine/design-issue-sanitary-pads-india.html.

Bilgic, Hatice, Erdogan Hakki, Anamika Pandey, Mohd Kamran Khan, and Mahinur Akkaya. "Ancient DNA from 8400 Year-Old

Çatalhöyük Wheat: Implications for the Origin of Neolithic Agriculture." *PloS One* 11, no. 3 (2016). https://www.ncbi.nlm.nih.gov/pmc/articles/PMC4801371/.

Bostrom, Nick. "Existential Risks: Analyzing Human Extinction Scenarios and Related Hazards." *Journal of Evolution and Technology* 9, no. 1 (2002). www.nickbostrom.com/existential/risks.html.

Brock, David. "How Gordon Moore Made 'Moore's Law.'" *Wired*, April 16, 2015. https://www.wired.com/2015/04/how-gordon-moore-made-moores-law/.

Bronner, Stephen J. "With $72 Million in Funding, the Entrepreneur behind Beyond Meat Pursues Innovation over Profit." *Entrepreneur*, January 22, 2018. https://www.entrepreneur.com/article/307715.

Business Today (India). "China Conducts Biological Test on Its Army to Create Super-Soldiers, Says Top US Official." December 9, 2020. https://www.businesstoday.in/current/world/china-conducts-biological-test-on-its-army-to-create-super-soldiers-says-top-us-official/story/424341.html.

Canada Press. "Shell Quest Carbon Capture and Storage Project Reaches Milestone of 4M Tonnes." May 23, 2019. https://www.cbc.ca/news/canada/calgary/carbon-storage-cnrl-shell-quest-1.5146857.

Carey, Bjorn, and Elizabeth Svoboda. "Live Reports from the 2005 Darpa Grand Challenge." *Popular Science*, October 8, 2005. https://www.popsci.com/scitech/article/2005-10/live-reports-2005-darpa-grand-challenge/.

Carrington, Damian. "England Could Run Short of Water within 25 years." *The Guardian*, March 18, 2019. https://www.theguardian.com/environment/2019/mar/18/england-to-run-short-of-water-within-25-years-environment-agency.

Center for Consumer Freedom. "Ad: Fake Meat Grows in Factories,

Not on Vines." August 13, 2019. www.consumerfreedom.com/2019/08/ad-fake-meat-grows-in-factories-not-on-vines/.

Cision PRNewswire. "Sweet Earth Debuts New Plant-Based Culinary Innovations, Awesome Burger & Awesome Grounds." September 25, 2019. www.prnewswire.com/news-releases/sweet-earth-debuts-new-plant-based-culinary-innovations-awesome-burger--awesome-grounds-300924874.html.

Clifford, Catherine. "Elon Musk: 'Mark My Words – A.I. Is Far More Dangerous than Nukes." CNBC, March 14, 2018. www.cnbc.com/2018/03/13/elon-musk-at-sxsw-a-i-is-more-dangerous-than-nuclear-weapons.html.

CNBC International TV. "Sebastian Thrun, Founder of Udacity." *The Brave Ones,* May 15, 2017. https://www.youtube.com/watch?v=DtO3pcNnEuA.

CNN. "Estonia: Obama on Obamacare Website." July 21, 2016. www.youtube.com/watch?v=v6vF_ARdlBA.

Cohen, Jon. "The Untold Story of the 'Circle of Trust' behind the World's First Gene-Edited Babies." *Science,* August 1, 2019. www.sciencemag.org/news/2019/08/untold-story-circle-trust-behind-world-s-first-gene-edited-babies.

Cong, Le, F. Ann Ran, David Cox, Shuailiang Lin, et al. "Multiplex Genome Engineering Using CRISPR/Cas Systems. *Nature,* February 15, 2013. http://science.sciencemag.org/content/339/6121/819.long.

Consultative Group to Assist the Poor. "CGAP Launches Global Microfinance Survey to Inform COVID-19 Response." June 1, 2020. www.cgap.org/news/cgap-launches-global-microfinance-survey-inform-covid-19-response.

Cookson, Clive. "Africa to Propel World's Population towards 10bn by 2050." *Financial Times,* June 17, 2019. https://www.ft.com/content/868e20d0-90ec-11e9-b7ea-60e35ef678d2.

Corder, Mike. "Water Wizards: Dutch Flood Expertise Is Big Export

Business." Associated Press, November 12, 2017. https://apnews.com/article/1bc1f137cb134fdca67afdff08c242bc.

Crider, Johnna. "Over 730,000 Tesla Vehicles with Autopilot 2 & 3 on the Road." *Cleantechnica*, January 5, 2020. https://cleantechnica.com/2020/01/05/over-730000-tesla-vehicles-with-autopilot-2-3-on-the-road/.

Daily Bhaskar. "This Man Is a Sanitary Pad Revolutionary. Watch Him Take On the Age-Old Taboo" [in Hindi]. March 3, 2016. https://daily.bhaskar.com/news/TOP-sanitary-pad-revolutionary-arunachalam-muruganantham-menstruation-man-5265202-PHO.html.

Darmiento, Laurence. "Ethan Brown Went Vegan but Missed Fast Food. So He Started a Revolution." *Los Angeles Times*, January 20, 2020. https://www.latimes.com/business/story/2020-01-08/beyond-meat-founder-ethan-brown.

Davis, Joshua. "Hackers Take Down the Most Wired Country in Europe." *Wired*, August 21, 2007. https://www.wired.com/2007/08/ff-estonia/.

de Bertodano, Helena. "Khan Academy: The Man Who Wants to Teach the World." *The Guardian*, September 28, 2018. www.telegraph.co.uk/education/educationnews/9568850/Khan-Academy-The-man-who-wants-to-teach-the-world.html.

Dilanian, Ken. "China Has Done Human Testing to Create Biologically Enhanced Super Soldiers, Says Top U.S. Official." NBC News, December 4, 2020. https://www.nbcnews.com/politics/national-security/china-has-done-human-testing-create-biologically-enhanced-super-soldiers-n1249914.

Disaster Preparedness Tokyo. "Disaster Prevention Information." 2019. https://www.bousai.metro.tokyo.lg.jp/book/pdf/en/01_Simulation_of_a_Major_Earthquake.pdf.

———. "Manga Comic: 'Tokyo "X" Day.'" 2019. https://www.bousai.metro.tokyo.lg.jp/book/pdf/en/comic.pdf.

Dorcas, Adeline. "Pad Man Makes Nationwide Campaign for Use

of Sanitary Napkins." *Medindia,* May 11, 2018. https://www.medindia.net/news/healthwatch/pad-man-makes-nationwide-campaign-for-use-of-sanitary-napkins-179368-1.htm.

Dorsey, David. "Expert Says Driverless Cars Will Disrupt Transportation Industry within 5 Years." *Fort Myers News-Press,* January 31, 2020. https://www.news-press.com/story/news/local/2020/01/31/driverless-cars-larry-burns-predicts-autonomous-cars-go-mainstream-within-five-years-disrupt-transpo/4625273002/.

Dreifus, Claudia. "It All Started with a 12-Year-Old Cousin." *New York Times,* January 28, 2014. http://www.nytimes.com/2014/01/28/science/salman-khan-turned-family-tutoring-into-khan-academy.html.

e-estonia. "Estonia to Open World's First Data Embassy in Luxembourg." June 2017. https://e-estonia.com/estonia-to-open-the-worlds-first-data-embassy-in-luxembourg/.

Economic Times. "Martin Eberhard Calls Tesla His 'Baby', Talks about Being Ousted from Company's Board." March 1, 2019, https://economictimes.indiatimes.com/magazines/panache/martin-eberhard-calls-tesla-his-baby-talks-about-being-ousted-from-companys-board/articleshow/68211761.cms.

Economist. "Safe Cities Index 2019." https://safecities.economist.com/safe-cities-index-2019/.

e-Governance Academy. "Toomas Hendrik Ilves – Estonia as a Think Tank of e-Government. The eGA/GISI Initative." May 16, 2013. https://www.youtube.com/watch?v=BBcUORzU3CA.

Elliott, Larry. "Climate Crisis Fills Top Five Places of World Economic Forum's Risk Report." *The Guardian,* January 15, 2020. https://www.theguardian.com/business/2020/jan/15/climate-crisis-environment-top-five-places-world-economic-forum-risks-report.

Estonian World. "Estonia Celebrates the Day of Restoration of

Independence." August 20, 2017. https://estonianworld.com/life/estonia-celebrates-the-day-of-restoration-of-independence/.

Etherington, Darrel, and Lora Kolodny. "Google's Self-Driving Car Unit Becomes Waymo." December 13, 2016. https://techcrunch.com/2016/12/13/googles-self-driving-car-unit-spins-out-as-waymo/.

European Press Agency (EPA). "Businessman Carlos Slim Will Invest More Than 317 US Dollar Millions in Education." January 19, 2013.

Ferziger, Jonathan. "Netanyahu Offers to Help Brown Manage California Drought." Bloomberg, March 5, 2014. www.bloomberg.com/news/articles/2014-03-05/netanyahu-offers-to-help-brown-manage-california-drought.

Financial Express (Bangladesh). "Sal Khan's Academy Lights the Way." December 26, 2010.

———. "Who Is Arunachalam Muruganantham?" February 8, 2018. www.financialexpress.com/entertainment/who-is-arunachalam-muruganantham-the-real-padman-who-was-the-inspiration-for-akshay-kumar-movie/1059638/.

Finkle, Colin. "Beyond Meat Brand Profile." Brand Marketing Blog, July 21, 2019. https://brandmarketingblog.com/articles/good-branding/beyond-meat-brand-profile/.

Foster, Tom. "Friends Thought Beyond Meat's Founder Was Crazy. His Billion-Dollar IPO Proved Them Wrong." *Inc.* August 2019. https://www.inc.com/magazine/201908/tom-foster/ethan-brown-beyond-meat-alternative-protein-plant-burger-ipo-public-whole-foods.html.

Fourcade, Marthe, and Corinne Gretler. "McDonald's Picks Beyond for Canada Trial, U.S. Prize Remains." Bloomberg, September 26, 2019. www.bloomberg.com/news/articles/2019-09-26/mcdonald-s-to-test-veggie-burger-from-beyond-meat-in-canada.

Free Press Journal (India). "Environment-Friendly Sanitary

Pads: Work in Progress." December 6, 2018. https://www.freepressjournal.in/cmcm/environment-friendly-sanitary-pads-work-in-progress.

Friedman, Thomas. "Moore's Law Turns 50." *New York Times*, May 13, 2015. https://www.nytimes.com/2015/05/13/opinion/thomas-friedman-moores-law-turns-50.html.

"From Rags to Riches." *The Nation* (Nigeria), September 20, 2014. https://thenationonlineng.net/from-rags-to-riches/.

Frostenson, Sarah. "America Has a Water Crisis No One Is Talking About." *Vox*, March 22, 2018. www.vox.com/science-and-health/2017/5/9/15183330/america-water-crisis-affordability-millions.

Gates, Bill. "Future of Food." *GatesNotes*, March 18, 2013. www.gatesnotes.com/About-Bill-Gates/Future-of-Food.

Geib, Claudia. "Genetic Engineering Can Treat Rare Diseases, but Could Also Give Wealthy Children a Biological Advantage." *Futurism*, August 9, 2017. https://futurism.com/neoscope/expert-argues-that-gene-editing-will-widen-economic-class-gap.

Geist, Edward, and Andrew J. Lohn. "How Might Artificial Intelligence Affect the Risk of Nuclear War?" Rand Corporation, 2018. www.rand.org/pubs/perspectives/PE296.html.

Gheorghe, Georgeta. "Estonia's Toomas Hendrik Ilves Talks Digitalisation, from X-Road to Digital Embassies and E-Residency." *Business Review*, August 11, 2017. http://business-review.eu/featured/estonias-toomas-hendrik-ilves-talks-digitalisation-between-x-road-digital-embassies-e-resdiency-152141.

Gilson, Dirk. "India's Menstruation Man." *Aljazeera*. https://interactive.aljazeera.com/aje/shorts/india-menstruation-man/.

Giridharadas, Anand, and Keith Bradsher. "Microloan Pioneer and His Bank Win Nobel Peace Prize." *New York Times*, October 13, 2006. https://www.nytimes.com/2006/10/13/business/14nobelcnd.html.

Goldstein, Dana. "Research Shows Students Falling Months Behind during Virus Disruptions." *New York Times,* June 5, 2020. www.nytimes.com/2020/06/05/us/coronavirus-education-lost-learning.html.

Golikeri, Priyanka. "An Enterprise Aimed at Boosting Women's Health." DNA, June 9, 2011. https://www.dnaindia.com/business/report-an-enterprise-aimed-at-boosting-women-s-health-1552766.

Goodman, Marc. "A Vision of Crimes in the Future." TEDGlobal 2012, June 2012. www.ted.com/talks/marc_goodman_a_vision_of_crimes_in_the_future?language=en.

Gupta, Arun. "Disaster Capitalism Hits New York." *In These Times,* February 2013. https://inthesetimes.com/article/14430/disaster_capitalism_hits_new_york.

Gupta, Surojit. " Salman's India Dream: Coaching 450m in 10 years," *Times of India,* December 5, 2015, https://timesofindia.indiatimes.com/india/salmans-india-dream-coaching-450m-in-10-years/articleshow/50036337.cms.

Hagstrom, Britt F. "The Man behind the Nobel Prize." *Daily Star,* December 10, 2006. https://archive.thedailystar.net/suppliments/2006/december/bangladeshstandingtall/index.htm

Handley, Lucy. "The Education of Sebastian Thrun." CNBC, January 1, 2017. https://www.cnbc.com/2017/06/01/sebastian-thrun-udacity-googlex.html.

Harney, Alexandra, and Kate Kelland. "China Orders Investigation after Scientist Claims First Gene-Edited Babies." Reuters, November 26, 2018. https://www.reuters.com/article/us-health-china-babies-genes/china-orders-investigation-after-scientist-claims-first-gene-edited-babies-idUSKCN1NV19T.

Harvey, Fiona. "Humanity under Threat from Perfect Storm of Crises – Study." *The Guardian,* February 6, 2020. https://www.theguardian.com/environment/2020/feb/06/humanity-under-threat-perfect-storm-crises-study-environment?fbclid

=IwAR3roRN91V7VtZY8vHsEQv3ScicNTSRJCr8hyay4u_Sm24J6DnB7lmsoyxQ.

Heller, Martin C., and Gregory A. Keoleian. "Beyond Meat's Beyond Burger Life Cycle Assessment: A Detailed Comparison between a Plant-Based and an Animal-Based Protein Source." Center for Sustainable Systems Report, University of Michigan, September 14, 2018. http://css.umich.edu/publication/beyond-meats-beyond-burger-life-cycle-assessment-detailed-comparison-between-plant-based.

Heller, Nathan. "Estonia, The Digital Republic." *New Yorker*, December 11, 2017. https://www.newyorker.com/magazine/2017/12/18/estonia-the-digital-republic.

Hindustan Times. "Tata Trusts to Support Khan Academy for Free Online Education" December 7, 2015. www.hindustantimes.com/education/tata-trusts-to-support-khan-academy-for-free-online-education/story-kpQu01Ph4QGi9HDBrT3SVP.html.

Hodson, Hal. "E-Citizens Unite." *NewScientist* 224, no. 2991 (October 18, 2014): 24.

Hoe, Wen. "E-Stonia: One Small Country's Digital Government Is Having a Big Impact." Harvard Government Innovators Network, June 7, 2017. https://www.innovations.harvard.edu/blog/estonia-one-small-country-digital-government-having-big-impact-x-road.

Houser, Kristin. "Many Self-Driving Car Accidents Have Been Caused by Humans." *Business Insider*, September 5, 2018. https://www.businessinsider.com/self-driving-car-accidents-caused-by-humans-2018-9.

Hunter-Hart, Monica. "8 Staggering Predictions from Ray Kurzweil." *Inverse*, May 29, 2017. https://www.inverse.com/article/30913-8-staggering-predictions-ray-kurzweil.

Hrynowski, Zach. "What Percentage of Americans are Vegetarian?" Gallup, September 27, 2019. https://news.gallup.com/poll/267074/percentage-americans-vegetarian.aspx.

Ilves, Toomas Hendrik. "e-Estonia and the New Europe." Välisministeerium, October 25, 2005. https://vm.ee/et/node/42610.

Insurance Institute for Highway Safety, Highway Loss Data Institute. "Fatality Facts 2018, State by State." https://www.iihs.org/topics/fatality-statistics/detail/state-by-state.

International Space Hall of Fame at the New Mexico Museum of Space History. "The Father of Astronautics and Rocket Dynamics." https://www.nmspacemuseum.org/inductee/konstantin-e-tsiolkovsky/.

Ip, Greg. "The Moonshot Mind-Set Once Came from the Government. No Longer," *Wall Street Journal*, July 14, 2019. www.wsj.com/articles/the-moonshot-mind-set-once-came-from-the-government-no-longer-11563152820. See also Casey Dreier, "Reconstructing the Cost of the One Giant Leap," *The Planetary Society*, June 16, 2019, https://www.planetary.org/blogs/casey-dreier/2019/reconstructing-the-price-of-apollo.html.

"ISS National Laboratory," NASA, https://www.nasa.gov/mission_pages/station/research/nlab/index.html.

ITUNews. "The Future Is Better Than You Think: Predictions on AI and Development from Ray Kurzweil." May 29, 2019. https://news.itu.int/the-future-is-better-than-you-think-predictions-on-ai-and-development-from-ray-kurzweil/.

Jinek, Martin, Krzysztof Chrlinski, et al. "A Programmable Dual-RNA-Guided DNA Endonuclease in Adaptive Bacterial Immunity." *Science*, August 17, 2012. https://science.sciencemag.org/content/337/6096/816.

Jones, Abigail. "The Silence That Still Surrounds Periods." *The Independent* (UK), April 25, 2016. https://www.independent.co.uk/life-style/health-and-families/silence-around-menstruation-a7000051.html.

Kagen, Julia. "Microfinance." *Investopedia*, April 21, 2020. www.investopedia.com/terms/m/microfinance.asp.

Kaljulaid, Kersti. "Tallinn Digital Summit – Opening Address by Estonia's President Kersti Kaljulaid." September 29, 2017. https://www.youtube.com/watch?v=h1I4LGaP6Io.

Kalra, Nidhi, and David Groves. "The Enemy of Good: Estimating the Cost of Waiting for Nearly Perfect Autonomous Vehicles." Rand Corporation, 2017. https://www.rand.org/pubs/research_reports/RR2150.html.

Kane, Mark. "Close to 1.18 Million Plug-In Electric Cars Were Sold in China in 2019." *InsideEVs*, January 22, 2020. https://insideevs.com/news/394229/plugin-electric-car-sales-china-2019/.

Kaplan, David A. "Bill Gates' Favorite Teacher." *Fortune*, August 24, 2010. https://archive.fortune.com/2010/08/23/technology/sal_khan_academy.fortune/index.htm.

Karami, Arash. "Iran Official Warns Water Crisis Could Lead to Mass Migration." *Al-Monitor*, April 28, 2015. www.al-monitor.com/pulse/originals/2015/04/iran-water-crisis-mass-migration.html.

Keen, Andrew. "E-Stonia: The Country Using Tech to Rebrand Itself as Anti-Russia." *The Guardian*, April 21, 2016. https://www.theguardian.com/world/2016/apr/21/e-stonia-country-using-technology-to-rebrand-itself-as-the-anti-russia.

Kennedy, Pagan. "William Gibson's Future Is Now." *New York Times*, January 13, 2012. https://www.nytimes.com/2012/01/15/books/review/distrust-that-particular-flavor-by-william-gibson-book-review.html.

Khan, Jennifer. "The Gene Drive Dilemma: We Can Alter Entire Species, but Should We?" *New York Times*, January 8, 2020. https://www.nytimes.com/2020/01/08/magazine/gene-drive-mosquitoes.html.

Khan Academy. "Bill Gates Talks about the Khan Academy at Aspen Ideas Festival 2010." July 19, 2010. www.youtube.com/watch?v=6A07Pj71TUA.

Khandker, Shahidur R., M. A. Baqui Khalily, and Hussain A.

Samad. *Beyond Ending Poverty: The Dynamics of Microfinance in Bangladesh*. World Bank Group, 2016. http://documents.worldbank.org/curated/en/366431468508832455/pdf/106672-PUB.pdf.

King, Michelle. "Dunkin' and Beyond Meat Accelerate Nationwide Launch of Beyond Sausage Sandwich." Dunkin' News Release, October 21, 2019. https://news.dunkindonuts.com/news/releases-20191021.

Kingdom, Bill, Roland Liemberger, and Philippe Marin. "The Challenge of Reducing Non-Revenue Water (NRW) in Developing Countries – How the Private Sector Can Help: A Look at Performance-Based Service Contracting," World Bank, *Water Supply and Sanitation Sector Board Discussion Paper Series*, no. 8, December 2006, https://openknowledge.worldbank.org/handle/10986/17238.

Kirka, Danica. "No More Paperwork: Estonia Edges toward Digital Government." Associated Press, December 26, 2018. https://apnews.com/article/da5e0dd3bf364bbda436fc16690f842c.

Kitman, Jamie. "President Ilves: The Man Who Made E-Stonia." *The Guardian*, November 3, 2011. www.theguardian.com/world/2011/nov/03/president-ilves-made-estonia.

Kolyako, Nina. "Estonia Ranked First Worldwide in Terms of Broadband Internet Speeds." *Baltic Course*, January 27, 2012. http://www.baltic-course.com/eng/good_for_business/?doc=52217.

Korjus, Kaspar. "Welcome to Our Digital Nation, Bill Gates." *Medium*, October 18, 2018. https://medium.com/e-residency-blog/welcome-to-our-digital-nation-bill-gates-a8c33edd79a8.

Kosik, Allison. "Experts: U.S. Water Infrastructure in Trouble." CNN, January 21, 2011. http://edition.cnn.com/2011/US/01/20/water.main.infrastructure/.

Koushan, Kasra. "Sal Khan on Creating a Better Future." The Signal, July 20, 2017. https://thesign.al/sal-khan/.

Kuropatwa, Rebeca. "Helping One Village at a Time." *Jewish Independent*, June 16, 2017. https://www.jewishindependent.ca/tag/sivan-yaari/.

Kurzgesagt – In a Nutshell. "A Selfish Argument for Making the World a Better Place – Egoistic Altruism." March 18, 2018. https://www.youtube.com/watch?v=rvskMHnosqQ&vl=en.

Kurzweil, Ray. "The Law of Accelerating Returns," Kurzweilai.net, March 7, 2001. https://www.kurzweilai.net/the-law-of-accelerating-returns.

La Monica, Paul R. "Impossible Whoppers Are a Huge Hit at Burger King, Fueling Its Best Quarter in Four Years." CNN Business, October 28, 2019. www.cnn.com/2019/10/28/investing/restaurant-brands-earnings-burger-king-popeyes/index.html.

Lambert, Fred. "Tesla Drops a Bunch of New Autopilot Data, 3 Billion Miles and More." *Electrek*, April 22, 2020. https://electrek.co/2020/04/22/tesla-autopilot-data-3-billion-miles/.

———. "Tesla Produces Its 1 Millionth Electric Car." *Electrek*, March 9, 2020. https://electrek.co/2020/03/09/tesla-produces-1000000th-electric-car/.

Lan, Shlomit. "Positive Energy" [in Hebrew]. *Globes*, May 17, 2010. https://www.globes.co.il/news/article.aspx?did=1000560436.

Lander, Eric, Françoise Baylis, Feng Zhang, Emmanuelle Charpentier, Paul Berg, et al. "Adopt a Moratorium on Heritable Genome Editing." *Nature*, March 13, 2019. https://www.nature.com/articles/d41586-019-00726-5.

Lanphier, Edward, Fyodor Urnov, Sarah Ehlen Haecker, et al. "Don't Edit the Human Germ Line." *Nature*, March 12, 2015. www.nature.com/news/don-t-edit-the-human-germ-line-1.17111.

Law School Admission Council. "Khan Academy Launches Free LSAT Prep for All." June 1, 2018. https://www.lsac.org/about/news/khan-academy-launches-free-lsat-test-prep-all.

Layne, Rachel. "Space Case: Why Reaching for the Stars Could Soon Be a $1 Trillion Industry." CBS News, July 16, 2019. https://www

.cbsnews.com/news/space-is-a-more-than-400-billion-market-and-getting-bigger/.

———. "Water Costs Are Rising across the US – Here's Why." CBS, August 27, 2019. https://www.cbsnews.com/news/water-bills-rising-cost-of-water-creating-big-utility-bills-for-americans/.

LETA (Latvian Information Agency). "Ministry Hopes Great Success for Estonian e-Residency Service." October 7, 2014.

Lollis, Kent D. "Living, Working, and Achieving While Black." Law School Admission Council, June 16, 2020. www.lsac.org/blog/living-working-and-achieving-while-black.

Love, Bruce. "Sal Khan: An Educational Entrepreneur." CampdenFB, May 7, 2015. https://www.campdenfb.com/article/sal-khan-educational-entrepreneur.

Lu, Denise, and Christopher Flavelle. "Rising Seas Will Erase More Cities by 2050, New Research Shows." *New York Times*, October 29, 2019. https://www.nytimes.com/interactive/2019/10/29/climate/coastal-cities-underwater.html.

Mac, Elizabeth. "A Burgeoning Crisis? A Nationwide Assessment of the Geography of Water Affordability in the United States." *PloS One* 12, no. 4 (January 11, 2017). https://journals.plos.org/plosone/article?id=10.1371/journal.pone.0169488.

Mansel, Tim. "How Estonia Became E-Stonia." BBC, May 16, 2013. www.bbc.com/news/business-22317297.

Marsters, Peter. "Electric Cars: The Drive for a Sustainable Solution in China." *Wilson Center*, August 2009. https://www.wilsoncenter.org/publication/electric-cars-the-drive-for-sustainable-solution-china.

Masr, Mada. "'We Woke Up in a Desert' – The Water Crisis Taking Hold across Egypt." *The Guardian*, August 4, 2015. www.theguardian.com/world/2015/aug/04/egypt-water-crisis-intensifies-scarcity.

Mckean, Cameron Allan. "Tokyo's Disaster Parks." *The Guardian*, August 19, 2014. www.theguardian.com/cities/2014/aug/19

/tokyo-disaster-parks-hi-tech-survival-bunkers-hidden-green-spaces-earthquake.

Mellgren, Doug. "Bangladeshi Economist Wins Nobel Peace Prize." *Seattle Times*, October 13, 2006. https://www.seattletimes.com/nation-world/bangladeshi-economist-wins-nobel-peace-prize/.

Menon, Nikhil. "How Sal Khan Is Exorcising the Math Spectre from the Minds of Millions of Students Worldwide." *The Economic Times*, August 10, 2012. https://economictimes.indiatimes.com/how-sal-khan-is-exorcising-the-math-spectre-from-the-minds-of-millions-of-students-worldwide/articleshow/15425169.cms.

Miller, Claire Cain. "Musk Unplugged: Tesla CEO Discusses Car Troubles." *New York Times*, October 24, 2008. https://bits.blogs.nytimes.com/2008/10/24/musk-unplugged-tesla-ceo-discusses-car-troubles/.

MIT Club of Northern California. "Pioneers in Clean Technology – Marc Tarpenning – Tesla Motors." March 6, 2014. https://www.youtube.com/watch?v=EDCYoAQmmAA.

Mixergy. "How Khan Academy Is Changing Education with Videos Made in a Closet." June 28, 2010. https://mixergy.com/interviews/salman-khan-academy-interview/.

MOJO Story. "Twinkle Khanna & the 'Pad-Man' with Barkha Dutt." December 11, 2017. www.youtube.com/watch?v=V4_MeS6SOwk.

Molteni, Megan. "The Wired Guide to CRISPR." *Wired*, March 12, 2019. https://www.wired.com/story/wired-guide-to-crispr/.

———. "The World Health Organization Says No More Gene-Edited Babies." *Wired*, July 30, 2019. www.wired.com/story/the-world-health-organization-says-no-more-gene-edited-babies/.

Morduch, Jonathan. "The Role of Subsidies in Microfinance: Evidence from the Grameen Bank." *Journal of Development Economics* 60 (1999). https://s18798.pcdn.co/jmorduch

/wp-content/uploads/sites/5267/2016/12/Morduch-Grameen-Subsidies-JDE-1999.pdf.

Mosa Meat. "Why We Need to Change How We Make Meat." July 5, 2018. https://www.youtube.com/watch?v=oSl9yZuPrdo.

Mosher, Dave. "Congress and Trump are Running out of Time to Fix a $100-Billion Investment in the Sky, NASA Auditor Says." *Business Insider*, May 18, 2018. www.businessinsider.com/space-station-trump-plan-problems-astronaut-risk-2018-5.

Muruganantham, Arunachalam. "How I Started a Sanitary Napkin Revolution." TED@Bangalore, May 2012. www.ted.com/talks/arunachalam_muruganantham_how_i_started_a_sanitary_napkin_revolution?language=en#t-70857.

Musgrove, Mike. "An Electric Car with Juice." *Washington Post*, July 22, 2006. http://www.washingtonpost.com/wp-dyn/content/article/2006/07/21/AR2006072101515.html.

Nair, Prashant. "QnAs with Rodolphe Barrangou." *PNAS*, July 11, 2017. www.pnas.org/content/114/28/7183.

Nandakumar, Sowmya. "Salman Khan: World's Most Influential Online Educator." *Indo-American News*, May 17, 2012, page 6. https://issuu.com/indoamericannews/docs/may18_pages_1-36.

NASA's Ames Research Center. "Jason Dunn – The Future of Making Things in Space." January 31, 2017, 25:00. www.youtube.com/watch?list=PLfnpkfDmrBqYgX1O3kRQnc5Sjzz_LMwDo&time_continue=3313&v=ZXYCKoTclCc.

"New Perspectives: Influencing Circumstances through Human Connection – Spotlight: Sivan Ya'ari '02." *Altitude*, Winter 2020, https://www.alumni.pace.edu/s/1655/bp20/interior.aspx?sid=1655&gid=2&pgid=2383.

Nha, Thu Ngo. "A Glimpse of Japan's Deadly Earthquakes." ArcGIS StoryMaps, August 18, 2019. https://storymaps.arcgis.com/stories/daoacfccaece4fc481353b1a01589259.

Nikel, David. "Electric Cars: Why Little Norway Leads the World

In EV Usage." Forbes, June 18, 2019. https://www.forbes.com/sites/davidnikel/2019/06/18/electric-cars-why-little-norway-leads-the-world-in-ev-usage/.

Nikkei Asian Review. "Osaka Quake Exposes Japan's Aging Infrastructure." June 19, 2018. https://asia.nikkei.com/Economy/Osaka-quake-exposes-Japan-s-aging-infrastructure.

Nobel Prize. "Muhammad Yunus – Nobel Lecture." December 10, 2006. www.nobelprize.org/prizes/peace/2006/yunus/26090-muhammad-yunus-nobel-lecture-2006-2/.

Noer, Michael. "One Man, One Computer, 10 Million Students: How Khan Academy Is Reinventing Education." Forbes, November 19, 2012. www.forbes.com/sites/michaelnoer/2012/11/02/one-man-one-computer-10-million-students-how-khan-academy-is-reinventing-education/#7e17495444e0.

Nolen, Stephanie. "The Absorbing Tale of One Man's Quest for Better Feminine Hygiene." *Globe and Mail* (Canada), October 4, 2012.

NPR Radio. "Live Episode! Beyond Meat: Ethan Brown." February 16, 2017. www.npr.org/2017/02/16/515420148/live-episode-beyond-meat-ethan-brown.

Nuttall, Chris. "Tale of a Supercharged Start-Up Entrepreneurship." *Financial Times,* April 12, 2007. https://www.ft.com/content/ab94e730-e83c-11db-b2c3-000b5df10621.

Ofek, Elie, and Matthew Preble. "Takadu." Harvard Business School Case 514-011, July 2013 (revised August 2017).

Oryszczuk, Stephen. "Africa Gains from Israel's Solar System." *Jewish News,* April 12, 2018. https://jewishnews.timesofisrael.com/africa-gains-from-israels-solar-system/.

Pakistan Today. "India's 'Menstruation Man' Reinvents Sanitary Pads." May 15, 2016. https://archive.pakistantoday.com.pk/2016/05/15/indias-menstruation-man-reinvents-sanitary-pads/.

Parekh, Par. "The Math of Khan: Salman Khan: Founder of the Kahn Academy." Bidoun. www.bidoun.org/articles/the-math-of-khan.

Park, Alice. "Why We Don't Need Animals to Keep Enjoying Meat." *Time*, June 6, 2019. https://time.com/5601980/beyond-meat-ceo-ethan-brown-interview/.

Passwaters, Arie. "President's Lecture Series to Present Educational Innovator Salman Khan." Rice University Office of Public Affairs, September 24, 2018. https://news.rice.edu/2018/09/24/presidents-lecture-series-to-present-educational-innovator-salman-khan/.

Patrizio, Andy. "IDC: Expect 175 Zettabytes of Data Worldwide by 2025." NetworkWorld, December 3, 2018. www.networkworld.com/article/3325397/idc-expect-175-zettabytes-of-data-worldwide-by-2025.html.

Pelley, Scott. "U.S., China, Russia, Elon Musk: Entrepreneur's 'Insane' Vision Becomes Reality." CBS News, May 22, 2012. https://www.cbsnews.com/news/us-china-russia-elon-musk-entrepreneurs-insane-vision-becomes-reality/.

Pilling, David. "Lessons Learnt from Kobe Quake." *Financial Times*, May 11, 2011. https://www.ft.com/content/5290f95e-4bc4-11e0-9705-00144feab49a.

Radhakrishnan, Manjusha. "Meet the Real 'Pad Man' Arunachalam Muruganatham." *Gulf News*, February 7, 2018.

Rajghatta, Chidanand. "Modi to Meet Salman Khan, the Educator not the Movie Star." *Times of India*, September 28, 2015. https://timesofindia.indiatimes.com/india/modi-to-meet-salman-khan-the-educator-not-the-movie-star/articleshow/49113900.cms.

Ramesh, Randeep. "Giving a Hand Up, Not a Handout." *The Guardian*, December 27, 2006. https://www.theguardian.com/business/2006/dec/27/money.

Rathi, Akshat, and Echo Huang. "More Than 100 Chinese Scientists Have Condemned the CRISPR Baby Experiment as 'Crazy.'" *Quartz*, November 26, 2018. https://qz.com/1474530/chinese-scientists-condemn-crispr-baby-experiment-as-crazy/.

Raz, Guy. "Sal Khan: Can Technology Help Create a Global Classroom." *NPR TED Radio Hour,* August 11, 2017.

Reed, John. "Elon Musk's Groundbreaking Electric Car." *Financial Times,* July 24, 2009. https://www.ft.com/content/e117987e-74eb-11de-9ed5-00144feabdc0.

Reedy, Christianna. "Ray Kurzweil Claims Singularity Will Happen by 2045." *Futurism,* May 10, 2017. https://futurism.com/kurzweil-claims-that-the-singularity-will-happen-by-2045.

Regalado, Antonio. "First Gene-Edited Dogs Reported in China." *MIT Technology Review,* October 19, 2015. https://www.geneticsandsociety.org/article/first-gene-edited-dogs-reported-china.

Ripon Society. "The Estonian Example – Q&A with Toomas Hendrick Ilves." *The Ripon Forum* 47, no. 1 (Winter 2013). https://www.riponsociety.org/article/the-estonian-example-qa-with-toomas-hendrik-ilves/.

Ritchie, Hannah. "You Want to Reduce the Carbon Footprint of Your Food? Focus on What You Eat, Not Whether Your Food Is Local." Our World in Data, January 24, 2020. https://ourworldindata.org/food-choice-vs-eating-local.

Rose, Gideon. "Google's Original X-Man: A Conversation with Sebastian Thrun." *Foreign Affairs,* November/December 2013. www.foreignaffairs.com/interviews/2013-10-15/googles-original-x-man.

Roser, Max, and Esteban Ortiz-Ospina. "Global Extreme Poverty." Our World in Data, 2013. https://ourworldindata.org/extreme-poverty.

Ruptly. "Russia: Putin Warns a GM Human Could Be 'Worse Than Nuclear Bomb.'" October 21, 2017. www.youtube.com/watch?v=E6pD96RRIUQ.

Salt Lake Tribune. "Yunus, Grameen Bank win Nobel Peace Prize for Loans to the Poor." October 13, 2006. https://archive.sltrib.com/article.php?id=4486938&itype=NGPSID.

Sampson, Zachary T. "First 3-D Printing in Space Had Bay Area Start." *Tampa Bay Times*, December 8, 2014. https://www.tampabay.com/archive/2014/12/08/first-3-d-printing-in-space-had-bay-area-start/.

Saval, Nikil. "Shigeru Ban." *New York Times*, October 18, 2019. www.nytimes.com/interactive/2019/10/15/t-magazine/shigeru-ban.html.

Schleifer, Leah. "7 Reasons We're Facing a Global Water Crisis." World Resources Institute, August 24, 2017. https://www.wri.org/blog/2017/08/7-reasons-were-facing-global-water-crisis.

Schreiber, Mark. "Hell on Earth in '23." *Japan Times*, August 26, 2001. www.japantimes.co.jp/community/2001/08/26/general/hell-on-earth-in-23/#.Xkww7S2ZORu.

Schultze, Elizabeth. "How a Tiny Country Bordering Russia Became One of the Most Tech-Savvy Societies in the World." CNBC, February 8, 2019. https://www.cnbc.com/2019/02/08/how-estonia-became-a-digital-society.html.

Sengupta, Somini. "Online Learning, Personalized." *New York Times*, December 4, 2011. https://www.nytimes.com/2011/12/05/technology/khan-academy-blends-its-youtube-approach-with-classrooms.html.

"Seven Wonders of the Modern World as Named by ASCE." Herff College of Engineering, University of Memphis. http://www.ce.memphis.edu/1101/interesting_stuff/7wonders.html.

Shemer, Simona. "On World Water Day, a Look at the Innovative Tech That Makes Israel a Water Superpower." No Camels, March 22, 2018. http://nocamels.com/2018/03/on-world-water-day-a-look-at-the-innovative-tech-that-makes-israel-a-water-superpower/.

Shnayerson, Michael. "Quiet Thunder." *Vanity Fair*, May 2007. https://www.vanityfair.com/news/2007/05/tesla200705.

Shorto, Russell. "How to Think Like the Dutch in a Post-Sandy World." *New York Times Magazine*, April 9, 2014. https://www

.nytimes.com/2014/04/13/magazine/how-to-think-like-the-dutch-in-a-post-sandy-world.html.

Shwartz, Mark. "Target, Delete, Repair." *Stanford Medicine,* Winter 2018. https://stanmed.stanford.edu/2018winter/CRISPR-for-gene-editing-is-revolutionary-but-it-comes-with-risks.html.

Simons, Marlise. "Dutch Floods Came in the Back Door." *New York Times,* February 5, 1995, https://www.nytimes.com/1995/02/05/world/dutch-floods-came-in-the-back-door.html.

Singularity University. "Singularity University Addresses the World's Greatest Challenges at SU Global Summit." August 9, 2016. https://su.org/press-room/press-releases/singularity-university-addresses-the-worlds-greatest-challenges-at-su-global-summit/.

Smith, Evan. "Sal Khan on Why Khan Academy Is a Non-Profit." *Overheard with Evan Smith.* Season 5, episode 18, aired April 20, 2015. www.youtube.com/watch?v=LI7UOuX9Qjs.

Solon, Olivia. "'It's about Expanding Earth': Could We Build Cities in Space?" *The Guardian,* April 21, 2018. https://www.theguardian.com/science/2018/apr/21/expanding-earth-could-we-build-cities-in-space.

Sorkin, Amy Davidson. "Japan: Earthquake, Water, Fire." *New Yorker,* March 11, 2011. https://www.newyorker.com/news/amy-davidson/japan-earthquake-water-fire.

Spencer, Stephan. "Exponential Medicine in a COVID-19 World with Dr. Daniel Kraft." *Get Yourself Optimized,* no. 246, May 7, 2020. www.getyourselfoptimized.com/exponential-medicine-in-a-covid-19-world-with-dr-daniel-kraft/.

Stanford Center for Professional Development. "Stanford Seminar: Electric Vehicles and Startups." January 5, 2017. https://www.youtube.com/watch?v=s47Cy8OUNcM.

Starostinetskaya, Anna. "Beyond Meat Is about to Go Public; Founder Ethan Brown Speaks about the Evolution of Meat." VegNews, April 4, 2019. https://vegnews.com/2019/4/beyond

-meat-is-about-to-go-public-founder-ethan-brown-speaks-about-the-evolution-of-meat.

Storr, Will. "Flooded Britain: How Can Holland Help?" *Daily Telegraph* (UK), April 26, 2014. https://www.telegraph.co.uk/culture/art/architecture/10769974/Flooded-Britain-how-can-Holland-help.html.

Stout, Kristie Lu. "India's 'Pad Man' Becomes Bollywood Blockbuster." CNN, February 8, 2018. http://edition.cnn.com/TRANSCRIPTS/1802/08/nwsm.01.html.

Straiton, Jenny. "Genetically Modified Humans: The X-Men of Scientific Research." *BioTechniques* 66, no. 6 (June 2019). https://www.future-science.com/doi/pdf/10.2144/btn-2019-0056.

Strauss, Valerie. "Khan Academy: The Revolution that Isn't." *Washington Post*, July 23, 2012. www.washingtonpost.com/blogs/answer-sheet/post/khan-academy-the-hype-and-the-reality/2012/07/23/gJQAuw4J3W_blog.html?utm_term=.36e13b742375.

Tabuchi, Hiroko. "$300 Billon War beneath the Street: Fighting to Replace America's Water Pipes." *New York Times*, November 10, 2017. https://www.nytimes.com/2017/11/10/climate/water-pipes-plastic-lead.html.

Tarpenning, Marc, and Shernaz Daver. "The Story of Building Tesla." Startup Grind, June 3, 2018. https://www.youtube.com/watch?v=pGf1tyPXBpA.

Telegram & Gazette. "Dutch Flood Expertise Is Big Export Business." November 13, 2017.

Teller, Astro, and Megan Smith. "What Is Your X? Amplifying Technology Moonshots." Google Official Blog, February 6, 2012. https://googleblog.blogspot.com/2012/02/whats-your-x-amplifying-technology.html.

Tesla. "Impact Report." https://www.tesla.com/ns_videos/tesla-impact-report-2019.pdf.

Teutsch, Betsy. "How Sanitary Pads Can Help Women Improve their

Health and Education." *The Atlantic,* April 22, 2014. https://www.theatlantic.com/business/archive/2014/04/sanitary-napkin-business/360297/.

Thompson, Clive. "How Khan Academy Is Changing the Rules of Education." *Wired,* July 15, 2011. www.wired.com/2011/07/ff_khan/.

Thrun, Sebastian. "Google's Driverless Car." Ted2011, March 2011. https://www.ted.com/talks/sebastian_thrun_google_s_driverless_car?referrer=playlist-planes_trains_and_automobiles.

———. "Leave the Driving to the Car, and Reap Benefits in Safety and Mobility." *New York Times,* December 5, 2011. https://www.nytimes.com/2011/12/06/science/sebastian-thrun-self-driving-cars-can-save-lives-and-parking-spaces.html.

———. "What We're Driving At." *Google Official Blog,* October 9, 2010. https://googleblog.blogspot.com/2010/10/what-were-driving-at.html.

Thrun, Sebastian, and Chris Anderson. "What AI Is – and Isn't." Ted2017, April 2017. https://www.ted.com/talks/sebastian_thrun_and_chris_anderson_what_ai_is_and_isn_t?language=en.

Ummelas, Ott. "World's Most High-Tech Voting System to Get Hacking Defenses." Bloomberg, July 18, 2017. https://www.bloomberg.com/news/articles/2017-07-17/world-s-most-high-tech-voting-system-to-get-new-hacking-defenses.

United Nations. "United Nations Dignitaries Hail Nobel Peace Prize Winners in Special Tribute to Muhammad Yunus, Grameen Bank." November 17, 2006. https://www.un.org/press/en/2006/dev2610.doc.htm.

University of Michigan. "Carbon Footprint Factsheet." Center for Sustainable Systems, Pub. No. CSS09-05, 2020. http://css.umich.edu/factsheets/carbon-footprint-factsheet.

University of Utah. "The Evolution of Corn." Genetics Learning

Center, July 2015. http://learn.genetics.utah.edu/content/selection/corn/.

Urstadt, Bryant. "Salman Khan: The Messiah of Math." *Bloomberg Businessweek*, May 19, 2011. www.bloomberg.com/news/articles/2011-05-19/salman-khan-the-messiah-of-math.

US Government Accountability Office. "Supply Concerns Continue, and Uncertainties Complicate Planning." May 20, 2014. www.gao.gov/products/GAO-14-430.

U.S. News and World Report. "Grace King High School." www.usnews.com/education/best-high-schools/louisiana/districts/jefferson-parish-public-school-system/grace-king-high-school-8639.

Valimised. "Statistics about Internet Voting in Estonia." www.valimised.ee/en/archive/statistics-about-internet-voting-estonia.

Virta. "The Global Electric Vehicle Market in 2020: Statistics & Forecasts." https://www.virta.global/global-electric-vehicle-market.

Wade, Andrew. "Made in Space: New Production Frontiers." *The Engineer*, February 8, 2016. https://www.theengineer.co.uk/made-in-space-new-production-frontiers/.

Wagner, Alex. "Can Sal Khan Reform Education in America?" *Huffington Post*, June 4, 2011. www.huffingtonpost.com/2011/04/04/the-khan-academy-and-educ_n_844390.html.

Wall, Mike. "In-Space Manufacturing Is about to Get a Big Test." Space.com, December 11, 2017. https://www.space.com/39039-made-in-space-off-earth-manufacturing-test.html.

Walters, Helen. "Two Giants of Online Learning Discuss the Future of Education." Ideas.Ted.com, January 28, 2014. https://ideas.ted.com/in-conversation-salman-khan-sebastian-thrun-talk-online-education/.

Wang, Brian. "Cheap Sanitary Pads Are Better Than Apple iPads for Poor Women in India by Improving Lives, Improving Health and

Providing Thousands of Jobs." *Next Big Future*, March 5, 2014. https://www.nextbigfuture.com/2014/03/cheap-sanitary-pads-are-better-than.html.

Wattles, Jackie. "NASA Says Moon Rocket Could Cost as Much as $1.6 Billion Per Launch." CNN, December 9, 2019. www.cnn.com/2019/12/09/tech/nasa-sls-price-cost-artemis-moon-rocket-scn/index.html.

We Build Value. "Tokyo, the Earthquake-Proof City." June 27, 2018. www.webuildvalue.com/en/reportage/tokyo-the-earthquake-proof-city.html.

Wee, Sui-Lee. "Chinese Scientist Who Genetically Edited Babies Gets 3 Years in Prison." *New York Times*, December 30, 2019. https://www.nytimes.com/2019/12/30/business/china-scientist-genetic-baby-prison.html.

Winegard, Timothy. "People v Mosquitos: What to Do about Our Biggest Killer." *The Guardian*, September 20, 2019. https://www.theguardian.com/environment/2019/sep/20/man-v-mosquito-biggest-killer-malaria-crispr.

Wolman, David. "Before the Levees Break: A Plan to Save the Netherlands." *Wired*, December 22, 2008. https://www.wired.com/2008/12/ff-dutch-delta/.

Wong, Julia Carrie. "Elon Musk Unveils Tesla Electric Truck – and a Surprise New Sports Car." *The Guardian*, November 17, 2017. https://www.theguardian.com/technology/2017/nov/17/elon-musk-tesla-electric-truck-sports-car-surprise.

World Bank. "Bangladesh Continues to Reduce Poverty but at a Slower Pace." October 24, 2017. www.worldbank.org/en/news/feature/2017/10/24/bangladesh-continues-to-reduce-poverty-but-at-slower-pace.

World Economic Forum White Paper. "Meat: The Future Series; Alternative Proteins." 2019. http://www3.weforum.org/docs/WEF_White_Paper_Alternative_Proteins.pdf.

World Health Organization. "Drinking Water." June 14, 2009. http://www.who.int/news-room/fact-sheets/detail/drinking-water.

———. "Mosquito Sterilization Offers New Opportunity to Control Chikungunya, Dengue, and Zika." November 14, 2019. https://www.who.int/news-room/detail/14-11-2019-mosquito-sterilization-offers-new-opportunity-to-control-chikungunya-dengue-and-zika.

World Watch. "Livestock and Climate Change." November/December 2009. https://awellfedworld.org/wp-content/uploads/Livestock-Climate-Change-Anhang-Goodland.pdf.

Wynne, Alexandra. "Special Report: Plugging the Floods Will Mean Going Dutch." *New Civil Engineer*, June 3, 2014. https://www.newcivilengineer.com/archive/special-report-plugging-the-floods-will-mean-going-dutch-03-06-2014/.

Xinhuanet. "The First Trial of 'Gene Editing Baby' Sentenced He Jiankui and Three Other Defendants to Be Held Criminally Responsible" [in Chinese]. December 30, 2019. http://www.xinhuanet.com/legal/2019-12/30/c_1125403802.htm.

Xolo. "10 Notable People Who Are E-Residents of Estonia." October 11, 2016. https://blog.leapin.eu/10-notable-people-who-are-e-residents-of-estonia-c3b7dcd3c7d6.

Ya'ari, Sivan. "Africa: Is Good Enough?" TEDx Tel Aviv University, February 15, 2016. https://www.youtube.com/watch?v=tfgAexR5Z2w.

Young, Jeffrey R. "College 2.0: A Self-Appointed Teacher Runs a One-Man 'Academy' on YouTube." *Chronicle of Higher Education*, June 6, 2010. www.chronicle.com/article/College-20-A-Self-Appointed/65793.

Yunus, Muhammad. "Grameen Bank at a Glance." Yunus Centre, March 12, 2008. https://www.muhammadyunus.org/post/370/grameen-bank-at-a-glance.

Zelaya, Ian. "Lighting a Continent." *Washington Jewish Week*,

November 13, 2013. http://washingtonjewishweek.com/7152/lighting-a-continent/news/world-news/.

Zimmer, Carl. "From Fearsome Predator to Man's Best Friend." *New York Times*, May 16, 2013. www.nytimes.com/2013/05/16/science/dogs-from-fearsome-predator-to-mans-best-friend.html.

Interviews

Abe, Yoshiko, PhD – Sustainability Strategist, Kokusai Koygyo Co. Phone, March 31, 2020.

Adriaenssens, Sigrid, PhD – Professor of Civil and Environmental Engineering, Princeton University. Phone, April 16, 2020.

Ahlin, Christian, PhD – Professor, Michigan State University. Phone, May 29, 2020.

Allen, Mark, MD – CEO and cofounder, Elevian. Phone, March 23, 2020.

Anand, Sashi – Vice President, Coimbatore Women's Welfare Organization. Phone, March 30, 2020.

Ani, Karim – founder, Mathalicious. Phone, May 19, 2020.

Arison, David – Former VP of global business relations, Miya. Phone, March 30, 2020.

Arnold, David, PhD – Professor, Umass-Amherst. Phone, May 28, 2020.

Ayala, Alberto, PhD – Air Pollution Control Officer, Sacramento Metropolitan Air Quality Management District. Phone, May 6, 2020.

Barrangou, Rodolphe, PhD – Professor, North Carolina State University. Phone, June 16, 2020; October 12, 2020.

Eli Beer – Founder, United Hatzalah. Phone, June 10, 2020.

Beena, Sujithra, MD – Physician, Coimbatore, India. Phone, March 28, 2020.

Berman, Bradley – Advanced automotive journalist. Phone, May 4, 2020.

Bouzari, Ali, PhD – culinary food scientist. Phone, May 28, 2020.

Bruysten, Saskia – Cofounder and CEO, Yunus Social Business. Phone, May 6, 2020.

Burke, Andrew, PhD – Research Engineer, University of California Davis, Institute of Transportation Studies. Phone, May 4, 2020.

Burns, Lawrence, PhD – author of *Autonomy*; current advisor to Waymo, Google's self-driving car initiative, and former executive with GM. Phone, April 27, 2020.

Cagle, Yvonne, MD – NASA astronaut. Phone, July 21, 2020.

Cahn, Amir – Executive Director, Smart Water Network Forum. Phone, April 2, 2020.

Campbell, Doug – CEO, Solid Power. Phone, May 5, 2020.

Charo, Alta, JD– Professor of law and bioethics, University of Wisconsin, Madison. Phone, September 1, 2020.

Chowdhury, Dilruba – Khan Academy, Bangla Program Director. Phone, May 27, 2020.

Colucci, Lamont, PhD – Associate Professor of Politics and Government, Ripon College. Phone, May 8, 2020.

Cummings, Missy, PhD – Director of Humans and Autonomy Laboratory and Professor in the Department of Electrical and Computer Engineering, Duke University. Phone, May 4, 2020.

Cunningham, Tshaka, PhD – Scientific Program Manager for aging and neurodegenerative diseases, US Department of Veterans Affairs. Phone, July 10, 2020; October 8, 2020.

Danko, Eric – Director of Federal Affairs, Cruise. Phone, June 11, 2020.

Daver, Shernaz – former Chief Marketing Officer, Udacity. Phone, November 16, 2020.

Diamond, Robbie – CEO, Securing America's Future Energy (SAFE). Phone, April 17, 2020.

Dijkman, Jos – Engineer, Deltares Water Research Institute, The Netherlands. Phone, April 11, 2020.

Distel, Oded – former director of the Israel NewTech program, Israeli Ministry of Economy and Industry. Phone, March 12, 2020.

Dunn, Jason – cofounder, Made in Space. Phone, July 20, 2020.
Eberhard, Martin – Founder and former CEO, Tesla. Phone, November 23, 2020.
Eisele, Stephen – VP of Business Development, Virgin Orbit. Phone, April 28, 2020.
Ejiri, Norihiro – Structural Engineer, EJIRI Structural Engineers. Written exchange, April 20, 2020.
Figenbaum, Erik – Chief research engineer, Institute of Transport Economics in Norway. Phone, May 15, 2020.
Finette, Pascal – Chair for Entrepreneurship and Open Innovation, Singularity University. Phone, November 12, 2018.
Foster, Andrew, PhD – Professor of Economics and Health Services, Policy, and Practice, Brown University. Phone, May 18, 2020.
Frank, Andrew, PhD – Emeritus Professor, Mechanical and Aerospace Engineering, University of California Davis, Institute of Transportation Studies. Phone, May 6, 2020.
Garretson, Peter – Senior Fellow, American Foreign Policy Council, and former US Air Force officer. Phone, May 12, 2020.
Gasson, Christopher – Publisher, Global Water Intelligence. Phone, March 30, 2020.
Glas, Peter – Delta Program Commissioner, The Netherlands. Phone, March 26, 2020.
Goldberg, Jodi, JD – Associate, Wiley. Phone, May 14, 2020.
Goldberg, Lev – CEO, Aquify. Phone, March 16, 2020.
Goldstein, Carol, JD – satellite and space law expert. Phone, April 19, 2020.
Goldstein, Nir – Managing Director, Good Food Institute, Israel. Phone, June 25, 2020.
Gondal, Vishal – CEO, GOQII. Phone, April 8, 2020.
Grey, Aubrey de, PhD – Chief Science Officer, SENS Research Foundation. Phone, March 27, 2020.
Griffin, JT – Chief Government Affairs Officer, Mothers Against Drunk Driving (MADD). Phone, April 20, 2020.

Grinstein, Gidi – founder, Reut Group. Phone, December 17, 2020.

Hardman, Scott, PhD – Researcher, Institute of Transportation Studies. Phone, May 7, 2020.

Haskel, Gil, Ambassador – Head of Mashav, Israel's Development Agency. Phone, March 10, 2020.

Hays, Keith – Managing Director, Bluefield Research. Phone, April 4, 2020.

Hsieh, Fu-Hung, PhD – Professor, University of Missouri. Written exchange, May 30, 2020.

Hurt, Brett – CEO, Data.World. Phone, June 4, 2020.

Hwang, Roland – Managing Director, Climate and Clean Energy Program, National Resources Defense Council. Phone, May 15, 2020.

Ilves, Toomas Hendrik – former President, Estonia. Phone, March 27, 2020; April 3, 2020; November 9, 2020.

Ishiwatari, Mikio, PhD – former Senior Advisor, Japan International Cooperation Agency (JICA). Phone, April 19, 2020.

James, Anthony, PhD – Distinguished Professor of molecular biology and biochemistry and of microbiology and molecular genetics, UC Irvine. Phone, June 12, 2020.

Jonkman, Bas, PhD – Professor of Hydraulic Engineering, Delft University of Technology. Phone, April 14, 2020.

Joyce, Meri – Board Member, Peace Boat Disaster Relief (PBV). Phone, April 7, 2020.

Jurado, Jennifer, PhD – Chief Resilience Officer, Broward County, FL. Phone, April 15, 2020.

Kaevats, Marten – National Digital Adviser, Estonia. Phone, April 6, 2020.

Kaljulaid, Kersti – President, Estonia. Interview, Washington, DC, March 3, 2020.

Kaner, Avi – CEO, Morton Williams, and philanthropist. Phone, March 24, 2020.

Kaner, Liz – Member of the Board, Innovation: Africa. Phone, March 12, 2020.

Kelly, Tony – former CEO, Yarra Valley Water in Melbourne, Australia. Phone, April 1, 2020.

Kemmer, Aaron – cofounder of Made in Space. Phone, June 19, 2020; July 3, 2020.

Khadem, Ramin, PhD – former Chairman of the Board of Trustees, International Space University. Phone, April 27, 2020.

Khandker, Shahidur R., PhD – former World Bank Economist. Phone, June 1, 2020.

Kinnan, Cynthia, PhD – Professor, Tufts University. Phone, May 21, 2020.

Kirsh, Natie, PhD (honorary) – philanthropist and billionaire businessman. Phone, March 25, 2020.

Kockelman, Kara, PhD – Professor of Transportation Engineering, University of Texas at Austin. Phone, April 27, 2020.

Komino, Takeshi – Disaster Management Practitioner, Japan. Phone, April 13, 2020.

Korjus, Kaspar – Managing Director, Estonia's e-residency program. Phone, April 29, 2020.

Kornhauser, Alain L., PhD – Professor of Operations Research and Financial Engineering Director of the Program in Transportation, Princeton University. Phone, April 29, 2020.

Koseff, Stephen – former CEO, Investec. Phone, March 19, 2020.

Kotka, Taavi – first digital ambassador and Chief Information Officer, Estonia. Phone, August 1, 2019.

Kraft, Daniel, MD – Chair for Medicine and Neuroscience, Singularity University; founder and Chair, Exponential Medicine. Phone, February 19, 2020.

Kramer, William, PhD – Lecturer, International Space University; retired official of the US Fish and Wildlife Service. Phone, April 28, 2020.

Krimmer, Robert, PhD – Professor of e-Governance, Talinn University of Technology. Phone, April 8, 2020.

Kutter, Bernard – Chief Science Officer, United Launch Alliance (ULA). Phone, May 20, 2020.

LaRocco, Phil – Adjunct Professor, Columbia University. Phone, March 25, 2020.

Lieu, Chan – Senior Safety Program Analyst, Uber. Phone, April 23, 2020.

Lindt, John W. van de, PhD – civil engineering professor, Colorado State University. Phone, April 23, 2020.

Loss, Chris, PhD – food scientist and lecturer at Cornell University. Phone, June 4, 2020.

Lubi, Viljar – Deputy Secretary General for Economic Development in the Ministry of Economic Affairs, Estonia. Phone, April 4, 2020; November 6, 2020.

Mackey, John – founder of Whole Foods. Phone, June 15, 2020.

Mahadevan, Venkatesh – Chief Information Officer, Dubai Investments. Phone, April 2, 2020.

Maniam, Aaron – Deputy Secretary, Industry and Information, Ministry of Communications and Information, Singapore. Phone, April 15, 2020.

Margolis, Aleta – Founder, Center for Inspired Teaching. Phone, May 26, 2020.

Martens, Tarvi – National Electoral Committee's head of e-voting, Estonia. Phone, July 30, 2019.

Martinez, Sylvia – coauthor, *Invent to Learn*. Phone, May 19, 2020.

May, Scott – Head of MISTA at Givaudan. Phone, June 8, 2020.

Meri, Joyce – International Coordinator and Board Member, Peace Boat. Phone, April 7, 2020.

Metz, Tracy – Author, *Sweet & Salt*. Phone, April 6, 2020.

MohanaSundari, J., PhD – Managing Director, Sharp Electrodes. Phone, March 28, 2020.

Molavi, Afshin, PhD – Senior Fellow at the Johns Hopkins SAIS Foreign Policy Institute. Phone, March 16, 2021.

Morduch, Jonathan, PhD – Professor, New York University. Phone, May 27, 2020.

Morris, Dale – Economist and Director of Strategic Partnerships Water Institute. Phone, April 2, 2020.

Murphy, Robert, PhD – Former Senior Researcher, Rand Corporation. Phone, May 27, 2020.

Muruganantham, Arunachalam – innovator of low-cost sanitary pad. Phone, February 21, 2020.

Nishikawa, Satoru, PhD – disaster resilience expert; former senior Japanese Cabinet official. Phone, April 6, 2020.

Ogden, Timothy – Managing Director, The Financial Access Initiative, a research center based at NYU. Phone, May 26, 2020.

Ono, Yuichi, PhD – Assistant Director and Professor, International Research Institute of Disaster Science (IRIDeS), Tohoku University. Phone, April 16, 2020.

Ovink, Henk – Special Envoy for International Water Affairs, Kingdom of the Netherlands. Phone, April 1, 2020; October 28, 2020.

Peleg, Amir – CEO, Takadu. Interview, Washington, DC, April 2, 2019; Phone, October 22, 2020.

Pitt, Mark, PhD – Professor of Population Studies and Professor Emeritus of Economics, Brown University. Phone, May 16, 2020.

Potter, Michael, PhD (Honorary) – founder, Geeks without Frontiers; Space Policy Expert. Phone, April 17, 2020.

Rahmstorf, Stefan, PhD – Professor of physics of the oceans, Potsdam University. Phone, March 26, 2020.

Ramesh, S R – Hand-loom weaver, Coimbatore, India. Phone, March 28, 2020.

Reis, Evan – cofounder, US Resiliency Council. Phone, April 23, 2020.

Sallaberger, Christian, PhD – CEO, Canadensys Aerospace

Corporation; Chairman of the Board, International Space University. Phone, May 7, 2020.

Sato, Jun, PhD – structural engineer and associate professor, University of Tokyo. Phone, May 4, 2020.

Schwartz, Rusty, JD – founder, Kitchentown. Phone, June 2, 2020.

Sena, Michael – Telematics Expert. Phone, May 12, 2020.

Shaw, Rajib, PhD – Professor of Media and Governance, Keio University. Phone, April 20, 2020; December 15, 2020.

Siegel, Seth M. – author, *Let There Be Water.* Interview, New York. October 28, 2017.

Sperling, Daniel, PhD – Founding Director, Institute of Transportation Studies at the University of California, Davis. Phone, May 11, 2020.

Stager, Gary, PhD – Veteran teacher of education. Phone, May 22, 2020.

Steenhuis, Marinke, PhD – founder, SteenhuisMeurs. Phone, April 14, 2020.

Stive, Marcel, PhD – Civil Engineering and Geosciences Professor, Delft University of Technology. Written exchange, March 28, 2020.

Stott, Chris – founder, MANSAT; and space entrepreneur. Phone, April 20, 2020; May 8, 2020.

Stott, Nicole – former astronaut, STEM advocate. Phone, May 11, 2020.

Takeuchi, Toru, PhD – Professor of Architecture and Structural Engineering, Tokyo Institute of Technology. Phone, April 17, 2020.

Tarpenning, Marc – cofounder, Tesla. Phone, November 17, 2020.

Tepper, Steve, JD – General Council, Selfhelp Community Services. Phone, March 16, 2020.

Theo, George – CEO, Unitywater of Queensland Australia. Phone, March 30, 2020.

Thirugnanam, Natarajan, MD – plastic surgeon, Sri Ramakrishna Hospital, Coimbatore, India. Phone, March 22, 2020.

Tripathi, Supriya – Hindi content creator at Khan Academy, Khan Academy India. Phone, May 23, 2020.

van de Lindt, John W., PhD – Professor of Civil and Environmental Engineering, Colorado State University. Phone, April 23, 2020.

Venkatesh, Rangaraj, MD – Chief Medical Officer, Aravind Eye Care System. Phone, March 18, 2020.

Viik, Linnar – Programme Director, Estonian e-Governance Academy. Phone, August 20, 2019.

Waggonner III, J. David – Founding Principal, Waggonner & Ball. Phone, April 13, 2020.

Wang, Yunshi – Director of the China Center for Energy and Transportation, UC Davis Institute of Transportation Studies. Phone, May 12, 2020.

Watkins, Todd – Economics Professor, Lehigh University. Phone, May 22, 2020.

Weeden, Brian, PhD – Director of Program Planning, Secure World Foundation. Phone, April 30, 2020.

Weintraub, Seth – Founder, Elektrek. Phone, May 12, 2020; May 18, 2020.

Whittaker, William (Red) L., PhD – Director of the Field Robotics Center, Carnegie Mellon University. Written exchange, April 28, 2020.

Williams, Peter – industry expert; IBM Distinguished Engineer (retired). Phone, March 11, 2020.

Winetraub, Yonatan – cofounder, SpaceIL. Phone, April 24, 2020.

Wydick, Bruce, PhD – author, *Shrewd Samaritan*; Program Director of International and Development Economics, University of San Francisco. Phone, May 14, 2020.

Yaari, David – Chairman, A.B. Yaari Holdings. Interview, Washington, DC, March 3, 2020.

Yaari, Sivan – Founder, Innovation: Africa. Interview, Herzliya. August 8, 2019; Phone, October 27, 2020.

Yogeswaran, Babu – director and filmmaker. Phone, March 26, 2020.

Zakai, Shira – Head of Communications, Beyond Meat. Phone, January 27, 2020.

Zellner, Debra, PhD – Montclair State University. Phone, June 4, 2020.

Zheng, Jenny – Khan Academy Simplified Chinese Text Translation Advocate. Written exchange, June 3, 2020.

Zingher, Tally, JD – CEO, Dawsat. Phone, February 27, 2020.

About the Author

Avi Jorisch is a Senior Fellow at the American Foreign Policy Council and bestselling author of *Thou Shalt Innovate: How Israeli Ingenuity Repairs the World,* which has been translated into over thirty languages. He holds a bachelor's degree in history from Binghamton University and a master's degree in Islamic history from the Hebrew University of Jerusalem. He also studied Arabic and Islamic philosophy at the American University in Cairo and al-Azhar University. He has lectured around the world, and his articles have appeared in influential outlets including the *New York Times,* the *Wall Street Journal, Foreign Affairs, Forbes,* and Al-Arabiya.net. He has also appeared on CNN, CSPAN, Fox News, and MSNBC. He is a member of the Council on Foreign Relations and the Young Presidents' Organization (YPO).

JAICO PUBLISHING HOUSE

Elevate Your Life. Transform Your World.

ESTABLISHED IN 1946, Jaico Publishing House is home to world-transforming authors such as Sri Sri Paramahansa Yogananda, Osho, the Dalai Lama, Sri Sri Ravi Shankar, Sadhguru, Robin Sharma, Deepak Chopra, Jack Canfield, Eknath Easwaran, Devdutt Pattanaik, Khushwant Singh, John Maxwell, Brian Tracy, and Stephen Hawking.

Our late founder Mr. Jaman Shah first established Jaico as a book distribution company. Sensing that independence was around the corner, he aptly named his company Jaico ('Jai' means victory in Hindi). In order to service the significant demand for affordable books in a developing nation, Mr. Shah initiated Jaico's own publications. Jaico was India's first publisher of paperback books in the English language.

While self-help, religion and philosophy, mind/body/spirit, and business titles form the cornerstone of our non-fiction list, we publish an exciting range of travel, current affairs, biography, and popular science books as well. Our renewed focus on popular fiction is evident in our new titles by a host of fresh young talent from India and abroad. Jaico's recently established translations division translates selected English content into nine regional languages.

Jaico distributes its own titles. With its headquarters in Mumbai, Jaico has branches in Ahmedabad, Bangalore, Chennai, Delhi, Hyderabad, and Kolkata.

SINCE 1946